160 914334 0

AF615379

Basin Modelling: Practice and Progress

Geological Society Special Publications

Series Editors A. J. FLEET

A. C. MORTON

A. M. ROBERTS

GEOLOGICAL SOCIETY SPECIAL PUBLICATION NO. 141

Basin Modelling: Practice and Progress

EDITED BY

S. J. DÜPPENBECKER
BP Exploration Inc., Houston, Texas, USA

&

J. E. ILIFFE
PGS Tigress (UK) Ltd, Aberdeen, UK
(Present address: Conoco Inc., Houston, Texas, USA)

1998
Published by
The Geological Society
London

THE GEOLOGICAL SOCIETY

The Society was founded in 1807 as The Geological Society of London and is the oldest geological society in the world. It received its Royal Charter in 1825 for the purpose of 'investigating the mineral structure of the Earth'. The Society is Britain's national society for geology with a membership of around 8500. It has countrywide coverage and approximately 1500 members reside overseas. The Society is responsible for all aspects of the geological sciences including professional matters. The Society has its own publishing house, which produces the Society's international journals, books and maps, and which acts as the European distributor for publications of the American Association of Petroleum Geologists, SEPM and the Geological Society of America.

Fellowship is open to those holding a recognized honours degree in geology or cognate subject and who have at least two years' relevant postgraduate experience, or who have not less than six years' relevant experience in geology or a cognate subject. A Fellow who has not less than five years' relevant postgraduate experience in the practice of geology may apply for validation and, subject to approval, may be able to use the designatory letters C Geol (Chartered Geologist).

Further information about the Society is available from the Membership Manager, The Geological Society, Burlington House, Piccadilly, London W1V 0JU, UK. The Society is a Registered Charity, No. 210161.

Published by The Geological Society from:
The Geological Society Publishing House
Unit 7, Brassmill Enterprise Centre
Brassmill Lane
Bath BA1 3JN
UK
(*Orders*: Tel. 01225 445046
Fax 01225 442836)

First published 1998

British Library Cataloguing in Publication Data
A catalogue record for this book is available from the British Library.

ISBN 1-86239-008-8
ISSN 0305-8719

Distributors

USA
AAPG Bookstore
PO Box 979
Tulsa
OK 74101-0979
USA
(*Orders*: Tel. (918) 584-2555
Fax (918) 560-2652)

Australia
Australian Mineral Foundation
63 Conyngham Street
Glenside
South Australia 5065
Australia
(*Orders*: Tel. (08) 379-0444
Fax (08) 379-4634)

India
Affiliated East-West Press PVT Ltd
G-1/16 Ansari Road
New Delhi 110 002
India
(*Orders*: Tel. (11) 327-9113
Fax (11) 326-0538)

Japan
Kanda Book Trading Co.
Cityhouse Tama 204
Tsurumaki 1-3-10
Tama-shi
Tokyo 206-0034
Japan
(*Orders*: Tel. (0423) 57-7650
Fax (0423) 57-7651)

Typeset by E & M Graphics, Midsomer Norton, Bath, UK.

Printed by The Alden Press, Osney Mead, Oxford, UK.

Contents

References to this volume

It is recommended that reference to all or part of this book should be made in one of the following ways:

DÜPPENBECKER, S. J. & ILIFFE, J. E. (eds) 1998. *Basin Modelling: Practice and Progress*. Geological Society, London, Special Publication, **141**.

GILES, M. R., INDRELID, S. L. & JAMES, D. M. D. 1998. Compaction: the great unknown in basin modelling. *In*: DÜPPENBECKER, S. J. & ILIFFE, J. E. (eds) *Basin Modelling: Practice and Progress*. Geological Society, London, Special Publications, **141**, 15–43.

Preface

Petroleum exploration in frontier to developing regions is model driven, whether the model is comprised of ideas in a geoscientist's head, or of results of numerical simulations investigating several scenarios that test the boundaries of possibility. Modelling is a valuable process of risk reduction in exploration prospectivity assessment, and an area which has significant room for risk reduction in new technologies, or in better practices with existing technologies.

The impact of basin modelling on the exploration business is on risking source effectiveness, and by providing the timing framework for petroleum generation, migration and reservoir/trap formation. The reconstruction of the geological evolution of a sedimentary basin, and numerical quantification of this information, is now considered to add value. Insight into the development and understanding of a basin that develops during a basin modelling study provides as much value as the final numerical results.

The objective of this volume is to focus for the first time on the application of basin modelling tools to real problems, and to share experiences as to which techniques have worked and which have not. With great regret we have had to reject several excellent contributions related to other themes in order to adhere to this concept. We have striven for a balance between how greater understanding of some of the processes and parameters could affect model results, and the application of software and techniques to solve particular questions posed about different basins. Although the contents of this volume do not reflect this as we would have wished, all the papers make a valuable contribution to the future of basin modelling.

In the vanguard, **Waples** rises to the unenviable task of providing a keynote paper entitled 'Basin modelling: how well have we done?'. This reflective analysis demonstrates the foundation of our present level of understanding, and outlines areas where we should venture forth.

Some of the basic concepts of basin modelling that have fallen into general acceptance over the past two decades are fundamentally challenged by the papers in this volume. **Giles *et al.*** revisit the concepts employed for calculating compaction, and their incorporation of physical, thermal and chemical processes into a more realistic behaviour is highly illuminating. Similarly, **Okui *et al.*** quantified hydrocarbon expulsion from shale source rocks as a function of porosity and permeability in the laboratory. From this they derived an expulsion model that is more representative of argillaceous rocks, rather than the typical values normally used for saturation thresholds, originally derived from reservoir rocks. These fundamental changes in our understanding of the real physical processes have important connotations for the results of our basin simulations in terms of pressure, temperature, maturity and phase of hydrocarbons produced from the source rock.

The enigma of overpressure formation is elucidated by **Waples & Couples**, who eloquently clarify the interactions of rock properties with the sensitive rate-determining processes that take place during compaction. **Tokunaga *et al.*** quantified these rate-controlling dynamic shale rock properties in the laboratory and investigated their effects on the physical processes by numerical modelling. The overpressure theme is continued with **Darby *et al.***'s paper on pressure simulation in 1D and 2D simulations. **Throndsen & Wangen**, also take up the question of our level of understanding of the 3D volume when our problems are simply constructed in one or two dimensions. Their three-dimensional simulator examples indicate the importance of modelling the system as a whole.

The importance of truly integrated multidisciplinary projects that represent the vital technical core of exploration decisions is demonstrated in the application papers by Symington *et al.*, Ho *et al.*, Hegre *et al.*, Shegg & Leu and Archard *et al.* **Archard *et al.*** assess uplift using a variety of techniques in a very complex basin history. The uplift and erosion analysis theme is shown to be crucial to assessing geothermal gradients and temperature histories in a compressive tectonic regime by **Schegg & Leu**. Both Archard *et al.* and Shegg & Leu describe pragmatic approaches that use various datasets to determine the same unknown parameters, and handle the uncertainties in their data and techniques. **Hegre *et al.*** present an attractive set of pyrolysis data that show an unexpected variability. Their problem is typical of that of the modeller in that several explanations might be valid, but which one is true? **Symington *et al.*** present an excellent example of an integrated multidisciplinary approach to date and quantify trap charge from the kitchen regions. The strength of their analysis is in the verification of the predictions from their 2D and 3D simulators using a wide variety of data types.

The reflective vein of the volume is continued by **Ho *et al.*** who report a comparative study of pre-drilling and post-drilling maturity predictions and results. They show how well they did in the early 1980s, using seismic velocities to predict thermal conductivity for their models, compared with the results using the more sophisticated software of the 1990s. Their approach shows the importance of information within the actual seismic data on which we base our models.

Rather than providing a unique solution, basin modelling allows investigation of a range of geological models and hypotheses and the selection of the most likely. The importance of sensitivity analysis and recognition of uncertainty in input datasets is highlighted by **Thomsen**, using a probabilistic technique, and by **Gallagher & Morrow** using numerical inversion techniques to constrain heat flow histories. So, how well can we do most of the time?

What remains clear from this volume is that modelling will continue to evolve as a joint effort between technological advances in the data acquisition, model building and simulation software, along with more rigorous quantification and understanding of the thermodynamics of the rock and fluid properties. We are rapidly approaching the age of true three-dimensional models, where we will likely discover new sets of interdependencies, new uncertainties to unravel – in particularly around fluid migration, and new calibrants for our models.

Acknowledgements

The editors gratefully acknowledge the work of the reviewers for the papers in this book, and we also wish to thank all those presenters who contributed to the success of the conference which gave rise to this volume. We reserve a special thanks to those who have made the time and effort to contribute to this stimulating volume.

Basin modelling: how well have we done?

D. W. WAPLES

Consultant, 9299 William Cody Drive, Evergreen, CO, 80439 USA

Abstract: Model developers have been highly successful in building usable, useful, and reliable one-dimensional models, and in convincing explorationists to use them. Development of convenient, universal software played an instrumental role in the acceptance of l-D modelling. The remaining weaknesses in technology and training, while serious, can be resolved through a few years of intensive work in certain specific areas. 2-D and 3-D fluid-flow modelling, in contrast, finds itself today largely in the position occupied by l-D modelling more than a decade ago. The concept of fluid-flow modelling has not yet been adequately sold to explorationists, mainly because existing software is often inadequate to meet the real needs of exploration personnel. Development of appropriate software will probably have to precede widespread popularization of fluid-flow modelling. To improve existing software, two major areas must be addressed: (1) making models more comprehensive by including all relevant geologic phenomena, and (2) making software easier to use.

The general topic of basin modelling includes both 1-D modelling (often called maturity modelling) and 2-D and 3-D fluid-flow modelling (sometimes itself called basin modelling). Both types of modelling are in some senses like the mythical monster Hydra, in which dangerous new heads grow as fast as old ones can be cut off. Over the past decade — but principally during the past five years — we have solved some important problems in maturity and fluid-flow modelling. However, the solutions to old problems have often uncovered additional hitherto-unrecognized issues, often of equal importance. This paper will discuss both those heads that we have succeeded in lopping off, and the new ones growing from the old roots.

The discussion is divided into four parts. The first two discuss the successes and the remaining problems in 1-D (maturity) modelling. The remaining two parts discuss 2-D or 3-D modelling carried out for the additional purpose of studying pressure and fluid flow (particularly hydrocarbon migration).

Maturity modelling and fluid-flow modelling share a common foundation, in that the kinetics of hydrocarbon generation, the expulsion model, and most of the geologic framework used in 1-D modelling are the same as those used in 2-D models. Furthermore, except for lateral heat transfer by convection or diffraction, the thermal regimes in 1-D and 2-D models are the same. Finally, the optimization sequence in 2-D modelling can depend heavily on 1-D optimization.

As a result of this common heritage, much of the technology within 2-D models can be discussed adequately using examples from 1-D models. I will therefore begin with 1-D (maturity) models, and will only discuss the same phenomena in 2-D or 3-D models when the change from 1-D to multiple dimensions makes a difference.

The views expressed in this paper should be taken as my personal opinions, rather than as dogma or absolute truth. Where my observations seem correct, I hope they will stimulate further research and development. Where they appear wrong, I hope they will stimulate discussion that will eventually clarify those issues for everyone.

1-D (maturity) modelling: successes

Development of maturity models

The development of Tissot's kinetic model (1969) and Lopatin's TTI model (1971), and the latter's popularization in the West and among explorationists (Waples 1980), represented important initial steps in uniting chemical and geochemical technology with geology. The ability to predict maturity and hydrocarbon generation provided geologists with powerful new tools for geologic analysis. For example, the requirements of both methods that the burial and thermal history of a section be properly reconstructed forced geologists to think of time as a dynamic quantity, and to address geologic history as a series of events from past to present.

Subsequent applications of maturity modelling have emphasized the need to use measured data such as temperatures and maturity values (e.g. Ro) to calibrate the thermal history in order to improve the quality of predictions from the model. The

WAPLES, D. W. 1998. Basin modelling: how well have we done?. *In*: DÜPPENBECKER, S. J. & ILIFFE, J. E. (eds) *Basin Modelling: Practice and Progress*. Geological Society, London, Special Publications, **141**, 1–14.

recognition that our first guesses about thermal and erosional history are likely to be highly uncertain — and even sometimes greatly in error — was an important step in learning how to reconstruct geologic histories with confidence. It also provided a sobering realization about the poor quality of our knowledge in most situations.

Development of sophisticated software

One of the most important advances in 1-D maturity modelling has been the development in the late 1980s of excellent PC-based software products that permit modelling to be carried out quickly and conveniently, even by non-specialists. In particular, computer modelling permits one to run multiple scenarios easily, thus promoting the use of sensitivity analysis and proper optimization. Computer-based modelling also permits use of sophisticated algorithms that would not be possible without a computer, such as kinetic modelling of vitrinite reflectance or hydrocarbon generation, or use of the heat flow/conductivity method instead of geothermal gradients. The importance for 1-D maturity modelling of computers and the software that drives them can hardly be over-estimated.

Replacement of TTI models with kinetic models

For many years TTI models (Lopatin 1971; Waples 1980) were more popular among explorationists than were kinetic models. The popularity was probably due to a number of factors, including:

(1) earlier promotion of the TTI method within exploration environments;
(2) the ability to execute TTI modelling by hand, in an era when computers and software were often not available for carrying out those calculations; and
(3) in the early days the more-graphical nature of input and output for the TTI models that made conceptualization easier for geologists.

Nevertheless, in spite of the great benefits accrued as geologists learned to use 1-D maturity models as a routine tool in geological evaluations, the technical details of the TTI method are weak. The kinetic function used in the TTI model is demonstrably inferior to that used in kinetic models. Furthermore, with the advent of personal computers and excellent software for maturity modelling, there are no longer any advantages in most companies in using the TTI method. The replacement of the TTI method in recent years by standard kinetic models for vitrinite-reflectance change and for hydrocarbon generation (e.g. those published by Tissot *et al.* 1987; Burnham 1989; Burnham & Braun 1990; Sweeney & Burnham 1990) has improved our ability to model Ro change and hydrocarbon generation. The recent emphasis on 'personalized kinetics' (discussed later) holds promise to make kinetic models even more powerful and accurate.

Growth in popularity of the heat flow/conductivity method

As 1-D maturity modelling has become more commonplace, the weaknesses of the geothermal-gradient method inherited from TTI modelling have became increasingly apparent. Before computerization, use of geothermal gradients was a necessity, but once computers took over the calculations, geothermal gradients had very few advantages. Real geothermal gradients are normally non-linear, a characteristic that cannot be captured easily using the geothermal-gradient method. Most workers today use heat flows and conductivities, which much more closely simulate a realistic cause-and-effect relationship between geological events (e.g. deposition, erosion, tectonism) and temperature.

Use of heat flows has allowed us to try to link thermal histories to tectonic events, since many such events are related to changes in heat flow. Thus the use of heat flow is fundamental in building models that mimic reality and are maximally predictive. Geothermal gradients, in contrast, are merely the result of the complex interaction of heat flow, thermal conductivity, and heat capacity, and thus can never serve in our models as a fundamental cause of subsurface temperatures.

Hydrocarbon composition

Until recently, the products of kerogen decomposition were divided simply into oil and gas, with no clear specification of whether the terms oil and gas refer to molecular size or to phase (and if to phase, under what conditions?) Recent work has led to the development of prototype versions of kinetic schemes that divide products of generation and cracking into several categories. The most common of these 'compositional-kinetics' schemes divide the hydrocarbon products into four or five fractions based on molecular size. Most workers agree that use of a few hydrocarbon fractions (especially dividing them by molecular size and by compound type, e.g. saturates vs aromatics) represents a good compromise between what we need to know in order to predict hydrocarbon composition and phase in the subsurface, and what we can hope to know based on our understanding of kerogen and its reactions.

The fundamental mathematical structure for carrying out relatively complex calculations of hydrocarbon composition has already been developed, and compositional-kinetics data for a few kerogen types have been made available for routine use. Detailed predictions of hydrocarbon composition should provide information of value in evaluating hydrocarbon phase behaviour under subsurface conditions.

Inclusion of expulsion calculations

Until recently, 1-D maturity models stopped after calculating hydrocarbon generation, and thus did not include expulsion. In recent years, however, most such models have begun to calculate expulsion. This development is logical, since maturity modelling essentially describes the behaviour of the source rock, and expulsion is the final step that occurs in the source rock. Inclusion of expulsion and generation together in a single model makes modelling of hydrocarbon generation more convenient and realistic, since the amount of hydrocarbons expelled is normally much more important to us than the amount generated.

Recognition of the importance of systematic optimization

Experienced modellers now recognize that measured data on, for example, porosity, subsurface temperature, sonic velocities for shales, thermal indicators (e.g. vitrinite reflectance), and pressures play a vital role in constraining the input data for our modelling simulations. The process of utilizing measured data to improve our geologic model is called 'optimization' or 'calibration.' Optimization can be slow and difficult, however. In order to make it as quick, painless. and effective as possible, it is important that the modeller follows a certain logical sequence and adheres to certain principles.

We begin the optimization process by creating a reasonable compaction history by calibrating our porosity-reduction function with measured porosity data. We then calibrate our matrix conductivities and present-day heat flows using measured temperatures and measured conductivities (either from our samples or from a data library). Next we calibrate our erosional history using a combination of measured thermal indicators (e.g. vitrinite reflectance and/or apatite fission-track data) and sonic velocities. (If we change our estimate of removal we may have to revise our previously optimized porosity-reduction function, conductivities, and present-day heat flow.) Finally, we calibrate the paleoheat flow using thermal indicators (e.g. vitrinite reflectance, apatite fission-track data, and/or fluid-inclusion temperatures).

The recognition of the need for good optimization procedures has been very important in the development of sophisticated, reliable models. As noted above, it has had the ancillary benefit of encouraging thinking and understanding about the geologic history of each model, and has thus increased the sophistication of geologists.

Recognition of the importance of measured calibration (optimization) data

As noted above, we have gradually become aware that our knowledge of geological history (especially thermal history and amounts and timing of erosion) is generally rather poor. Understanding the limitations of our knowledge in these important areas has itself represented a significant accomplishment. In addition, we now recognize that certain types of common data can be used to constrain or test our proposed geologic models. These data include surface and near-surface temperatures, downhole temperatures from logging runs and fluid tests, indicators of total thermal history (e.g. vitrinite reflectance), indicators of maximum burial depth (e.g. vitrinite reflectance, apatite fission-track data, fluid-inclusion temperatures, and sonic velocities), and indicators of instantaneous temperatures, such as apatite fission-track data and fluid inclusions. Considerable effort and expense now go into accumulating and using calibration data, with corresponding increases in the quality of our optimization, modelling, and overall understanding.

Recognizing the need for calibration data has in turn stimulated interest in the acquisition of those data. Consequently, the quality of calibration data has improved, and new sources of data have been discovered (e.g. fission-track data, FAMM, sonic velocities). Intensive use of calibration data to constrain models has in turn stimulated our geologic thinking, since in the course of optimization, we are required to find a geologic history that is compatible with the measured data.

Improved understanding of the strengths and weaknesses of thermal indicators

Although most of the weaknesses and uncertainties of the various thermal indicators have been recognized by specialists for many years, explorationists have historically tended to take measured data rather literally. and to use them uncritically. Today, however, we know that the process of modelling properly means not only fitting calculated Ro values to measured Ro values, but also critiquing the measured Ro data prior to attempting the fit. We have recognized that if measured and calculated values do not agree, part or all of the error could lie in the measured data instead of in our geologic model.

The general recognition of the weaknesses of all thermal indicators, coupled with the demand for reliable thermal indicators to constrain thermal histories for modelling, has led to two developments. One is an increasing tendency to use multiple thermal indicators, reasoning that this method provides internal checks and balances that can help identify problem data. The second development is some significant recent improvements in the quality of measured data, including the development of new technologies such as FAMM (Fluorescence Alteration of Multiple Macerals: e.g. Wilkins *et al.* 1992, 1995). Although still far from perfect, thermal indicators today are more reliable than ever before. This strength, coupled with our increased awareness of the possibility of errors. has improved the quality of our modelling.

Sensitivity analysis

Modellers have become well aware that some uncertainties are much more important than others. Sensitivity analysis has developed to a moderately sophisticated degree to address the question of which uncertainties in a given simulation are most critical. Recognition of the sources of greatest uncertainty pays two dividends: the modeller can then focus his or her attention on acquiring additional data to help decrease the greatest uncertainty, and the modeller can recognize more clearly the magnitude of the uncertainty in the modelling results.

Selling of 1-D maturity modelling to explorationists

Although the basic concepts of 1-D maturity modelling have been around since the late 1960s (Tissot 1969), maturity modelling was rarely used within the exploration community in the 1970s. Waples' 1980 paper on the TTI method seems in retrospect to have represented a critical point in the popularization of maturity modelling. However, the early 1980s was a period of slow growth of maturity modelling among explorationists, with use of this technology apparently limited by several factors:

(1) lack of recognition of its importance in exploration;
(2) lack of available software, requiring that calculations be done by hand; and
(3) lack of a 'critical mass' of explorationists who were using the technology, thus relegating it to a supporting role rather than a primary one.

Factors (2) and (3) go together, as they do so often in the development of many new technologies. Fax machines are wonderful if most people have one, but are of limited value if only a few people have access. Similarly, when it is considered normal for every farmout offer or acreage evaluation to include 1-D maturity modelling output, geologists will be doing more maturity modelling.

1-D maturity modelling did not become widespread among explorationists until the mid- to late 1980's. Today it is a standard technique that plays a major role in modern exploration.

There is no question that 1-D maturity modelling has now been sold successfully to explorationists. Model developers have played an important role in selling maturity modelling, because without good software that meets the needs of explorationists, the technique would not have achieved the level of acceptance and routine use it enjoys today. Model builders have been largely passive but very successful salesmen. They have built the proverbial 'better mousetrap,' in a mouse-infested world, and the world has beaten a path to their door. These lessons are important for the development and sales of other technologies, such as 2-D fluid-flow modelling.

1-D maturity modelling: unresolved problems

1-D maturity modelling today is very popular and highly successful in its exploration applications. However, despite the enormous advances that have been made in the last few years in improving the technology and making it more accessible and useful, a number of important problems remain to be solved. The following sections discuss briefly some of the most important problems.

Quality of measured subsurface temperature data

Hermanrud *et al.* (1990, 1991) and Waples & Mahadir Ramly (in press a), among others, have shown that the quality of measured subsurface temperature data is highly variable, and that most measured values, even after standard corrections are made, under-estimate true (virgin) subsurface temperatures rather badly. Many modellers are unaware of this problem, however, and thus will make major errors in reconstructing present-day temperatures, thermal histories, and maturation. Even when modellers are aware of these problems, proper corrections are currently difficult or impossible due to lack of data.

Hermanrud *et al.* (1990) have taken one approach, in attempting to refine correction methods using additional information about measurement conditions. Waples & Mahadir Ramly (in press a), in contrast, have developed statistical corrections based on a study of the Malay Basin. Preliminary reconnaissance suggests their statistical correction methods may be fairly applicable in other areas, but further research is needed.

Drill-stem tests and production tests are by far the most reliable subsurface temperature measurements, although errors do occur (Hermanrud *et al.* 1991). Log-derived temperatures, in contrast, are very poor (Hermanrud *et al.* 1990; Waples & Mahadir Ramly, in press a). Until we can base most of our subsurface temperature estimates on DSTs and production tests, or until reliable methods are developed for correcting log-derived temperatures (doubtful), present-day subsurface temperatures will remain uncertain in most modelling studies. Since much of our reconstruction of the palaeothermal environment depends on a correct assessment of present-day temperatures, our entire thermal edifice — so vital for maturity modelling — is built on a shakey foundation unless modern subsurface temperatures are accurately known.

Surface temperatures

Most workers consider both present and palaeosurface temperatures to be well constrained. Uncertainties in average present-day air temperatures are usually no more than a few degrees, and standard methods exist to estimate temperatures of surface rocks in both subaqueous and subaerial settings if one knows the air temperature.

Recently, however, Muller *et al.* (1998) have suggested that near-surface temperatures can be much higher than one would anticipate from average true surface temperatures. They attribute this effect to a combination of the inability of present software to subdivide the near-surface rock layers (where porosity and hence conductivity are changing very rapidly with depth) into thin-enough intervals to model porosity accurately, and a dominance of convective heat-transfer processes in the near surface which are not taken into account in most maturity models. They believe it is better to use a pseudo-surface temperature by extrapolating the subsurface trend to the surface, rather than to use a realistic surface temperature.

Since this idea is new, it is difficult to judge its validity, universality, or importance. My own experience, particularly in the Malay Basin does not show such an effect, although our near-surface-temperature data base was smaller than that of Muller *et al.* This topic will be of interest for future research.

Thermal conductivities

Several problems with thermal conductivities remain to be solved. One is the question of accurately estimating porosity, since total conductivities of rocks represent a weighted average of the conductivities of the mineral grains and the pore fluids. Methods of estimating porosity are still somewhat crude, especially in overpressured (undercompacted) regimes.

A second problem is how to calculate an average conductivity, even when the mineral (or lithological) and fluid composition is precisely known. Arithmetic, geometric, and harmonic averages are each appropriate under certain conditions, but many software programs do not use the most appropriate averaging method in each case. Interbedded lithologies, for example, should be averaged differently from the same gross lithologies mixed in a homogeneous fashion. Luo *et al.* (1994) have suggested an interesting method using fabric theory.

A third problem is scaling. If one has conductivity data for a single sample from a formation, is one justified in using that single conductivity as the conductivity of an entire formation? How can one scale up from the size of samples on which measurements are made to the size of our simulation grids? This difficult question remains largely unaddressed today.

A fourth problem is the lack of a large, accurate, and public base of conductivities of different lithologies. Existing databases are fragmentary and often contradictory. A comprehensive and reliable database for pure lithologies and for other common mixed lithologies needs to be developed.

The better the quality of the temperature data, the less important are conductivity databases, because if temperatures are known, a reasonable estimate of both heat flow and conductivity can usually be made. However, in many modelling studies, where temperature data are fragmentary or unreliable, the ability to predict conductivities accurately is very important.

Conceptual models for palaeoheat flow

Conceptual models today for palaeoheat flow are rather weak. Extensive work has been done on rift basins and failed rift basins, linking the magnitude of increase in heat flow with the degree of crustal attenuation. Recently, however, new conceptual models (e.g. two-layer models including a brittle upper part and a ductile lower part) for crustal thinning have led some workers to postulate lower Beta factors and hence lesser increases in heat flow than was previously believed.

Because basin types other than rift basins have received much less attention, our conceptual models for their heat-flow histories are even weaker. Fortunately, most other styles of basin are not associated with large changes in conductive heat flow, and thus our uncertainties are usually not huge.

Finally, our models for the heat flow prior to basin development are weak, because there is so much natural variation in heat flow of cratons of various ages and compositions. Age seems to play an important role for continental crust (younger crust has higher heat flow), but a wide range of heat flows is seen for any given age (Cull & Denham 1979; Cull & Conley 1983). Radiogenic heat from oceanic crust always seems to be very small (Sclater *et al.* 1980). Radiogenic heat originating within continental crust, in contrast, is highly variable, with low values comparable to those for oceanic crust in a few areas, and much higher values in other areas (Sclater *et al.* 1980). Without knowing the heat flow prior to the basin-forming event, it is difficult to estimate the magnitude of any tectonically induced change in heat flow.

Convective heat flow

1-D models cannot directly handle convective heat transfer, because much of the convective movement is non-vertical. This limitation is intrinsic, and cannot be overcome with better software; we must simply live with it.

Moreover, we are becoming increasingly aware that fluid flow and the consequent convective heat transfer can play important roles in the thermal history of basins (e.g. Burrus *et al.* 1994; Deming 1994; Hulen *et al.* 1994). Convective heat transfer can either raise or lower temperatures, depending upon the source and temperature of the moving fluid, if we can somehow estimate the magnitude of the convective heat flux, we may be able to build those effects crudely into our 1-D model by treating convective heat flow as though it were conductive.

Thermal indicators

The critical role played by thermal indicators in estimating amounts of erosion and in optimizing palaeoheat flow makes it essential that these indicators be accurate. Vitrinite reflectance (Ro) has historically been the most important thermal indicator, although its weaknesses have been recognized (and largely downplayed) for a number of years. Recently, however, the problem of Ro suppression has gained considerable attention, ironically because a solution has finally been developed. Measurement of equivalent Ro values using FAMM technology now allows us to correct for Ro suppression or even to replace direct Ro measurements (e.g. Lo 1993; Wilkins *et al.* 1992, 1995). FAMM appears to be useful and valid, but only time will tell whether it provides a significant improvement over Ro measurements.

Other thermal indicators have been much more disappointing. T_{max} is plagued by its kerogen-type dependence, but can be very useful in cases where that hurdle is overcome. Biomarkers have developed a rather bad reputation in recent years, as we have learned about facies effects and complex reaction pathways that preclude simple interpretations. Fission-track data are difficult to interpret and somewhat controversial. Fluid inclusions have never been utilized as much as is perhaps warranted.

Much more research will be required to provide us with thermal indicators that are reliable enough to use confidently for modelling. This important area represents one of the main challenges of the next few years in maturity modelling.

Personalized kinetics

Over the past few years there has been a strong movement to acquire kinetic parameters for individual kerogens for use in maturity modelling. The theory behind personalized kinetics is that since there are so many varieties of kerogens, the standard kinetics for Type I, II, and III kerogens are inadequate to describe all kerogens, particularly those with unusual characteristics such as high sulfur contents. Therefore, by measuring kinetic parameters directly for any desired kerogen, one should be able to improve modelling of hydrocarbon generation.

Space does not permit a detailed analysis of all the problems associated with personalized kinetics. There is some potential for analytical error in the measurements themselves, although most laboratory analyses seem to be very reliable. The main problem occurs in the mathematical reduction of the raw pyrolysis data to derive the kinetic parameters. The kinetic parameters thus obtained provide excellent mathematical fits to the pyrolysis data, but are often not physically reasonable, since the combinations of activation energies and frequency factors thus obtained violate the laws of thermodynamics applied to chemical reactions (Waples 1996; unpublished data). Use of these incorrect personalized-kinetic data can lead to major errors in modelling of hydrocarbon generation (Waples & Mahadir Ramly in press b). Thus personalized kinetics as often applied today represents a step backward rather than forward. Personalized kinetics properly applied have a good future, but in order to work well, the mathematical data-reduction process will have to be linked closely with reaction thermodynamics.

Hydrocarbon composition

In spite of the advent of prototype models for compositional kinetics, as discussed earlier, most modellers still use simple kinetic schemes that classify maturation products into three categories: oil, gas, or residue. More training of modellers in the proper use of compositional kinetics is therefore necessary to improve predictions of hydrocarbon composition. Moreover, software programs need to build in compositional networks, because for most modellers the task of creating the complex reaction networks necessary to use compositional kinetics is too daunting and time consuming.

Furthermore, as noted earlier, there are very few data available today on compositional kinetics, and until more data become available for a range of kerogen types, compositional kinetics will be of limited value. Finally, in order for compositional kinetics to represent a significant advance over simple oil/gas/residue systems, we will have to have better models for expulsion and for phase behaviour in the subsurface. Only then can we take advantage of the detailed compositional information provided by compositional kinetics in modelling expulsion.

Kinetics of cracking

The kinetics of hydrocarbon cracking have received very little attention over the years, and thus should be considered a weak area in maturity models. Most software programs still use a very simple model in which oil is cracked to gas plus residue in some proportion (usually *c.* 50% of each). The kinetics employed are normally very simple, often consisting of a single activation energy. Some of the commonly used frequency factors and activation energies seem to me to be too low when thermodynamic constraints are taken into consideration. In many cases the same kinetics are used for all types of oils.

Compositional kinetics (see above) holds some promise for providing more-realistic kinetic parameters for oil cracking. Some work has suggested that kinetic cracking parameters should be different for different oil types (e.g. Behar *et al.* 1988). However, the variations between analyses for a single oil sample can be greater than the differences among oil samples (compare Behar *et al.* 1988 and Ungerer *et al.* 1988). More recently, Kuo & Michael (1994) have proposed a kinetic scheme for oil cracking, but their kinetic parameters appear to violate the laws of thermodynamics, probably for different reasons than in the case of personalized kerogen kinetics (Waples 1996).

Finally, some workers (e.g. Price 1993) disputed the accepted ideas on cracking. They generally believe that liquid hydrocarbons are much more stable than the accepted kinetics indicate, and that cracking is much less important than most workers currently believe. Further work, including both experiments and theory, in this area is necessary to settle this dispute and provide reliable kinetics for hydrocarbon cracking.

Expulsion models

Expulsion models included in software programs today are, in my opinion, more crude than our present state of knowledge could support. Expulsion models can generally be divided into five groups:

(1) those where expulsion is some direct function of maturity (e.g. vitrinite reflectance) without any reference to actual hydrocarbon generation;
(2) those where expulsion is a direct function of fractional hydrocarbon generation (e.g. transformation ratio) without regard to volumes of hydrocarbons generated;
(3) overflow models (e.g. no expulsion occurs until a certain threshold volume of hydrocarbons has been generated, but after that threshold has been reached, all additional hydrocarbons are expelled);
(4) saturation models, in which the amount of expulsion is a function of a number of factors that determine the hydrocarbon saturation in the pore space of the source rock and relative permeability to hydrocarbons; and
(5) diffusion models, in which hydrocarbons depart the source rock by diffusion through a continuous organic network.

Of these models, all except the saturation model and the diffusion model are far too crude to be regarded as mechanistic models, and should be abandoned as quickly as possible. They can be useful, but always require empirical calibration with data that are often not available. Present saturation models seem to be reasonably complete conceptually, but many of the quantitative details are still poorly known. The main problems with existing saturation models are:

(1) in most models only two fluid phases are considered, rather than three (although many

workers argue that in a source rock a separate gas phase is unlikely to exist);

(2) porosities of source rocks are usually poorly known, lending a high degree of uncertainty to the calculated pore volume and hence to calculated saturations;

(3) adsorption of hydrocarbons on kerogen is not always considered; and

(4) relative-permeability curves used for source rocks are poorly known and in many models are probably not correct.

Adsorption of hydrocarbons on kerogen seems to be very important (e.g. Sandvik & Mercer 1990; Sandvik *et al.* 1992). Further work is needed to resolve such issues as:

(1) total adsorptive capacities of various types of kerogen and of kerogens at different maturity levels; and

(2) differences in adsorptive capacity for different oil and gas molecules, including effects of molecular size and polarity.

Relative-permeability curves for source rocks in most models are based on analogy with reservoir rocks. However, these concepts are probably not appropriate for rocks with much smaller pore spaces, and thus may greatly over-estimate the saturation threshold required to initiate hydrocarbon expulsion (see Okui & Waples 1993 and Pepper & Corvi 1995). Scaling issues should also be considered: we must use average values for organic richness in our simulations, whereas the most-effective source rocks are likely to be richer than average and their behaviour may be significantly different from the average.

The diffusion model for expulsion has been recently given an indirect boost by the work of Vernik & Landis (1996), who using measurements of acoustic velocities and anisotropies concluded that kerogen networks are continuous in both horizontal and vertical directions. This conclusion, if correct, solves a major difficulty with the diffusion model, since many workers have been sceptical about the continuity of kerogen in many moderately rich rocks. Further work on this model is needed, however.

Optimization

All serious modellers today recognize the need to carry out good optimization as a preparatory step to performing final simulations. The basic logical scheme for 1-D optimization is agreed upon by most workers: porosity (and sometimes pressure), present-day temperatures, erosion, thermal history, and a final pressure optimization. The main remaining problems are to make sure that optimization is actually carried out properly, since even experienced modellers are liable to make errors or to forget steps. Properly designed software could provide valuable aids in optimization, but to date no software developer has concerned itself with actively aiding modellers in optimizing their input data.

When only a single well is available, optimization is relatively simple; the modeller need only find a set of input values that yields calculated results in as close agreement as possible with the most-reliable measured data. The main problems to be confronted are:

(1) deciding what the true or target values are within the set of measured data, and

(2) making sure that all input values are geologically, geophysically, and geochemically reasonable.

Fitting calculated curves to measured data should never simply be mindless curve fitting.

When multiple wells are available, however, an additional issue must be addressed. Optimization must be carried out consistently from one well to another. Simply fitting each well in isolation is not enough; the input values used in the optimization must vary in geologically reasonable ways from well to well. In some cases one must be willing to sacrifice, to some degree, the quality of the fit in one well in order to achieve a greater overall harmony in the input values (e.g. trends of heat flow across a basin). These compromises are usually justifiable because the quality of our measured data is often mediocre to poor. However, modellers should be careful not to compromise data they truly believe in.

I hope that in the near future software designers will pay more attention to optimization. Aid in optimization could be included in the software in one or more of the following forms:

(1) providing a checklist so that the modeller does not inadvertently forget an optimization step or perform the steps out of the proper order;

(2) providing semiautomated optimization that finds an optimal value for a given parameter within a range of plausible values preselected by the modeller; or

(3) performing the entire optimization automatically.

I personally recommend option (2). Option (1) would be very easy for software developers to include, but is of limited value. Option (3) is too likely to lead to mathematically correct but scientifically unreasonable solutions, even where the modeller provides guidelines. Furthermore, option (3) would deprive modellers of many of the intangible benefits of performing optimization themselves. Option (2) will relieve modellers of much of the burden of number crunching in optimizing each parameter, but allows them to intervene after optimization of each variable. They will therefore retain most of the benefits of the manual optimization process.

Structure of models: coupled vs modular

Some 1-D maturity models perform their calculations in a fully coupled mode, whereas others use a modular structure. The theory behind fully coupled models is that in nature all processes are coupled, and it is necessary to take all possible feedback relationships into consideration at each calculation step in order to get the most accurate answers possible. The philosophy behind modular models, in contrast, is that most of these feedback relationships are of very minor importance, and can be ignored for practical purposes. This simplification leads to much faster simulations.

The question of coupled versus modular models is much more than academic, and has ramifications for 2-D modelling as well. Modular systems are much better suited to optimization, both because simulations are faster (optimization requires making many runs by trial and error), and because in a well-designed modular system any modules that are unaffected by a particular change in an input parameter (e.g. during optimization) are not recalculated, leading to a further saving of time.

Neither type of model is perfect, and perhaps it is not necessary for the industry to decide finally on one type or the other for 1-D modelling. Nevertheless, it is worth reconsidering whether the benefits of each type of model are worth the sacrifices. The ramifications of these questions for 2-D modelling will be discussed later.

Training for modellers

Performing 1-D maturity modelling with modern software is easy, but performing it well is very difficult. One must master the art of optimization, which itself requires that one be knowledgeable about subsurface temperatures, porosity reduction, thermal indicators, petrophysics, and theoretical models for heat flow, among others. One must also be conversant in the various issues related to kerogen type, kinetics of hydrocarbon generation and destruction, and expulsion. It is not easy to master these various disciplines. especially if one is attempting to educate oneself.

It is therefore important to establish good training programmes for modellers that will show them the correct approaches and methods in modelling, especially in optimization. Such courses must include hands-on practice as well as theory. Moreover, such training should be followed by testing of the candidate modellers. Although the idea may seem snobbish to some, I believe that some sort of certification examination for modellers would greatly improve the quality of modelling, as well as safeguard the technology from abuses.

Sensitivity analysis

Although the basic principles of sensitivity analysis are well known to experienced modellers, the concepts are not familiar to many casual users. Moreover, even experienced modellers often do not carry out thorough sensitivity analysis routinely. Greater effort needs to be expended in the future in making sensitivity analysis routine.

Fluid-flow modelling: successes

The technical successes include:

(1) the building of 2-D and 3-D models that show the flow of one, two, or even three phases of fluids;
(2) the inclusion in some models of fluid flow through user-specified faults;
(3) the ability to transfer heat non-vertically via moving fluids or by focusing due to variations in conductivity;
(4) the ability to trap fluids in both structural and stratigraphic traps;
(5) the ability to handle minor tectonic complexity (mostly in the form of normal faulting);
(6) the ability to link up with seismic work stations as a source of input data; and
(7) the successful testing and calibration of these models in several areas.

Nontechnical successes include the building of such models on a proprietary or commercial basis by several companies, and the successful selling of these models at least locally within some exploration companies.

Fluid-flow modelling: unresolved problems

Despite the successes listed above, fluid-flow modelling today is, in my opinion, characterized more by unfulfilled promises, remaining questions, lingering doubts, and a sceptical base of potential clients than by its successes. There seem to have been relatively few successful exploration applications of fluid-flow modelling, and most applications seem to be more in the form of special or *ex post facto* studies than for routine exploration purposes. A major problem with fluid-flow models, in my opinion, is that the existing ones have been characterized as much by what they don't do as what they do. The sections below outline some of the technical and non-technical hurdles to be cleared in the future if fluid-flow modelling is to make a significant impact on exploration.

Selling the technology to explorationists

In contrast to 1-D maturity modelling, 2-D fluid-flow modelling has been very poorly sold to the exploration industry. Consequently, it is at most only a minor contributor to exploration in most areas. Fluid-flow modelling is mainly utilized by specialists for special studies, rather than routinely. It is used fairly intensively in certain countries like Norway that actively promote 2-D modelling, and occasionally in mature exploration areas like the North Sea or Mahakam Delta, where calibration data are abundant. A common perception within the exploration community is that the technology is of little value in frontier areas, due to the lack of data and resulting large uncertainties in input parameters. In any case, a number of important technical improvements must be made before fluid-flow modelling can be sold successfully within the exploration community. Some of these issues are discussed briefly below.

Tectonic complexity

A major weakness of nearly all 2-D and 3-D models is their inability to handle tectonic complexity in a realistic way. Most 2-D models handle normal faulting in at least a rudimentary fashion, but some do not deal at all with reverse faulting. Diapirism, intrusions, and other similar events are not handled. Sections being modelled cannot normally be lengthened or compressed. Faults are only emplaced where the modeller designates, and their permeability characteristics must be specified by the modeller. Features below seismic resolution, such as minor faults or fractures, are normally not simulated, since their presence is usually unknown to the modeller. In short, any tectonic complexity that can be included in fluid-flow models must be entered manually by the modeller; models do not include any predictive capabilities in this regard.

In order for fluid-flow models to be of great value for exploration, it is absolutely necessary that future 2-D and 3-D models include at least two types of tectonic complexity. First, they must predict tectonic fracturing and fracture permeability (Larson *et al.* 1993). Second, they must be able to accommodate geometric changes associated with both normal and reverse faulting, including extension and compression of sections.

Fluid flow through non-matrix permeability

Existing fluid-flow models depend almost exclusively on matrix permeability to transport fluids. In reality, we know that a large fraction of total fluid flow, including both heat transfer and hydrocarbon migration, occurs through non-matrix permeability in the form of fractures and, less commonly, faults. By not considering non-matrix permeability, existing fluid-flow models may grossly misinterpret patterns of fluid flow (see Larson *et al.* 1993).

Our main problems in modelling fluid flow through non-matrix permeability are knowing where such permeability exists, and what values the permeability has through time. Larson *et al.* (1993) have proposed a way to predict the occurrence of tectonic fracturing, as well as to estimate fracture permeability through time. Improvements on their simple scheme can undoubtedly be made, but inclusion of some model for predicting fracturing and fracture permeability is essential for any fluid-flow model that will be used in disturbed areas.

System permeabilities

Most of the data we use in assigning permeabilities to rock units in fluid-flow modelling come from measurements of hand samples, and most of those measurements are made on reservoir rocks. We therefore have a dearth of data on non-reservoir formations, including source rocks and seals. Moreover, the data we do have for a given formation are usually biased toward high-permeability samples. Finally, permeabilities measured on hand samples invariably exclude non-matrix permeability.

These weaknesses are very serious. Most fluid flows through high-permeability pathways, so we must know the maximum permeabilities in a system in order to judge the system permeability. Measurements made on hand samples will seldom give us this information (Waples 1991). Secondly, system permeabilities are very dependent on the geometrical relationships among the various lithologies. Data from hand samples normally do not give us this information. Finally, permeability anisotropies in the subsurface may be huge, with differences between lateral and vertical permeabilities up to at least five orders of magnitude, largely due to lithological anisotropies (e.g. Burrus *et al.* 1994; Massonat & Manisse 1994; Norris *et al.* 1994).

If we are to be successful in predicting pressures and directions and magnitudes of fluid flow, we must be able to predict system permeabilities much more accurately. We will have to learn more about scaling effects in matrix permeability in going from hand-size specimens to formation scale. We will have to find ways to consider non-matrix permeability and better ways to characterize the internal character of rock units in order to determine the magnitude of permeability anisotropies.

If we have enough pressure data, we can determine system permeabilities for our models empirically, and some of the weaknesses discussed here can be overcome. However, pressure data may be lacking or inadequate in many areas where we may wish to execute modelling studies. In such cases we will have to estimate system permeabilities based on an analysis of geological factors such as depositional facies, and on scaling theory.

Pressure

A major selling point of fluid-flow models is their ability to model known overpressure and predict pressures in undrilled areas. Obviously, the ability to consistently and correctly predict pressures is critically dependent upon including all processes relevant to pressure in the model, and providing accurate values of the critical parameters. The previous sections have made it clear that I am sceptical as to whether existing fluid-flow models meet these criteria. While it is true that where pressure data are available, a fluid-flow model can always be optimized to give the observed pressures, this success does not prove that the optimized model is correct. For example, if the main source of permeability is fractures, but the model does not consider fracture permeability, the correct pressures can only be achieved by assigning an inordinately high matrix permeability. While this answer may satisfy our present-day data, it may give very erroneous answers in two related situations:

(1) in calculating pressures and fluid flow at other times, especially prior to the fracturing, when the erroneous model will probably under-estimate pressure and over-estimate fluid flow; and
(2) in other areas today where the lithology may be similar but the tectonic history (and fracture permeability) may be different.

Data entry and manipulation

A major barrier to the execution of multi-dimensional modelling studies is the time, knowledge. and money required to obtain, enter, and edit all the geologic data necessary to run 2-D or especially 3-D simulations. If fluid-flow modelling is to have a significant impact on exploration, it is obvious that data entry and manipulation must be made easier.

Data entry can be made easier principally by linking fluid-flow models to databases that already exist. Since geological data come in a variety of formats, fluid-flow models will have to be able to read data in tabular form, from maps of various kinds, from cross-sections, and from various well-profile formats. An alternative approach would be for some company to develop a universal data manipulator that is able to reduce data from a wide variety of sources to a standard format which could be read by any modelling software program.

Data manipulation within the program is also important. If a model is acquiring data from a variety of sources in a variety of formats, the specific data will have to be properly stored in a 2-D or 3-D network. In addition, data from different sources will inevitably have internal inconsistencies. Deterministic models require one and only one value for each parameter at each node at each time. If a point is over-constrained or if the value is impossible, incorrect answers will be obtained. A software program must therefore be able to recognize illegal data and alert the modeller to the problem. The program should ideally prompt the modeller toward the correct value by displaying such information as nearby values of the same or relevant parameters (in tabular form or as maps or cross-sections, as requested by the modeller). Automatic correction of errors is probably not feasible, since it could lead to serious mistakes. In the distant future we may even see intelligent programs that recognize not only impossible values, but also unlikely ones that should be scrutinized by the modeller.

Optimization

The problems with optimization noted for l-D modelling are also relevant to 2-D modelling. However, optimization in 2-D or 3-D is usually considerably more difficult and time consuming than for 1-D modelling, not only because one is working with many more data points and additional types of data (e.g. permeabilities and pressures), but also because achieving a fully optimized and geologically reasonable 2-D or 3-D model is conceptually and logistically more difficult.

I personally believe that optimization of 2-D or 3-D models should be based heavily on the philosophy and technology of 1-D optimization. In such a scheme the dataset would first be optimized in a coarse fashion using a series of 1-D simulations where optimizing data exist. Once a satisfactory regional model has been developed, the final optimization would be performed in 2-D or 3-D.

This point, however, is controversial. If one is going to perform much of the 2-D optimization in

1-D. several requirements must be fulfilled. First, the 2-D-modelling software must include the capability of functioning in 1-D, or else must be fully compatible with a 1-D program. Fully compatible means not only in terms of importing and exporting data, but also in the mathematical algorithms for physical processes (e.g. porosity reduction, thermal calculations) and in the structure of the calculations (modular or fully coupled). If the 1-D and 2-D models are not fully compatible, the discrepancies between results for the two systems may be large enough to negate the benefits of performing 1-D optimization.

Moreover, the greatest benefits of using 1-D optimization will accrue if the 1-D model is modular, since optimization is faster. Here we encounter a major problem, however. Although modular models appear to me to be superior for 1-D modelling, they are clearly inferior for 2-D or 3-D modelling, where the feedback interactions (e.g. fluid flow linked with pressure and temperature) are much more important. Linking a modular 1-D system with a fully coupled multidimensional model is probably not a satisfactory solution. Therefore, one option is to link a fully coupled 1-D model with a fully coupled multidimensional model.

Another option, which I personally prefer but which is not currently under consideration by model developers, is to build models based on explicit calculation methods. Although such calculation methods have been largely rejected by model developers, I believe that there is a way to overcome the stability problems that have been associated with them (see Waples & Couples, 1998). Such models would combine the best aspects of fully coupled models (rapid and comprehensive feedback among the various processes) with those of modular models (calculation speed). The calculation structure would probably be slightly different for the 1-D and multi-dimensional versions, but they would be compatible.

Regardless of which model structure one chooses, 1-D modelling will never give precisely the same results as 2-D modelling, because 2-D models permit lateral flow of fluids and heat. Therefore, 1-D optimization does not give a final answer about input values for 2-D modelling, but rather suggests values that, in most cases, will be close to the correct ones. Using these values obtained from 1-D modelling, one can then fine-tune the 2-D model until a fully optimized 2-D model is obtained. Thus in my opinion proper 2-D optimization should use 1-D modelling as the main tool for coarse tuning, with final optimization performed in 2-D.

Simulation speed

Simulation speed is an issue that raises its head in two distinct areas. Speed is particularly important during optimization, where numerous runs must be made. We have already discussed two possible ways of solving this problem: by relying heavily on the much-faster 1-D optimization for the coarse tuning, and by building models that utilize an explicit calculation method and that may be much faster.

Speed is considerably less important during final simulations once optimization is complete. If we know we must perform only one more simulation, we can be more tolerant of slow simulation speeds. However, I think the maximum simulation times should still be no more than a few hours, and preferably considerably less.

A major hurdle to performing rapid simulations in fluid-flow models is dealing with multiphase flow. Maintaining mathematical stability in such systems using conventional implicit calculation methods requires short time steps, and hence requires long simulation times. Limiting the number of fluid phases may mean sacrificing important scientific information for the sake of speed, and thus is not a very acceptable solution. The explicit calculation scheme mentioned above may be able to solve these problems.

While it is true that computer speeds are increasing at a fantastic rate, and that some issues of calculation speed that were important only a few years ago have now disappeared, speed remains an issue. As computer speeds increase our appetites increase, so that we will always remain somewhat dissatisfied. We will want to include new types of calculations, as well as refining our calculation intervals (shorter time steps and/or smaller calculation cells). Thus computer speed will always remain to some degree an issue of concern to fluid-flow modellers.

The issue of simulation speed is not simply one of how long a given run takes, or how long it takes to perform a study. If simulations are rapid, fluid-flow modelling will have much greater appeal for explorationists. I therefore consider improving simulation speed a crucial step in selling fluid-flow modelling in the exploration community.

Communication with reservoir models

There has been considerable discussion within the software and modelling industries about the possibility of linking fluid-flow models with reservoir simulators. To me, however, this seems like an odd marriage: the objectives of exploration and production are quite different, and many of the scientific and technical issues in the two types of models are also different. A vendor who provides a fully integrated modelling system for exploration and production may derive a competitive advantage, although in my opinion this assumption remains unproved. It is likely that such link-ups will be established in the future, but their significance, both for applications and for developments of new technologies necessary to make the linkup successful, is still speculative.

Summary

We have been highly successful in building usable, useful, and reliable 1-D maturity models, and in convincing explorationists to use them. Development of convenient, universal software played an instrumental role in the acceptance of 1-D maturity modelling by the exploration community. The remaining weaknesses in most 1-D models, while serious, can be resolved through a few years of intensive work in certain specific areas. Of particular importance are improvements in training for modellers, linkage with geological theory on such phenomena as thermal histories of basins, quality of data used in optimization, quality of kinetic parameters for non-standard kerogens and for hydrocarbon cracking, models for expulsion, and ways of performing optimization.

2-D and 3-D fluid-flow modelling, in contrast, finds itself today largely in the position occupied by 1-D maturity modelling more than a decade ago. The concept of fluid-flow modelling has not yet been adequately sold to explorationists, largely because existing software is inadequate to meet the real needs of exploration personnel. As was the case with 1-D modelling, development of appropriate software will probably have to precede widespread popularization of fluid-flow modelling. To improve existing 2-D software, two major areas must be addressed: making models more comprehensive by including all relevant geological phenomena, and making software easier to use.

Geological phenomena that need further attention in 2-D models include tectonic complexity, predicting the occurrence of non-matrix permeability and fluid flow through it, the relationship between values for various parameters measured on hand samples and those appropriate for modelling systems (scaling effects), and issues of anisotropy and heterogeneity. Issues related to the software itself are the structure of the computational system and its relationship to optimization, data entry and manipulation, simulation speed, and selling fluid-flow modelling successfully to explorationists.

I thank J. Burrus and S. Larter for helpful and encouraging comments that improved the manuscript. They do not necessarily agree with all my comments, however.

References

BEHAR, F., UNGERER, P., AUDIBERT, A. & VILLALBA, M. 1988. Experimental study and kinetic modelling of crude oil pyrolysis in relation with enhanced oil recovery processes. *In:* MEYER, R. F. & WIGGINS, E. J. (eds) *Fourth Unitar/UNDP International Conference on Heavy Crude and Tar Sands.* 747–759.

BURNHAM, A. K. 1989. A simple kinetic model of petroleum formation and cracking. *Lawrence Livermore Laboratory Report UCID-21665.* Lawrence Livermore National Laboratory.

—— & BRAUN, R. L. 1990. Development of a detailed model of petroleum formation, destruction, and expulsion from lacustrine and marine source rocks. *In:* DURAND, B. & BEHAR, F. (eds) *Advances in Organic Geochemistry 1989.* Pergamon, Oxford, 27–39.

BURRUS, J., BROSSE, E., DE CHOPPIN, J. & GROSJEAN, Y. 1994. Interactions between tectonism, thermal history, and paleohydrology in the Mahakam Delta, Indonesia: model results, petroleum consequences. *Bulletin of the American Association of Petroleum Geologists*, **78**, 1136.

CULL, J. P. & CONLEY, D. 1983. Geothermal gradients and heat flow in Australian sedimentary basins. *BMR Journal of Australian Geology and Geophysics*, **8**, 329–337.

—— & DENHAM, D. 1979. Regional variations in Australian heat flow. *BMR Journal of Australian Geology and Geophysics*, **4**, 1–13.

DEMING, D. 1994. Overburden rock, temperature, and heat flow. *In:* MAGOON, L. B. & DOW, W. G. (eds) *The Petroleum System — from Source to Trap.* AAPG Memoir 60, 165–186.

HERMANRUD, C., CAO, S. & LERCHE, I. 1990. Estimates of virgin rock temperature derived from BHT measurements: bias and error. *Geophysics*, **55**, 924–931.

——, LERCHE, I. & MEISINGSET, K. K. 1991. Determination of virgin rock temperature from drillstem tests. *Journal of Petroleum Technology*, **43**, 1126–1131.

HULEN, J. B., GOFF, F., ROSS, J. R., BORTZ, L. C. & BERESKIN, S. R. 1994. Geology and geothermal origin of Grant Canyon and Bacon flat oil fields, Railroad Valley, Nevada. *Bulletin of the American Association of Petroleum Geologists*, **78**, 596–623.

KUO, L.-C. & MICHAEL, G. E. 1994. A multicomponent oil-cracking kinetics model for modelling preservation and composition of reservoired oils. *Organic Geochemistry*, **21**, 911–925.

LARSON, K. W., WAPLES, D. W., HAN FU & KODAMA, K. 1993. Predicting tectonic fractures and fluid flow through fractures in basin modelling *In:* DORÉ, A. G. *et al.* (eds) *Basin Modelling: Advances and Applications.* Elsevier, Amsterdam, 373–383.

LO, H. B. 1993. Correction criteria for the suppression of vitrinite reflectance in hydrogen-rich kerogens: preliminary guidelines. *Organic Geochemistry*, **20**, 653–657.

LOPATIN, N. V. 1971. Time and temperature as factors in coalification. *Izvestiya Akademii Nauk USSR, Seriya Geologicheskaya*, **3**, 95–106 [in Russian].

LUO, M., WOOD, J. R. & CATHLES, L. M., 1994. Prediction of thermal conductivity in reservoir rocks using fabric theory. *Journal of Applied Geophysics*, **32**, 321–334.

MASSONAT, G. J. & MANISSE, E. 1994. Evaluation of vertical permeability anisotropy in fractured reservoirs. *Bulletin of the American Association of Petroleum Geologists*, **78**, 1154.

MULLER, D. S., LEROY, R. A., MARZI, R. W, PASLEY, W. A., STONER, D. J. & WAGNER, B. E. 1998. The upper boundary condition in basin modelling. The case for using surface intercept temperature rather than surface temperature for modelling maturity history. Poster presented at the Geological Society Petroleum Group Conference 'The Application of Basin Modelling Techniques to Hydrocarbon Exploration: What Have we Learnt', 1–2 November 1994, London.

NORRIS, R. J., MASSONAT, G. J. & GOMMARD, D. 1994. Horizontal permeability anisotropy developed from the up scaling of anisotropic facies patterns. *Bulletin of the American Association of Petroleum Geogists*, **78**, 1158.

OKUI, A. & WAPLES, D. W. 1993. Relative permeabilities and hydrocarbon expulsion from source rocks. *In:* DORÉ, A. G. *et al.* (eds) *Basin Modelling: Advances and Applications.* Elsevier, Amsterdam, 293–301.

PEPPER, A. S. & CORVI, P. J. 1995. Simple kinetic models of petroleum formation. Part III: Modelling an open system. *Marine and Petroleum Geology*, **12**, 417–452.

PRICE, L. C. 1993. Thermal stability of hvdrocarbons in nature: Limits, evidence, characteristics, and possible controls. *Geochimica et Cosmochimica Acta*, **57**, 3261–3280.

SANDVIK, E. I. & MERCER, J. N. 1990. Primary migration by bulk hydrocarbon flow. *Organic Geochemistry*, **16**, 83–89.

——, YOUNG, W. A. & CURRY, D. J. 1992. Expulsion from hydrocarbon sources: the role of organic absorption. *Organic Geochemistry*, **19**, 77–87.

SCLATER, J. G., JAUPART, C. & GALSON, D. 1980. The heat flow through oceanic and continental crust and the heat loss of the earth. *Reviews of Geophysics and Space Physics*, **18** (1), 269–311.

SWEENEY, J. J. & BURNHAM, A. K. 1990. Evaluation of a simple model of vitrinite reflectance based on chemical kinetics. *Bulletin of the American Association of Petroleum Geologists*, **74**, 1559–1570.

TISSOT, B. 1969. Premières données sur les mécanismes et la cinétique de la formation du pétrole dans les sédiments. *Revue de l'Institut Français du Pétrole*, **24**, 470–501.

TISSOT, B. P., PELET, R. & UNGERER, PH. 1987. Thermal history of sedimentary basins, maturation indices, and kinetics of oil and gas generation. *Bulletin of the American Association of Petroleum Geologists*, **71**, 1445–1466.

UNGERER, P., BEHAR, F., VILLALBA, M., HEUM, O. R. & AUDIBERT, A. 1988. Kinetic modelling of oil cracking. *Organic Geochemistry*, **13**, 857–868.

VERNIK, L. & LANDIS, C. 1996. Elastic anisotropy of source rocks: implications for hydrocarbon generation and primary migration. *Bulletin of the American Association of Petroleum Geologists*, **80**, 531–544.

WAPLES, D. W. 1980. Time and temperature in petroleum formation: application of Lopatin's method to petroleum exploration. *Bulletin of the American Association of Petroleum Geologists*, **64**, 916–926.

—— 1991. Innovative methods for quantifying poorly known parameters necessary for numerical basin modelling. *Journal of the Japanese Association for Petroleum Technology*, **56**, 96–107.

—— 1996. Comments on 'A multicomponent oil-cracking kinetics model for modelling preservation and composition of reservoired oils' by Kuo, L.-C. and Michael, G. E., *Organic Geochemistry*, **24**, 323–325.

—— & COUPLES, G. D. 1998. Some thoughts on porosity reduction — rock mechanics, fluid flow, and pressure. *This volume.*

—— & MAHADIR RAMLY. In press a. A simple statistical method for correcting and standardizing heat flows and subsurface temperatures derived from log and test data. *In:* TEH, G. H. (ed.) *Proceedings Volume Kuala Lumpur 1994 AAPG International Conference — Southeast Asian Basins: Oil and Gas for the 21st Century,* Kuala Lumpur, Geological Society of Malaysia.

—— & MAHADIR RAMLY. In press b. Developing consistent and reasonable kerogen kinetics using Rock-Eval-type pyrolysis. *Geological Society of Malaysia Bulletin.*

WILKINS, R. W. T., WILMSHURST, J. R., RUSSELL, N. J., HLADKY, G., ELLACOFF, M. V. & BUCKINGHAM, C. 1992. Fluorescence alteration and the suppression of vitrinite reflectance. *Organic Geochemistry*, **18**, 629–640.

——, ——, HLADKY, G., ELLACOTT, M. V. & BUCKINGHAM, C. P. 1995. Should fluorescence alteration replace vitrinite reflectance as a major tool for thermal maturity determination in oil exploration? *Organic Geochemistry,* **22**, 191–209.

Compaction — the great unknown in basin modelling

M. R. GILES, S. L. INDRELID & D. M. D. JAMES[1]

Shell International Exploration & Production BV, RTS Rijswijk, Volmerlaan 8, 2280 AB Rijswijk, The Netherlands

[1] *Present address: 3 Finedon Hall, Finedon NN9 5NL, UK*

Abstract: Decompaction routines are used in basin modelling packages to calculate sediment thickness and material properties such as thermal conductivity. However, compaction in nature is dependent on initial porosity, composition, and effective stress, and a considerable range of porosity–depth trends exists. Simple thermal modelling demonstrates that significant uncertainties (up to VRE ± 0.5) arise in predicted maturities due to this variation. Furthermore, the validity of using porosity loss as a measure of compaction is questionable because changes in solid volume can occur. Chemical reaction may increase or decrease porosity without changing sediment thickness, although an apparently smooth transition occurs from dominantly mechanical processes of porosity loss (e.g. grain rearrangement) at shallow levels to dominantly chemical processes (e.g. grain dissolution/cementation) at depth. That compaction and porosity loss are processes dependent upon effective stress, time and temperature is illustrated by the observation of overpressuring in the subsurface, comparison of experimental and natural compaction rates and analysis of porosity–depth trends for sediments of different ages. Mechanistic models of the processes involved in compaction (e.g. pressure solution) also indicate time dependency. Time-dependent models of compaction can be constructed, but these are difficult to incorporate into basin models as they cannot be run in a simple backstripping mode.

Basin modelling technologies are now widely used to reconstruct thermal history (Royden & Keen 1980; Sclater & Christie 1980; Falvey & Middleton 1981), the evolution of pore fluid pressure (Schneider *et al.* 1993), stratigraphic thickness and geometry (e.g. Lawrence *et al.* 1990), and the generation and migration of hydrocarbons (Hermanrud 1993; Sylta 1993). Common to all these applications is a need to calculate the thickness of a sedimentary layer through time and to estimate its porosity. The former is necessary so that the depth of an individual horizon is known through time, whereas the latter is needed because porosity is a basic property used to calculate other variables such as thermal conductivity, bulk sediment heat capacity, permeability etc. Thus porosity-loss models impact virtually all areas of basin modelling from subsidence analysis through to fluid migration models.

Here we associate the term 'compaction' with change in sediment dimensions during burial. Commonly, compaction is considered to be the largely irreversible reduction in sediment thickness as a result of vertical load. However, in compressive regimes it is also necessary to account for lateral compaction (i.e. bulk shortening). We thus prefer to define compaction as the three-dimensional strain, rather than the vertical change in sediment thickness. Compaction defined in this way is not the same as porosity loss, which may be due to increase in solid volume (e.g. by introduction of cement) as well as to volume strains.

Theoretical background

Porosity can be defined as the volume fraction of fluids present in a bulk volume of sediment,

$$\phi = \frac{V_f}{V_b} = 1 - \frac{V_s}{V_b} \tag{1}$$

where V_f is the volume of fluid, V_s is the volume of solids, and V_b is the bulk volume. Compaction, defined above as the change in the bulk volume (dV_b), is then given by:

$$dV_b = dV_f + dV_s \quad . \tag{2}$$

By assuming that solid volume remains constant, the change in porosity can be related to the bulk volumetric strain $\frac{dV_b}{V_b}$ by differentiating and rearranging equation (1):

$$d\phi = (1 - \phi)\left(\frac{dV_b}{V_b}\right)_{V_s} \quad . \tag{3}$$

GILES, M. R., INDRELID, S. L. & JAMES, D. M. D. 1998. Compaction — the great unknown in basin modelling. *In*: DÜPPENBECKER, S. J. & ILIFFE, J. E. (eds) *Basin Modelling: Practice and Progress*. Geological Society, London, Special Publications, **141**, 15–43.

In sedimentary basins, sediment loading initially results in a vertical reduction in sediment thickness. Vertical strain continues during burial, but in some cases the principal stress axis later rotates (for example in a compressive regime) and hence horizontal shortening can also occur. If the solid volume (V_s) remains constant the change in porosity is dependent only on the change in sediment thickness (vertical strain). The validity of this important assumption will be explored later.

Decompaction (Watts & Ryan 1976; van Hinte 1978; Gallagher 1989) — the reconstruction of sediment thickness back through time — relies totally on the assumptions of constant solid volume and vertical compaction. The following methodology is applied (Fig. 1): The thickness of solids (H_s) can be defined as the thickness of sediment layer ($z_{base}-z_{top}$) minus the volume of porosity in a unit area. Hence, at present day ($t = 0$):

$$H_s = z_{base} - z_{top} - \int_{z_{top}}^{z_{base}} \phi_{(z,\, t=0)} dz \quad . \tag{4}$$

For a given time ($t = i$) between the deposition of the layer and the present day, the thickness of the layer ($z_{base}{}^{i} - z_{top}{}^{i}$) is calculated by adding the pore volume (for a unit area) predicted from the porosity – depth law to the (constant) thickness of solids (H_s):

$$z_{base}{}^{i} - z_{top}{}^{i} = H_s + \int_{z_{top}{}^{i}}^{z_{base}{}^{i}} \phi_{(z,\, t=i)} dz \quad . \tag{5}$$

This equation can be solved by iteration to find the depth to the base of the layer at time i ($z_{base}{}^{i}$) if the depth at the top ($z_{top}{}^{i}$) is known. Hence to decompact a stratigraphic column the top layer ($z_{top}{}^{i} = 0$) must be decompacted first, and then the base of this layer can be input as the top of the underlying unit, and so on sequentially down through the sediment pile.

To apply the above method to compute the decompacted sediment thicknesses for a sequence of sediments, we need to be able to specify how the porosity changes with depth. Note that the porosity need not be dependent on depth *per se*, but instead can be a function of a number of variables each of which varies smoothly with depth. Because it is incorrect to predict porosity from depth at a given time in the past without first decompacting to find the true depth at that time, the two have to be solved simultaneously. This can be done iteratively, using any (necessarily non-unique) porosity–depth function which is invariant with time and satisfies the present day observations. In practice the porosity–depth relationship needs to be relatively

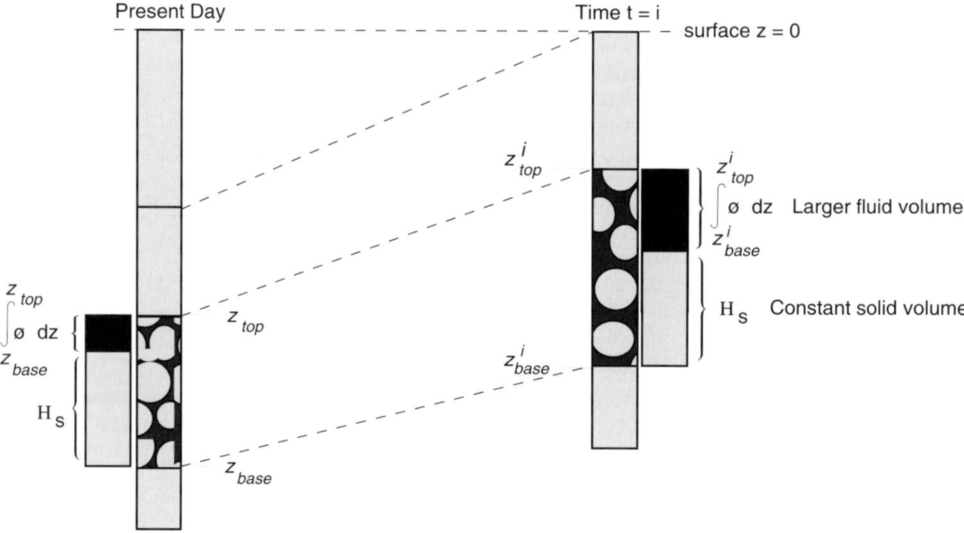

Fig. 1. Standard decompaction methodology. The volume of grains (which can be simply derived from the present day layer thickness and porosity) is assumed to remain constant. The porosity at a given time in the past is added on to give the total thickness of the layer at that time. Past porosity is most commonly determined using a porosity–depth curve. Decompaction is generally performed simultaneously with back-stripping to obtain corrected sediment thicknesses, and hence depths, in the past.

simple in order for a solution to be obtained. In contrast the use of a time variant porosity–depth model would require repeated forward runs of the whole burial process to find a (non-unique) match between the present day predicted and observed thicknesses.

If porosity loss is modelled as a more complex process, involving variables such as temperature or effective stress which themselves depend on the porosity, then the decompaction of sediments cannot be treated as a separate exercise. Instead it is necessary to solve for the sediment thicknesses simultaneously with the other variables which control porosity. For example, if porosity loss is modelled as a function of temperature, then the temperature history of the sediments is also required in order to calculate the sediment thicknesses, but the temperature is a function of the thermal conductivity, which is in turn dependent on the porosity. Thus the heat flow and compaction equations would have to be solved simultaneously. As such equations are generally solved forward in time, there is no guarantee that the calculated sediment thicknesses and porosities would match those actually measured. Hence multiple iterations would be necessary.

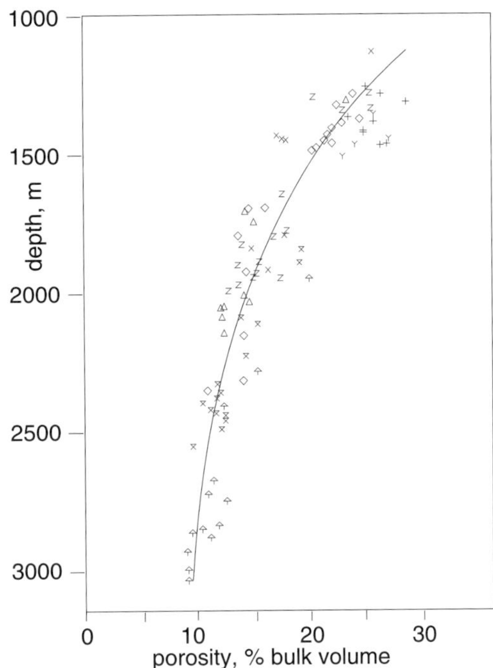

Fig. 2. Example of a well calibrated porosity–depth curve from Sarawak. The curve was constructed using porosity data from logs from channel sandstones in a group of eight neighbouring wells. Each well is indicated by a different symbol.

Porosity–depth relationships

Empirical observations

The observation that porosity commonly decreases smoothly with depth in continuously subsiding basins (Fig. 2) permits the construction of simple porosity–depth functions. Such functions are widely used, both in decompaction algorithms, and as a first step in estimating rock properties such as permeability and thermal conductivity. These functions are purely empirical: they have no physical foundation because no compaction mechanism links porosity and depth simply, directly and universally. Furthermore, a large number of these curves are available (Fig. 3).

Porosity loss with increasing depth is largely irreversible (Lang 1978; Serra 1984; Giles 1996), with the exception of the creation of enhanced porosity by grain dissolution (Giles & de Boer 1990), and hence is commonly determined by the maximum burial depth rather than the present day depth: uplifted areas tend to exhibit anomalously low porosities at a given depth compared to continuously buried areas.

Construction of a reliable porosity–depth trend thus requires a large number of porosity measurements from sediments which are now at their maximum depth of burial, distributed over a wide depth range. Furthermore, porosity is also dependent on initial composition (Nagtegaal 1978), grain size, shape and sorting (Beard & Weyl 1973; Atkins & McBride 1992), and these variables should also be taken into account when selecting porosity values for inclusion in the analysis. Even clean, quartz-rich sands from the same area can show differing porosity–depth trends, if their depositional environments, and hence sorting, are not identical (Curves 7 & 8, Fig. 3a). Furthermore, strained quartz, such as that sourced from an accretionary prism, is more susceptible to dissolution, and hence chemical compaction, than an unstrained quartz sourced from a stable craton (Kamb, 1961; Paterson, 1973).

A number of porosity–depth functions have been proposed for fitting field data. The simplest of these is a linear decrease in porosity with depth. Simple linear trends can fit porosity–depth data for sandstones over a limited depth interval (see Fig. 3a, Etive trend). However, a linear trend implies negative porosities below a certain depth and does not adequately represent the curved trends seen at shallow depths (Fig. 2).

The observation that porosity decreases more gradually at greater depth (e.g. Fig. 2) led Rubey &

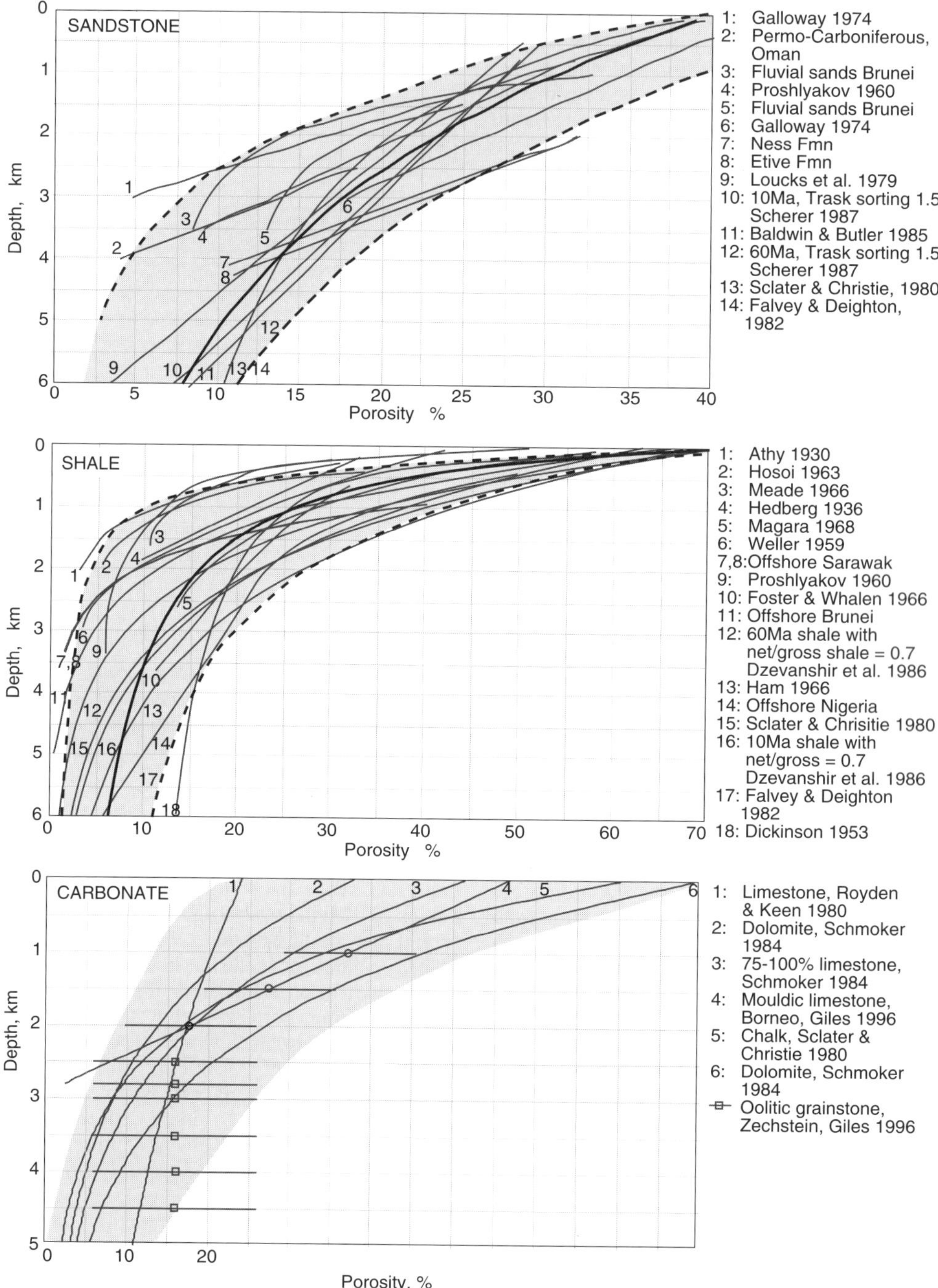

Fig. 3. Porosity versus depth trends for (a) sandstones, (b) shales and (c) carbonates. A broad range of porosity–depth behaviours exist, depending on the initial porosity (sorting, grain size), composition, age, pressure regime and temperature gradient. In general, porosities are higher than average in overpressured areas, and lower than average in areas where uplift has occurred, or where there is a high geothermal gradient and/or unstable mineralogy, although both pressure and temperature may have varied through time. The maximum, minimum, and average curves, used to investigate the effect of the choice of porosity model on thermal modelling results are in upper and lower dotted curves and heavy solid line, respectively.

Hubbert (1959) to propose an exponential relationship:

$$\phi = \phi_o e^{-\lambda_{exp} Z} \qquad (6)$$

where λ_{exp} is a constant, generally known as the compaction coefficient, and ϕ_o is the initial (surface) porosity. Sclater & Christie (1980) used this exponential function to fit porosity data from the North Sea. This model and the parameters used by Sclater & Christie (1980) have subsequently become the most commonly used, both in textbooks and as defaults in many basin modelling programs.

Falvey & Middleton (1981) noted that an exponential porosity–depth relationship does not provide a good fit to the data at high porosities. As an alternative, they proposed a function based on the assumption that a differential change in porosity is proportional to the differential change in load:

$$\phi = \frac{\phi_o}{(1 = \phi_o \lambda_{fm} z)} \qquad (7)$$

where λ_{fm} is again a constant of proportionality. This function reproduces the general form of the porosity–depth curve, but also predicts that some porosity is maintained down to depths of greater than 10 km. A similar approach was taken by Baldwin & Butler (1985) who chose to represent solidity $(1 - \phi)$ purely empirically as a power-law function of depth.

The importance of calibrating the porosity–depth curve for a given lithology in a given area (i.e. finding the parameters for the function which best fit the data) can be illustrated by comparing the range of porosity–depth trends exhibited by sandstones and shales (Fig. 3) from a variety of settings around the world. Different data sets may exhibit differences in porosity of more than 20% at the same depth. A large number of possible causes may explain the differences between porosity - depth curves for the same lithology, including differences in composition (Nagtegaal 1978), age (Maxwell 1960), geothermal gradient (Galloway, 1974), overpressure (Dickinson 1953), and initial porosity and packing (Beard & Weyl 1973, Atkins & McBride 1992). In general, porosity is preserved better at a given depth in areas with mature mineralogies, a low thermal gradient, rapid sedimentation rates, a young age, and where overpressures are developed. Conversely, porosity is generally destroyed rapidly in hot basins with an immature mineraolgy. Some data sets are also derived from areas where the sedimentary column has experienced uplift, so that the sediments are not presently at their maximum depth of burial. However, even within the same basin significant differences may exist. For example, comparing the exponential relationship which Sclater & Christie (1980) derived from North Sea sandstone data (Curve 13, Fig. 3a) with another North Sea data set (Curve 7) shows that the porosity–depth trends for sandstones from the same basin with a similar provenance may diverge significantly. The potential cause(s) of the difference between any two porosity - depth curves are discussed below. However, the influence of many of the listed factors is currently unquantified, so that it is not possible to predict the best porosity–depth curve to be used in modelling a given formation in a given location without constraint from local data. Furthermore, although such functions adequately reproduce the general trends in porosity with depth at *some* depths, none of them can reproduce porosity–depth data at *all* depths.

Estimates for the initial porosity (ϕ_o) of the rock prior to burial are well constrained for sandstones from the work of Tickell & Hiatt 1938; Krumbein & Monk 1942; Pryor 1973; Beard & Weyl 1973; Schenk 1983; and, most recently, Atkins & McBride 1992. Sufficient data exist to allow for differences in the initial porosity and packing of sandstones due to facies variations. The same is not true of other lithologies: there are few compilations available for carbonates (see Enos & Sawatsky 1981) and only a few general references for mudrocks (Burst 1969; Greensmith 1978). Porosities from shallow well data are notoriously unreliable because of the effect of large borehole size and rugosity on porosity estimates from wireline logs, and the lack of core material.

The alternative to using an assumed value for the initial porosity of these lithologies is to extrapolate the trend from the calibration data set to the surface. However, the initial porosity estimate is then dependent on the choice of porosity–depth function: consider the range of initial porosities which would be extrapolated from the curves shown in Fig. 3. Similarly, modelling of deep kitchen areas requires downward extrapolation of the function as there is seldom much data from below 4 km with which to calibrate the deeper part of the curves.

Measurement of porosity

Porosity data for sandstones and porous carbonates are relatively easily obtained either from direct measurements on core plugs or from well logs. For those lithologies, the large amount of core data available means that the relationships between porosity–depth, porosity–permeability etc. are often quite well calibrated, at least in individual wells.

For mudstones and muddy carbonates this is not the case. Core samples are usually fewer in number and the interpretation of petrophysical measurements may be ambiguous. Mudrocks contain free water in the pore space, but also contain bound water within the clay lattice and as more loosely electrostatically bound water associated with clay particles. Different petrophysical tools respond to this water in different ways. For instance, the neutron tool responds primarily to the hydrogen ion content, whereas resistivity tools record electrical conduction by the pore-fluid and grain matrix, and are thus affected by salinity variations. Porosity estimation from a resistivity log requires assumptions about the shale mineralogy, the water composition, and the amount of clay-bound water, as well as knowledge of the temperature. In general an empirical calibration between the porosity and the resistivity is made. Hence, depending on the underlying assumptions and method of calibration, the resistivity log will indicate different porosities.

Similar problems exist with sonic velocity–porosity correlations, which are the most widely used method for deriving the porosity of mudrocks. The Wyllie time average equation (Wyllie *et al.* 1956) is the simplest and most commonly used method for calculating mudstone porosities from sonic velocity. Unfortunately, it is based on laboratory measurements for sandstones, and is known to be significantly in error for shallow sandstones (porosities above 30%, Serra 1984). More complex formulas, such as those of Raymer *et al.* (1984) need to be employed. In deriving porosity from sonic velocity, estimates of critical parameters such as the matrix velocity must be found either by directly correlating porosity with sonic velocity, where such measurements exist, or by estimating parameters from the sonic velocity when the rock has a near zero porosity, or by assuming default values. This last method does not take account of lithological differences.

Caution is thus necessary in constructing porosity trends from petrophysical logs and in comparing trends derived from different data sources. In particular, porosity values obtained by different methods should not be mixed together to form a single calibration curve.

Construction of porosity–depth curves

Porosity–depth or porosity–effective stress relationships are used in basin modelling to calculate subsidence history and properties such as heat capacity, thermal conductivity, and permeability. Therefore, errors in these derived properties can be reduced by providing better calibrated porosity models. Suitable data sources for the construction of porosity–depth curves are provided by porosity data from core plugs or porosity/density measurements from wireline logs. Because many factors may affect the porosity loss behaviour, these sources of variation should be isolated as far as possible. An idealized recipe for the construction of these curves is outlined below.

(1) A large number (100s–1000s) of samples is ideally required. These should be distributed over a wide depth range.
(2) Only facies with similar initial porosities and porosity/depth or effective stress behaviour should be included in the construction of the curve. Other facies should be excluded. Similarly, samples with unusual compositions should be excluded (e.g. do not mix arenites with wackes).
(3) Only samples currently at the maximum effective stress they have experienced should be included (i.e. discard samples which have been uplifted).
(4) Samples from areas with markedly different geothermal gradients should be excluded. A separate trend could be constructed for these.
(5) The samples should be derived from a narrow stratigraphic range to rule out any possible age effects.
(6) Samples from overpressured areas should either be treated separately (as these may have retained porosity), or the model should be formulated in terms of effective stress rather than depth (see below).

A best fit to the data (calibrated curve) can then be obtained. At this stage the outliers should be checked to see if they are assigned to the correct facies, or exhibit signs of an alternative process (e.g. leaching, localized mineralization etc.) which explains their anomalous porosity. If there is sufficient reason, they can be discarded from the analysis.

Finally, after checking the anomalous data, the best fitting porosity–depth curve can be re-derived.

Unfortunately, such porosity–depth trends rely on large amounts of data, and are thus unlikely to be available in the early stages of basin evaluation. Note also that such statistical representations of the data are valid only over the effective stress–depth range of the data, and extrapolation beyond these limits can be dangerous.

Effect on basin modelling results

The ability to reconstruct subsidence through time for the basin fill is fundamental to modelling the time variant profiles of temperature and pore-fluid pressure that are needed to estimate the generation and migration of hydrocarbons. Many basin modelling applications require the use of porosity–depth models at several stages, but the impact of the choice of porosity–depth function on the model output is rarely analysed. In the following section a simple one-dimensional steady - state thermal model has been used to estimate the uncertainty introduced into maturity calculations as a result of the porosity–depth functions employed.

We adopt a pragmatic approach to the wide range of data, in that the porosity–depth curves in Fig. 3 have been used to define both an 'average' porosity–depth trend, and an envelope bounding the range of values. A sufficiently wide range of curves is represented that we may assume that this is a fair summary of the likely variability in porosity–depth functions. These values and uncertainties can then be used to evaluate the potential error which could propagate into a maturity model if the porosity–depth curve used for decompaction and determination of thermal parameters is assumed rather than derived using local calibration data. We model initially the simple case of a 4 km thick shale section continuously deposited from 200 Ma until the present day (Fig. 4).

When the various porosity loss curves are used for the decompaction, there is only a small range in the derived decompacted burial history. This is because the depth at the present day, and the depth at the time of deposition are both constrained. However, a depth difference of several hundred metres may occur in the intervening time: even for the two most common defaults (values from Falvey & Middleton 1981; Sclater & Christie 1980) a difference in depth of around 190 m is predicted, and there is 400 m difference between the depths predicted using the maximum, and minimum shale porosity–depth curves. This difference has little impact in the case where back-stripped tectonic subsidence is used to constrain the beta stretching factor, and hence heat flow history, for a basin with a single phase of extension (Wooller *et al.* 1992), but may be significant in cases where there are multiple phases of extension, or where extension continued over a long period.

Thermal conductivity is generally derived from the modelled porosity according to an equation of the form (Woodside and Messmer, 1961):

$$K_{bulk} = K_{matrix}^{(1-\phi)} K_{fluid}^{\phi} \quad . \tag{8}$$

It is thus highly dependent on the porosity model. Figure 4c shows the dependency of the conductivity on the porosity–depth model. The range in the calculated conductivity is greatest (± 40%) at burial depths of 0.5–1.5 km, and decreases with increasing depth. This error in conductivity is enough to give a significant error in the predicted temperature at depth (a range of ±20 °C at 4 km) for a given heat flow, *if there is no constraint from temperature data available*. For accurate forward modelling of temperatures at depth (for example for use in well planning) where the regional heat flow is known, it is thus important that a locally calibrated porosity–depth curve should be applied, rather than a default model.

More commonly, temperature measurements are available to calibrate the model, and the modeller is trying to create a picture of regional heat-flow history before predicting maturity in a hydrocarbon kitchen area. Even in this case, where well data can be used to calibrate the model results, the present day heat flow is not uniquely determined unless the conductivity and porosity–depth profiles are also calibrated locally. For the case of a 4 km shale section and a measured temperature of 150 °C at 4 km depth, the present day heat flow derived by fitting the measured temperature varies between 50.6 mWm^{-2} (for the most porous shale curve) to 67.5 mWm^{-2} (for the least porous shale curve). Similarly, for a sand section (steadily deposited over 200 Ma) with a measured temperature of 135 °C at 5 km depth, the optimized present day heat flow varies between 58.7 and 78.4 mWm^{-2}.

When the regional heat flow model is applied to an undrilled kitchen area the uncertainties in the heat flow history will propagate through into the predicted maturity of the sediments, despite the fact that the maturity is known at the wells used to calibrate the model. Figure 5 shows that for both the sandstone and shale model sections the uncertainty in the depth to the top of the oil window due to the range of porosity–depth functions applied is approximately ±500 m, and the uncertainty in the depth to the top of the gas window is even larger. Similarly, the error in predicted maturity (VRE) at a given depth (e.g. a source rock horizon) is as much as ±20%. This large uncertainty in the maturity calculation occurs because the effect of the porosity on the depth history (from decompaction) and its effect on the thermal properties work in the same direction: not only is a more porous shale buried more deeply over much of its history, but it is also hotter at the same depth. Naturally, the uncertainty in maturity affects the timing and magnitude of predicted hydrocarbon generation — the error in the timing of peak generation may be as much as 75 Ma (Fig. 4f). Even at a well with reliable temperature data available to calibrate the thermal properties, the timing of peak generation still exhibits a significant uncertainty (±15 Ma) because the generation history depends directly on the decompacted depth of burial as well as temperature history.

In thermal modelling studies, the modeller should be aware of the potential error due to the choice of porosity–depth curve. Such errors could be vital when assessing the hydrocarbon prospectivity of a basin. This is highlighted in an example from Sarawak (Fig. 6) where a real heat flow model

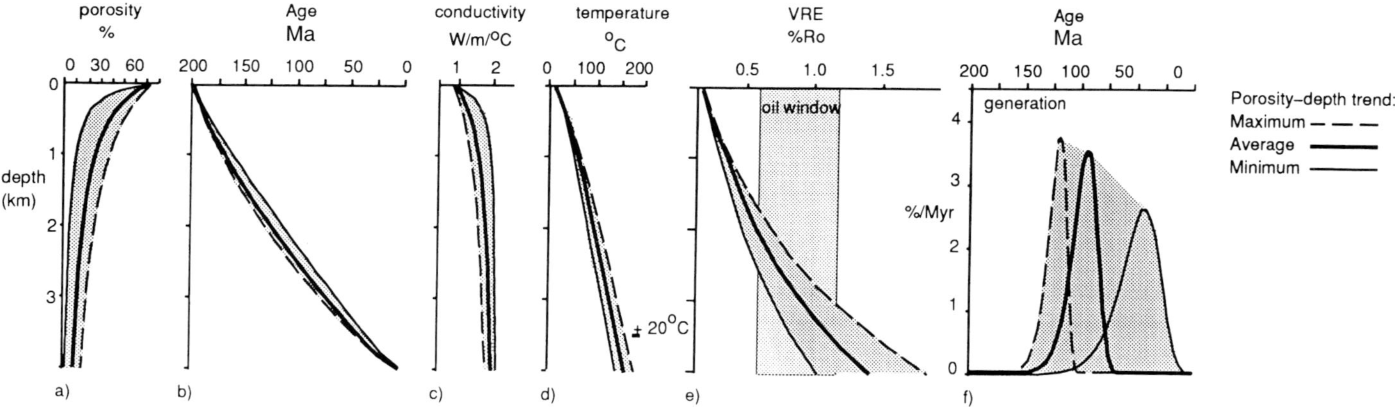

Fig. 4. Effect of the choice of porosity model on thermal modelling. The model is based on a simple stratigraphy: 4 km of shale deposited steadily from 200 Ma to present day. (a) Range of porosity versus depth curves for shale, as in Fig. 3a. (b) Decompacted depth to the deepest horizon, versus time. There is a range of about 400 m in the depth during the middle part of the burial history due to the choice of porosity model. This range could be crucial in determining the time of entering the oil window versus trap formation. (c) Thermal conductivity versus depth. Conductivity increases as the porosity decreases; hence the conductivity varies significantly with the choice of porosity–depth model. (d) Predicted temperatures versus depth (assuming no calibration with measured formation temperatures). The range of thermal conductivities results in a range of predicted temperatures with a total range of 40 °C at 4 km depth. (e) Predicted maturity (VRE) versus depth. The range of solution for the temperature history and depth result in an uncertainty of ± 450 m in the depth to the oil window (VRE = 0.6 %Ro). (f) Predicted generation rate versus time. The timing of peak generation is shifted by the choice of porosity model, again due to the range in decompacted depth and temperature history.

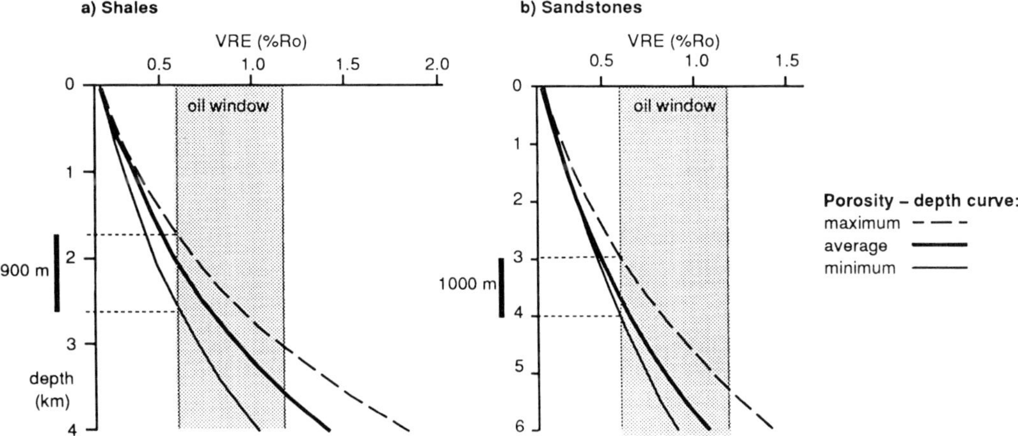

Fig. 5. Range of possible uncertainty in the depth to the oil window resulting from the range of porosity–depth curves. (a) For a shale section — total uncertainty = ± 450 m. (b) For a sandstone section — total uncertainty = ± 500 m.

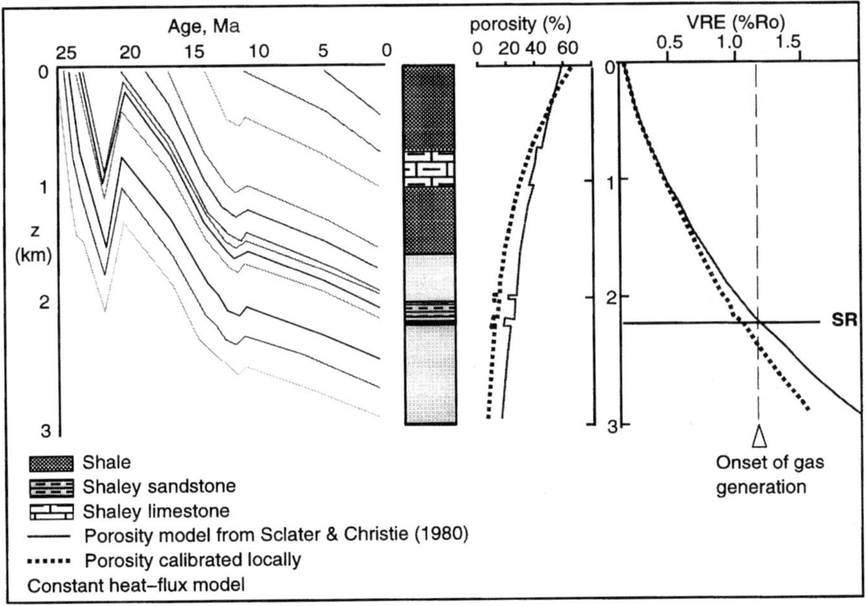

Fig. 6. Example of the effect of the choice of porosity–depth model on predicted maturity for a well from Sarawak. A regional heat flow derived by matching temperature data at near by wells was applied to this stratigraphy in order to assess the hydrocarbon generation potential of an undrilled kitchen area. The coal source rock (SR) at 2.2 km depth is just gas mature if default porosity coefficients (values from Sclater & Christie 1980) are used, but it is not gas mature if locally calibrated curves for both sandstone and shale porosity loss are applied. If this model were for a real well and could be calibrated against temperature, both models would predict the same temperature history and maturity, although different heat flow values would be assumed in doing this. However, in areas where much older sediments are present and the heat flow may have varied through time, fitting present day temperature does not necessarily constrain the present day maturity.

calibrated at a number of wells has been applied to an undrilled kitchen area. Using the typical defaults for most basin modelling packages (i.e. values from Sclater & Christie 1980), a coal at 2.2 km depth would be predicted to be just gas mature. However, using a locally calibrated porosity–depth curve suggests that the coal is not gas mature.

Thermal modelling thus illustrates that a significant error is possible in assessing maturation if the porosity model is incorrect. Although the errors in deriving the heat flow can, of course, be minimized by calibrating the model against measured temperature and maturity (VR) data at the well locations, the errors from extrapolating this derived heat flow over the area as a whole cannot be reduced. For the case where the lithologies and depths in the kitchen area are not significantly different to those at the well location this error will be minimal as long as the same porosity–depth model is applied in the kitchen as at the calibration well(s). However, if the kitchen area is significantly deeper (as is often the case) and/or the facies present in this area vary from those in the calibration wells, errors from using defaults can be significant. This error can be reduced by using a locally calibrated porosity–depth curve. However, even the use of a locally calibrated curve cannot guarantee that the error due to the porosity model is zero, as the porosity–depth model itself may be inadequate for the area: for example, in areas with a history of overpressuring (see below), changes in pressure through time mean that the present day porosity–depth curve is not representative of the whole history. In areas where there is no constraint from either temperature and/or maturity data, and no locally calibrated porosity–depth curve is available, it is important to quantify the uncertainty in predicted maturity due to the porosity model: for example by running low-side and high-side sensitivity tests.

The choice of porosity model gives rise to another potential error where thermal modelling leads into hydrocarbon-migration modelling or pore fluid pressure modelling, because the porosity is used to calculate the permeability. The simplest, and most commonly applied, relation between porosity and permeability is the Kozeny–Carmen equation (Scheidegger 1974) in which permeability is a function of the cube of porosity and the square of the specific surface. Thus if the porosity estimate is in error, there will be an even greater error in the permeability estimate, which will feed through into calculations of migration rate and fluid pressure.

Porosity–effective stress relationships

Experimental evidence

Triaxial compaction experiments (de Boer, 1975; Giles, unpublished data) aimed at studying the porosity loss behaviour of unconsolidated sands show that porosity loss is primarily a function of the applied effective stress (defined below) and the composition of the grains making up the sand. Figure 7 shows that, for a given lithology there is an approximately linear trend between the log of porosity loss (ϕ/ϕ_o) and the effective stress, which suggests that porosity loss behaviour conforms to an exponential function of effective stress. Exceptions to this rule occur where the sediment contains a high proportion of ductile grains, although this deviation may cease at high effective stresses. Both the presence of clay (the most common ductile grains) and an increase in temperature (Fig. 7b), increase the mechanical compaction rate. The temperature-dependence of the compaction rate (and therefore the compaction coefficient) is relatively high, and thus increased compaction rates in areas with high geothermal gradients may help to explain the scatter observed in Fig. 3. Similarly, abnormally high porosities (for a given depth) are observed in many overpressured zones, where the effective stress is reduced by the high pore fluid pressure.

Experimental and natural data thus clearly illustrate that load rather than depth *per se* is the driving force for compaction. At any point in the subsurface the vertical stress generally depends on the weight of the overlying porous rock, which is partially offset by the isotropic stress or pressure exerted by the pore fluid. These competing stresses lead to the concept of effective stress (Fig. 8), which in its simplest form, commonly called Terzaghi's law (Terzaghi & Peck 1948), may be written as:

$$\sigma_{\text{eff}} = \sigma - P_{\text{f}} \quad (9)$$

where σ is the imposed stress and P_{f} the fluid pressure. The validity of this equation will be discussed in detail later.

Modifications of the exponential function (eqn 6), where depth is replaced by Terzaghi effective stress can be fitted to natural data extremely well. Furthermore, by making porosity reduction a function of effective stress rather than depth, the effects of overpressures on porosity reduction are automatically taken into consideration (Fig. 9). Sclater & Christie (1980) rewrote the porosity–depth equation in terms of Terzaghi's effective stress by simply substituting depth (in eqn 6) by the stress divided by the average density and the acceleration due to gravity ($\boldsymbol{g}$):

$$\phi = \phi_o e^{-\sigma_{\text{eff}}/c} \text{ where } c = \lambda(\bar{\rho}_{\text{b}} - \rho_{\text{w}})\boldsymbol{g} \quad . \quad (10)$$

One of the major problems with this equation is because the average bulk density of the water saturated sediment ($\bar{\rho}_{\text{b}}$) is dominantly a function of

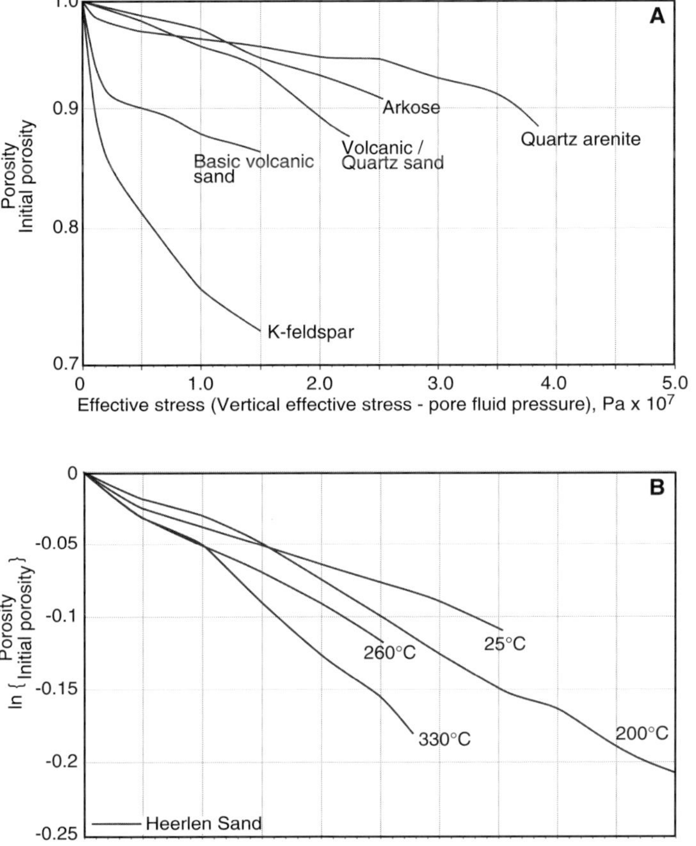

Fig. 7. Experimental compaction of unconsolidated sands. The approximately linear trends indicate an exponential relation between porosity-loss and effective-stress. Data from de Boer (1975) and Giles (unpublished data). (a). Effect of the composition of the sand. (b) Effect of temperature. At higher temperatures porosity loss is more severe for a given level of effective stress.

porosity, and the density of water (ρ_w) is influenced by temperature and salinity; both are variable rather than constant. Furthermore, experimental (de Waal 1986) and natural data suggest that porosity loss and vertical strain are largely irreversible functions of the applied stress, which results in hysteresis (Giles 1996) during loading and unloading cycles (Fig. 10). Although modifications of eqn 6 can model the loading portion of the porosity loss curves, they cannot represent the unloading portion. Thus the use of simple equations such as eqn 10 in decompaction requires that any uplift can be accurately quantified and corrected for, and that the pressures have been hydrostatic for the entire burial history.

The concept of effective stress stems from a consideration of the competing effects of the imposed overburden load forcing the grains together and the pore fluid pressure forcing them apart. Experimental evidence and natural data show that porosity loss can be considered a function of effective stress. However, Terzaghi's law implies that the pore fluid is continuous, and therefore exerts a pressure uniformly over the complete surface of the grains. In consolidated rocks this clearly is not the case. Let us therefore re-examine this law.

Theoretical considerations

The capacity for bulk volume change can be considered to be a function of two variables, the confining pressure (P_c) and the fluid pressure (P_f):

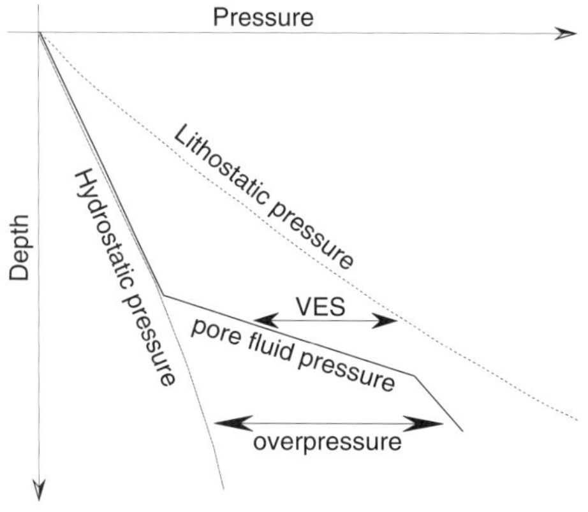

Fig. 8. Definition of vertical effective stress.

$$V_b = f(P_c, P_f) \tag{11}$$

where the confining pressure can be defined in terms of the trace of the stress tensor (σ_{ij}). A differential change in the bulk volume can therefore be written:

$$dV_b = \left(\frac{\partial V_b}{\partial P_c}\right)_{P_f} \left[dP_c - \alpha dP_f\right] \quad ; \tag{12}$$

with:

$$\alpha = \left\{ \left(\frac{\partial V_b}{\partial P_f}\right)_{P_c} \Big/ \left(\frac{\partial V_b}{\partial P_c}\right)_{P_f} \right\} \quad . \tag{13}$$

The term α is commonly known as the poro-elastic coefficient (Berryman 1992) although elasticity is not implied in the above definition. This coefficient is the ratio of the rate of change of bulk volume with fluid pressure at constant confining pressure and the rate of change of bulk volume with confining pressure at constant fluid pressure. Note that α can also be expressed in terms of compressibilities (e.g. Appendix 1). Equation 12 defines a differential effective stress law for bulk volume changes of the form:

$$d\sigma_{eff} = dP_c - \alpha dP_f \quad . \tag{14}$$

Integration of this equation with α equal to 1 gives Terzaghi's law (eqn 9) which has been found experimentally to describe the compaction behaviour of soils and unconsolidated rocks (Skempton 1970). However, experimental data on consolidated rocks imply a smaller poro-elastic coefficient (Berryman 1992). Berryman (1992) was able to show that the poro-elastic coefficient for

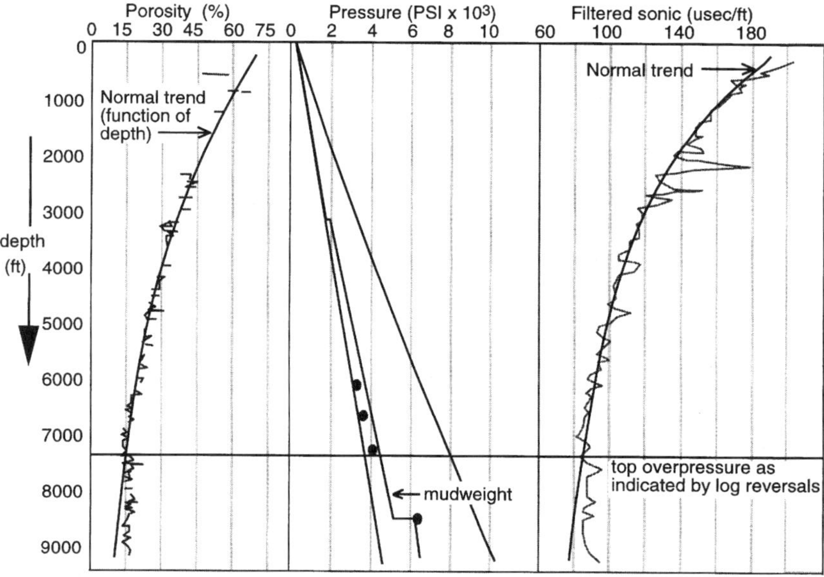

Fig. 9. Porosity - effective stress correlation. Porosity derived from sonic log velocity is shown on the left in grey. Note that this trend exhibits a high frequency lithology effect. A locally derived porosity–depth trend derived from nearby hydrostaticlly pressured wells is shown for comparison (heavy line). Note that, as the top of overpressures is encountered the effective stress starts to fall, and hence the porosity deviates from the hydrostatically pressured trend. A porosity–depth model cannot take account of this deviation.

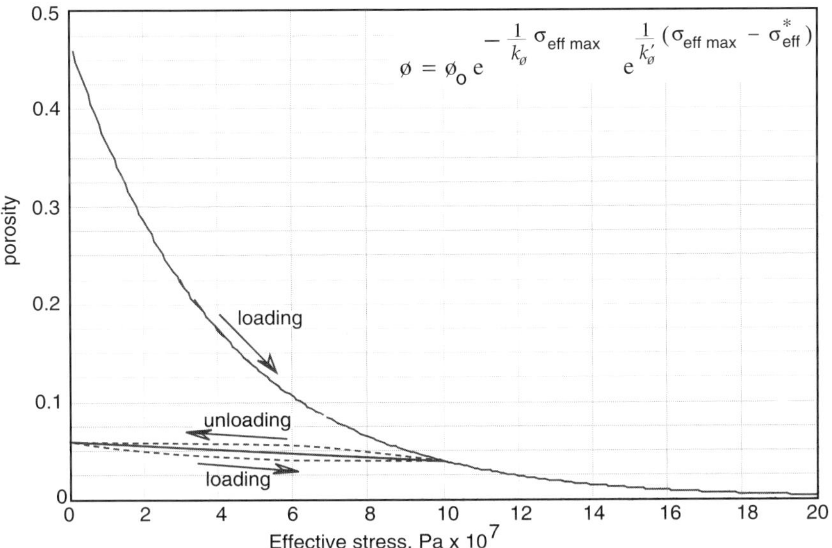

Fig. 10. Porosity hysteresis. During experimental compaction, a small amount of porosity is recovered on unloading (dotted lines). This behaviour is approximated by an equation which has a large term dependent on the maximum effective stress ($\sigma_{eff\text{-}max}$), and a small term dependent on the present day effective stress ($\sigma_{eff}{}^*$).

bulk volume changes lies in the range $\phi \le \alpha \le 1$. Clearly $\alpha = 1$ when $\phi = \phi_o$ (when fluid pressure = confining pressure) and $\alpha = 0$ when $\phi = 0$ (when the lack of pore-space means that pore-fluids cannot influence the stress). Note that the effective stress law for one property (e.g. bulk volume) is different from that of another property (e.g. porosity) although the coefficients may be related (Berryman 1992), as demonstrated in Appendix 1.

Assume for the moment that:

(a) The confining and pore pressures are independent.
(b) The poro-elastic coefficient for porosity change is a constant.
(c) The rate of porosity loss with maximum confining pressure is a constant.
(d) The rate of porosity rebound with decreasing confining stress (unloading) is also a constant.

It is then possible to show that porosity loss may be described by an equation of the form (see derivation in Appendix 1):

$$\phi = \phi_o e^{-\frac{1}{k}\sigma_{\text{max-eff}}} \; e^{\frac{\sigma_{\text{max-eff}} - \sigma_{\text{eff}}}{k'}} \qquad (15)$$

where the effective stress is given by: $\sigma_{eff} = P_c - \alpha' P_f$, k is the compressibility coefficient for the rate of porosity loss with increasing confining pressure, and k' is the compressibility coefficient describing the elastic rebound caused by a reduction in confining pressure (see Appendix 1) (ks are used in preference to λs to emphasise that these are physically meaningful constants, rather than empirically derived values, as in eqs 6 and 7.) The validity of the assumptions used to derive this equation will be examined later.

Note that in this equation the maximum effective stress ($\sigma_{\text{max-eff}}$) is a function of time because the burial path and fluid pressures followed can themselves be expressed as functions of time. When the load is greater than at any time previously (i.e. $\sigma_{\text{max-eff (t)}} = \sigma_{\text{eff}}$) porosity will decrease. During unloading ($\sigma_{\text{max-eff}} > \sigma_{\text{eff}}$) the effective stress is less than the maximum and hence porosity will increase marginally as a result of elastic recovery. As shown in Fig. 10, this equation adequately describes porosity hysteresis during loading and unloading cycles, with the constant k' controlling the amount of elastic recovery of porosity.

The time-dependence of compaction behaviour is not explicitly indicated in this equation. Despite this omission, eqn 15 is a major improvement over porosity–depth functions. It is more physically realistic and it embodies the maximum effective stress history. Furthermore, if any of the parameters within the equation are time variant then this equation can describe the porosity loss through time. For example, because it is a function of effective stress, variations due to overpressure are eliminated.

However, because the effective stress depends on

the load, which in turn depends on the porosity of the overlying sediments, it is impossible to use such an equation for simple (non-iterative) decompaction without knowing the stress history. Note that, in the case where lithostatic and hydrostatic gradients are approximately constant (as they commonly are, very crudely, below ≈1 km in a continuously subsiding basin) eqn 15 can be rewritten in terms of depth and is then analogous to eqn 6.

In deriving eqn 15 the assumption was made that both the poro-elastic coefficient and the compressibility moduli are spatially and temporally constant. Experimental (Fig. 11) and theoretical (Laurent *et al.* 1990; Zimmerman 1991; Berryman 1992) show that this is not the case for the poro-elastic coefficient. The bulk moduli at constant effective stress depend on the consolidation state of the sample, notably the porosity, degree of cementation, and overall composition. The poro-elastic coefficient for bulk volume change varies between 1 (when $\phi = \phi_0$) and 0 (when $\phi = 0$) (Berryman 1992). Thus different porosity loss curves would be derived for different values of the poro-elastic coefficient (Fig. 12a). Rocks with moderate porosity commonly exhibit a value of α of about 0.7 (Berryman 1992). Thus eqn 15 (and also the simplified exponential version, eqn 6) is inappropriate for fitting data with very low porosity or porosity close to the depositional porosity, where α is different from the value at typical subsurface porosities. This explains why many low porosity data deviate significantly from the values predicted by a simple exponential function (Fig. 12b). We know of no rigorously calibrated studies of the porosity loss behaviour at high effective stress from which to derive the correct form of behaviour at low porosities.

The poro-elastic coefficient for bulk volume change (a) has been correlated with porosity for individual rock units (Laurent *et al.* 1990; 1993). However, this leads to equations specific to a given rock. As the poro-elastic coefficient is 1 when the porosity equals the depositional porosity (Fig. 11) and 0 when the rock contains no porosity, a more general correlation can be obtained by considering the poro-elastic coefficient as a function of the relative change in porosity:

$$\alpha = \alpha_0 \left(\frac{\phi}{\phi_0}\right)^n . \qquad (16)$$

Available data (Fig. 11) suggest that the poro-elastic coefficient for a variety of rock types can be adequately fitted over the complete range of porosities if $\alpha_0 = 1$ and $n = 1/3$. The reason for the dependence of the poro-elastic coefficient on the one third power of the porosity ratio may reflect its dependence on the change in pore radius and therefore the interconnectivity and packing co-ordination of the solid framework with increased stress.

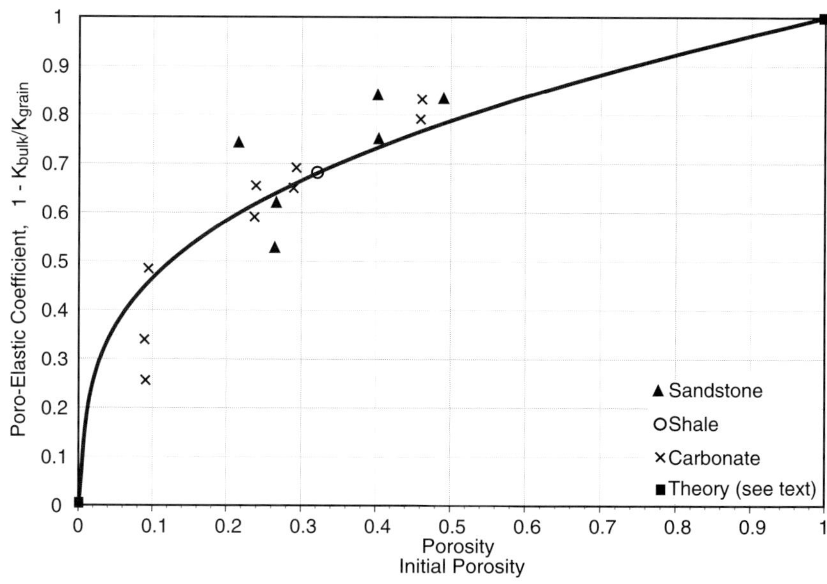

Fig. 11. Variation of the static poro-elastic coefficient with porosity. The theoretical line (which assumes that the poro-elastic coefficient is related to the one-third power of porosity) provides a good approximation to the (sparse) data.

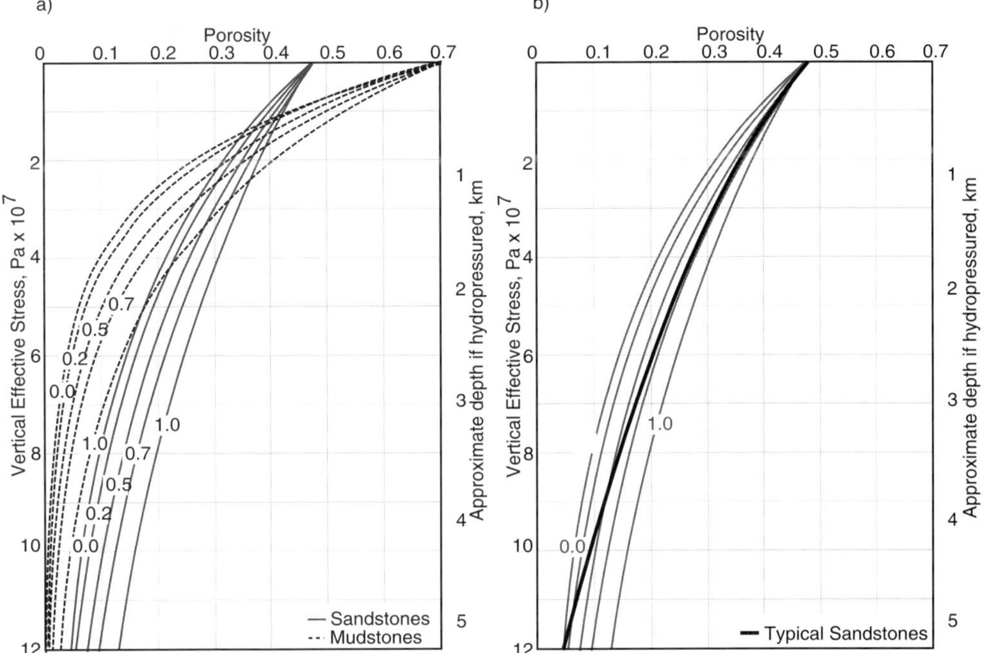

Fig. 12. Dependence of the porosity loss with effective stress (or depth) on the value of the poro-elastic coefficient. (a) The various curves were calculated using eqn 15, with standard initial porosities and compaction coefficients, and a number of different values for α. (b) For a typical sandstone, the feedback between porosity loss and the change in poro-elastic coefficient with porosity results in more rapid porosity loss at large depths than would be predicted by extrapolation of a simple exponential function (i.e. constant α).

Unfortunately, the dependence of the poro-elastic coefficient on porosity means that it is impossible to integrate eqn 14 simply. Hence no simple effective stress law analogous to eqn 15 can be derived. The observation that rocks with a wide range of porosities commonly have poro-elastic coefficients for bulk volume change of around 0.7 suggests that the effective stress law represented by eqn 15 may be a valid approximation, but at the range of depths applicable to most hydrocarbon exploration, evaluation of the porosity evolution would require a model which accounted for the spatial and temporal variability in the poro-elastic coefficient.

Practical considerations

Ideally we wish to be able to derive calibration data for porosity–depth or porosity–effective stress curves from well data, since these provide a continuous point sampling of the subsurface.

The effective stress can be derived relatively easily by integrating the density log to derive lithostatic pressure and using fluid pressure measurements (such as RFTs). Fluid pressure may also be crudely estimated from the mudweight used during drilling.

As discussed above, porosity measurements from cores are ideal for use in calibrating porosity–effective stress curves, provided that the core samples are representative of the lithologies over the whole depth range required. This situation is rare: cores are often taken only from the reservoir section, with little data being available for other lithologies. Similarly, a method is required by which porosity estimates can be made for undrilled regions such as kitchen areas or previously unexplored basins.

Seismic velocities (which are generally available) provide a large data set from which it should be possible to invert porosity data for all the lithologies in a given area. Seismic P-wave velocity (V_p) is a function of the bulk density (ρ_b), the shear modulus (μ), and the compressibility, or the bulk modulus (k_b). These parameters can be related to the porosity as follows: bulk density can clearly be expressed in terms of the porosity, the grain density (ρ_g) and the fluid density (ρ_f). Standard elastic relationships allow the shear modulus to be expressed in terms of the bulk modulus and

Poisson's ratio (ν). If the bulk modulus is related to the grain modulus via the poro-elastic coefficient for bulk volume change α (which relates the bulk compressibility to the grain compressibility), then:

$$V_p = f(k_b, \mu, \rho_b) = f(k_g, \alpha, \nu, \phi, \rho_g, \rho_f) \quad . \quad (17)$$

Given that the poro-elastic coefficient correlates with relative porosity change, the poro-elastic coefficient can be expressed as a function of porosity and eliminated. Similarly, Poisson's ratio could also be expressed as a function of porosity, as it falls predictably from 0.5 (i.e. fluid) under depositional conditions to values around 0.15 for silicate minerals with negligible porosity at large burial depths. It is then necessary to make the simplifying assumption that within the sedimentary column the pore fluid is generally aqueous, so that the fluid density does not change significantly with depth. Thus the relationship between P-wave velocity and porosity is a function of only two other variables:

$$V_p = (k_g, \phi, \rho_g) \quad \text{or} \quad \phi = f'(V_p, k_g, \rho_g) \quad . \quad (18)$$

Thus, knowing k_g and ρ_g, it is possible to derive estimates of porosity from either seismic or sonic log velocities. Furthermore, the low frequency content of seismic waves results in an averaging effect which subdues the dependence of P-wave velocities on small-scale lithology effects. Seismic and sonic velocities therefore provide a potential avenue by which realistic average porosities may be derived for local calibration of porosity loss models.

Compaction and porosity loss

Is solid volume constant?

Fundamental to the discussions so far has been the assumption that the solid volume remains constant, so that vertical strain can be equated to porosity loss. This assumption is particularly relevant in the case of decompaction (Fig. 13), but is the assumption of constant solid volume reasonable?

Most sedimentary basins can be considered as isochemical systems, in that there is little overall transport of components into or out of the system. However, the distribution of the components will vary as a function of temperature and time, with different phases being stable under different conditions.

Dehydration reactions are particularly important as a means of changing solid volume, because they do not require mass transport into the system, but do result in a net mass transport out of the system in the form of water. This water from dehydration is not usually considered as part of the pore fluid. Clay mineral dehydration reactions in shales are

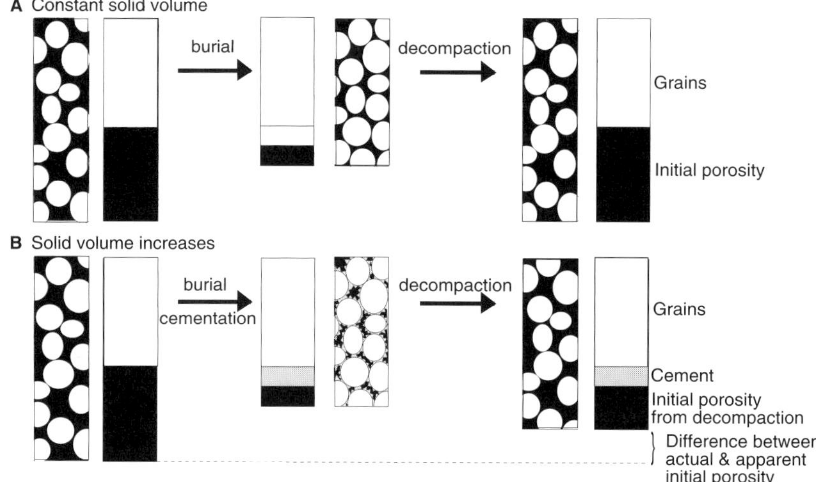

Fig. 13. Errors introduced into decompacted thicknesses by changes in solid volume. (A) Decompaction recovers the original porosity if solid volume is conserved. (B) If the solid volume changes, and this is not recognized, the amount of porosity predicted from the porosity–depth curve will be incorrect. In this example the solid volume has increased between the time of deposition and the present day, hence the calculated initial porosity is an underestimate. This error will pass through to the decompacted depths and all the underlying strata will also be placed at incorrect depth.

particularly important, because mudrocks are the most abundant lithology in the sedimentary record. Significant deviation from the assumption of conservation of solid volume may give rise to serious problems in understanding compaction (and hence decompaction) in sedimentary basins. Consider a fully compacted shale, composed of 25 % Quartz, 25 % K-feldspar, and 50 % kaolinite by volume. If all the K-feldspar and most of the kaolinite is converted to illite (see Table 1) the solid volume lost can be calculated using the molar volumes of these minerals. This calculation indicates that 43 cm^3 of solid volume is lost per mole (molar volume = 108.7 cm^3) of K-feldspar dissolved, which is equivalent to a 9.6 % reduction in volume! Similarly, in some mudrocks the conversion of kerogen to hydrocarbons may result in a volume decrease.

However, the most spectacular dehydration reaction, in terms of volume change, is the gypsum to anhydrite reaction which results in a reduction in solid volume of more than 37 % (Table 1). This reaction has been investigated extensively by Jowett *et al.* (1993) who found that the dehydration reaction is controlled by the temperature, activity of water and the pore fluid pressure. The largely temperature dependent transformation may occur at depths as little as 400 m when overlain by a poor heat conductor, but could theoretically occur at depths in excess of 4 km when overlain by salt. These authors concluded that the reaction was of considerable potential importance to basin modellers as a result of its impact on subsidence analysis, the four-fold increase in thermal conductivity, and the fact that the reaction, being endothermic, acts as a heat sink, reducing the temperatures below those which would be predicted by a simple conductive heat flow model..

Transformation reactions may also be an important mechanism for changing the solid volume within a sedimentary sequence. This is particularly important for carbonates in the shallow subsurface, which undergo transformation of aragonite to calcite. This transformation (on a mole for mole basis) results in an 8 % increase in volume (Table 1) despite the fact that mass is conserved.

Solid volume change can also be important in sandstones, in the form of cementation in the pore spaces, and dissolution of framework grains. For example, many reservoir sandstones below depths of 1–2 km contain significant volumes of quartz cement (McBride 1989). However, introduction of cement does not necessarily have an adverse effect on the porosity: consider Fig. 14 which shows the porosity–depth and quartz cementation–depth curves for Oligo–Miocene sandstones from offshore Borneo. It seems unreasonable to assume that up to 16 % bulk volume of new authigenic quartz could have been introduced into these sandstones without affecting the shape of the porosity–depth curve. If instead, the volume of quartz remains constant, this implies that changes in solid volume are included as part of the compaction process, and should not be treated separately.

Dissolution reactions can also cause significant changes in volume. For example, in the Brent Group of the North Sea up to 25 %B.V. (by volume) of feldspars may have been dissolved from deeply buried Brent Group sandstones (Giles *et al.* 1992). Even if the ions liberated from this reaction

Table 1. *Change in volume associated with transformation/dehydration reactions*

		molar volume (cm^3)	change in volume
gypsum to anhydrite	$CaSO_4.2H_2O$ to $CaSO_4$	73.7 46.15 27.6	 –37.5 %
aragonite to calcite	$CaCO_3$ to $CaCO_3$	34.15 36.93 2.78	 +8 %
K-feldspar + 2 kaolinite to illite + 2 water	$KAlSi_3O_8 + 2\ Al_2(Si_2O_5)(OH)_4$ to $KAl_4(Si_7Al)O_{20}(OH)_2 + 2\ H_2O$	108.7 + (2 x 99.5) 139.6 + (2x 18.1) 131.9	 –42.9 %
2K-feldspar + water + 2H$^+$ to kaolinite + 4 quartz + 2 K$^+$	$2\ KAlSi_3O_8 + H_2O + 2\ H^+$ $Al_2(Si_2O_5)(OH)_4 + 4\ SiO_2 + 2\ K^+$	(2 × 108.7) + 18.1 99.5 + (4 x 22.7) 45.2	 –19.2 %

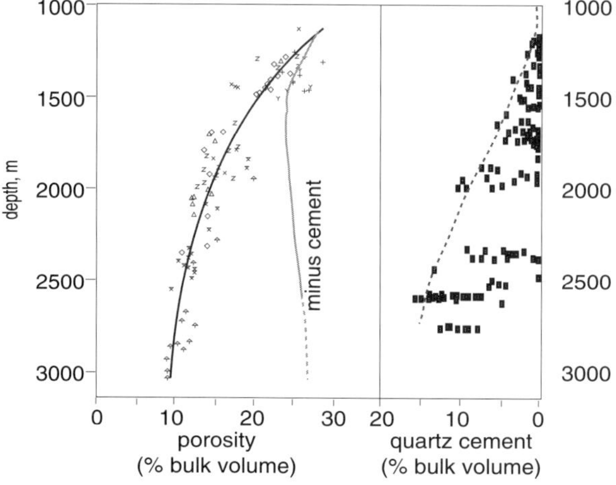

Fig. 14. Porosity and quartz cement versus depth, Sarawak. Up to 16 % quartz cement by volume (BV) has been precipitated, and thus the 'minus cement porosity' (the total volume not taken up by grains) has apparently not decreased between 1500 m and 3000 m depth. Therefore, either compaction has been halted completely despite continued sediment loading, or compaction and cementation are not independent.

had been re-precipitated as other minerals, so that no mass left the system in solution, over 3 % additional porosity would have been created due to the different molar volumes of the minerals. Thus the use of the constant solid volume assumption is fundamentally flawed.

It is possible to correct for changes of solid volume using the following development. The solid may be considered to be made up of J minerals so that the total solid volume is given by:

$$V_s = \sum_{j=1}^{j=J} V_{sm,j} n_j \tag{19}$$

where n_j is the number of moles of component j, and $V_{sm,j}$ is the molar volume of component j. Thus the rate of change of solid volume is:

$$\frac{dV_s}{dt} = \sum_{j=1}^{j=J} \left(V_{sm,j} \left(\frac{dn_j}{dt} \right) - n_j \left(\frac{dV_{sm,j}}{} \right) \right) . \tag{20}$$

The rates of change of mineral composition $(\frac{\partial n_j}{\partial_t})$ may be modelled using kinetic expressions for the reaction rate. Rate expressions are known for some reactions, such as the conversion of smectite to illite (Eberl & Hower 1976, Vasseur & Velde 1992).

The temporal variation in molar volume can be described in terms of the variations in effective stress and temperature through time, and the relations between molar volume and temperature and effective stress:

$$\frac{dV_{sm,j}}{dt} = \left(\frac{\partial V_{sm,j}}{\partial T} \right)_{\sigma_{\text{eff}}} \left(\frac{dT}{dt} \right) + \left(\frac{\partial V_{sm,j}}{\partial \sigma_{\text{eff}}} \right)_T \left(\frac{d\sigma_{\text{eff}}}{dt} \right)$$

$$= \mathrm{v}_i \left(\frac{dT}{dt} \right) + \beta_j \left(\frac{d\sigma_{\text{eff}}}{dt} \right) \tag{21}$$

where v_j and β_j are the expansivitiy and compressibility of mineral j, respectively. Hence the total rate of change of solid volume per unit solid is:

$$\frac{1}{V_s} \frac{dV_s}{dt} = \frac{1}{V_s} \left\{ \sum_{j=1}^{j=J} \left(V_{sm,j} \left(\frac{\partial n_j}{\partial t} \right) - n_j \left[v_j \left(\frac{dT}{dt} \right) - \beta_j \left(\frac{d\sigma_{\text{eff}}}{dt} \right) \right] \right) \right\} - (R_{w,e} - R_{w,p}) \tag{22}$$

where $R_{w,p}$ is the rate of water production from dehydration reactions and $R_{w,e}$ is the rate at which water can escape from the sediment.

Thus, in principle, changes in solid volume can be accommodated within compaction models. Dehydration reactions, and other essentially closed system reactions which are time or time–temperature dependent may be incorportated relatively easily into this framework, but for cementation reactions this is very difficult because of the need to know the fluxes of solid material as a function of time.

Time dependency

Time has been suggested to play a role in compaction by many authors including Dallmus

(1958), Weyl (1959), Boswell (1961), Maxwell (1960; 1964), Griffiths (1964), Dzevanshir *et al.* (1986) and Scherer (1987) but without general acceptance. We examine the evidence for time-dependent compaction processes below. Its importance can be demonstrated using a simple porosity–effective stress models such as equation 15. As an example, we consider the change in porosity caused by overpressures (P_{ex}).

Taking into consideration the rate of fluid escape from a compacting rock will ultimately result in a time-dependence to compaction. By combining a porosity–effective stress model (such as equation AP8, Appendix 1) with mass conservation equations and Darcy's law for fluid movement a general expression for the rate of overpressure generation can be derived (Giles 1996):

$$\frac{dP_{ex}}{dt} = \frac{(1-\phi)}{\phi k} \nabla \cdot (P_{ex}\mathrm{k}/\mu) + \frac{\upsilon(1-\phi)}{k}\frac{dT}{dt} - \frac{1}{k}\left(\frac{d\phi}{dt}\right)_{cem} + \frac{(1-\phi)R_{w,p}}{\rho_f \phi k} - \frac{\beta(1-\phi)}{k}\frac{dP_f}{dt} + g(\rho_R - \rho_f)(1-\phi_o)S(t) \quad . \tag{23}$$

This equation (given in terms of the material derivatives) relates the rate of overpressure generation to the principal driving mechanisms provided by the loading rate (i.e. sedimentation rate ($S(t)$) multiplied by density, where where ρ_R is the average grain density), aquathermal (temperature-dependent) effects, and rate of fluid generation ($R_{w,p}$). Of prime importance is the rate of fluid dissipation ($\nabla \cdot (P_{ex}\mathrm{k}/\mu)$) as a result of the excess pressure generated, which in turn largely depends on the permeability (**k**) of the rock and the specific storage. Solution of this equation shows that dissipation of excess pressure will only be geologically instantaneous if the permeability of the rock is high enough. In rocks of low permeability, such as shales, water can only escape at a finite rate during burial. Overpressures can therefore be generated simply by a process of rapid burial of low permeability sediments — a process known as undercompaction (Fig. 15a). High pore-fluid pressures mean that the effective stress on the rock is reduced, and hence the porosity is higher than the porosity at the same depth in a normally (hydrostatically) pressured system. Modelling of the relationship between flow rate, permeability and excess fluid potential (driving the flow) shows that if there is no further loading by sedimentation the overpressures will decay as the pore fluid gradually escapes (Fig. 15b). The porosity shows a similar decay to progressively lower values as the effective stress increases due to decreasing fluid pressure. Such compaction is clearly a time-dependent process.

Thus in thick sedimentary successions the variance in porosity loss curves can be explained to a large degree by the variable effective stress conditions which affect the mudrocks, commonly linked to time variant drainage. Decompaction is therefore intrinsically inexact in all but the simplest cases. Plotting porosity loss against depth focuses attention on vertical variation of hydraulic conductivity but in practice drainage is as much determined by lateral communication along dipping aquifers or up faults. Both of these routes evolve (often discontinuously) through time (Fig. 16). The lateral path lies essentially in sand only, whereas the vertical path lies across a layered sequence whose permeability is essentially determined by the shales. As sand permeabilities exceed those of shale by 4–6 orders of magnitude over the range of effective stresses encountered in most basins (Hanor 1987, p 155), the relative flux transmitted via lateral bleed-off along aquifers is, in principle, 3–4 orders of magnitude greater than that via vertical bleed-off, despite the path length commonly being 1–2 orders of magnitude greater. In practice, however, the relative mean permeabilities for lateral and vertical bleed-off are not quite as disparate as might be imagined from the above two-dimensional reasoning. Erosional relief at sandstone bases, particularly valley fills cut at sea-level low-stands, commonly ensures a three-dimensional vertical connectivity (albeit tortuous) not apparent from a single vertical profile. Such pathways can often explain the lack of correlation between pressure change across many shales and their thickness (e.g. Weedman *et al.* 1992).

Clearly, the dominant bleed-off flux for many sand/shale sections will be directed laterally up-dip (Bredehoft *et al.* 1988) unless hindered by faults. By corollary, only relatively minor flux will be vertically transmitted, unless assisted by faults. It is thus worth considering the impact of faulting on fluid flow, and hence on compaction.

Faulting induces a variety of fabrics (Knipe 1992) which may be grouped as either shaley (smear) or sandy (particulate and cataclastic flow) gouges (Watts 1987), each of which has locally variable continuity within the fault plane. Each type of gouge is of markedly lower permeability than sand, but sandy gouge (particularly in the direction parallel to the fault plane) is of markedly higher permeability than shale. The effect of faulting is thus to decrease the mean lateral permeability of the sand/shale section and to increase its mean vertical permeability. The lateral decrease results predominantly from the juxtaposition of shale against sand across the fault and the generation of shaley gouge within the fault plane, the severity of

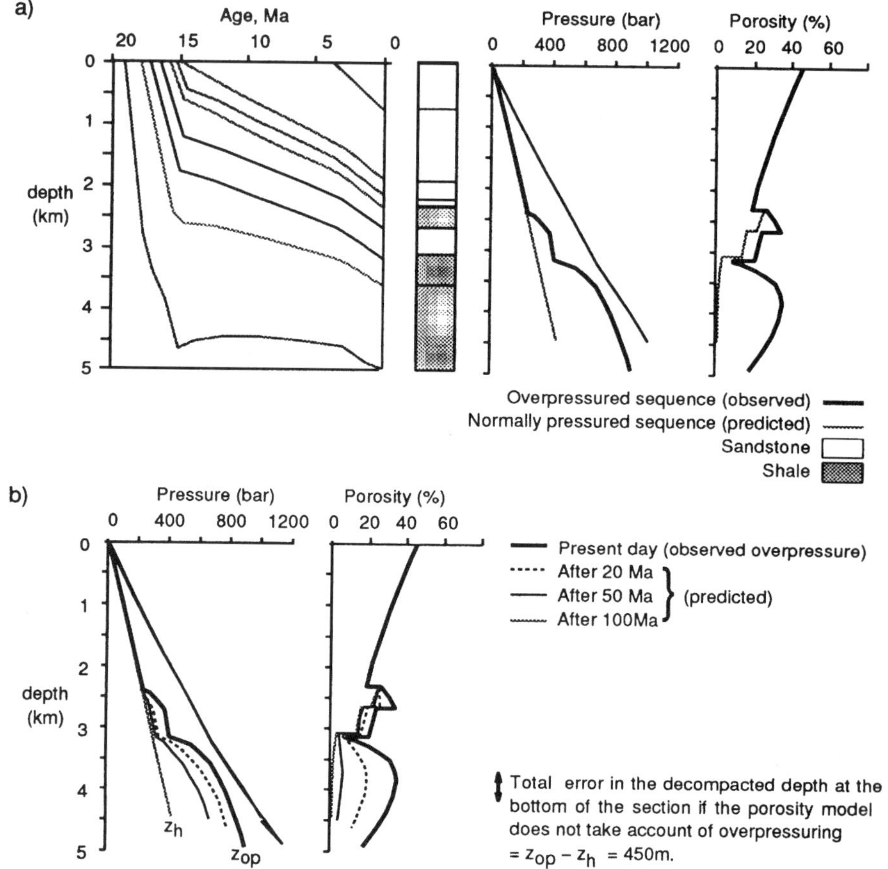

Fig. 15. Time-dependence of porosity: undercompaction case. (a) A well from Nigeria, where rapid sedimentation has led to the generation of an over-pressured section with anomalously high porosity in the low permeability shales. In contrast, the overlying sandstones are normally pressured because water could easily escape to the surface during compaction. Note that subsidence is plotted relative to the sediment surface, with the result that the basement appears to be uplifted due to compaction. In reality this compaction would result in surface subsidence and hence increased water depths. (b) If rapid sedimentation was now switched off, the slow flow through the shales would result in the gradual reduction of the overpressures, until the whole section was normally pressured. As the overpressure decreases, so the effective stress increases, and thus porosity is reduced from the undercompacted state observed in (a) to an equilibrium state ($\sigma_{eff} = \sigma_{eff\ max}$). Thus the porosity changes through time, despite the fact that no further loading by sedimentation has occurred.

both processes being statistically an inverse function of the sand/shale ratio. The vertical increase results from the generation of sandy gouge within the fault plane and three-dimensional fluid flow across the fault where footwall and hanging-wall geometries are not congruent and sand-on-sand contacts exist (Allan 1989).

Faulting may thus re-orient much of the fluid flux from lateral flow toward a largely vertical bleed-off. Whether this vertical flux is dominantly via the fault plane or lies within the fault blocks depends largely on the ratio of fault-zone transmissibility to that of the sedimentary section in the blocks (where vertical transmissibility is dominated by the major shale horizons). Where fault-zone transmissibility is relatively high, the fluid flux in sandstones will be largely lateral (towards faults) and hydrocarbon–water contacts will be strongly tilted. The common occurrence of sub-horizontal hydrocarbon–water contacts in oil and gas fields supports the idea that fluid flux in the sedimentary section is often sub-vertical. Thus because porosity is dependent on fluid drainage, the waxing and waning of pressure cells within the deforming

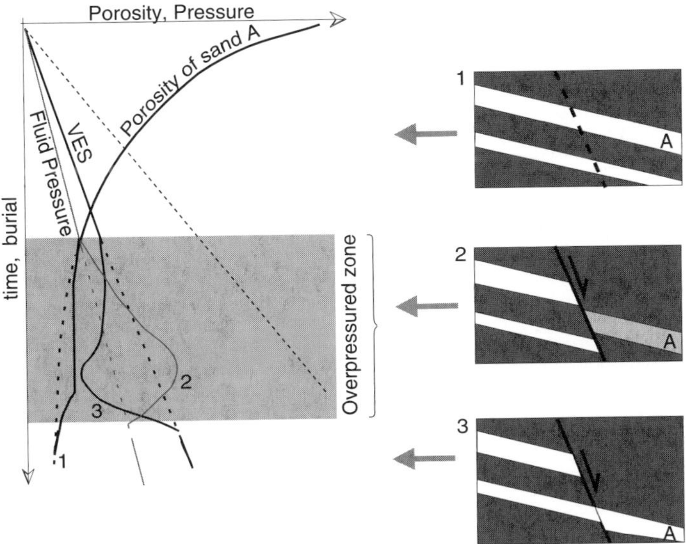

Fig. 16. Time-dependence of porosity — inflation case. Faulting causes temporary sealing of sand A. During this time the sand becomes overpressured. The reduced vertical effective stress (VES) means that compaction is halted (the porosity cannot increase again). When continued faulting juxtaposes sand against sand, pressure is released by up-dip flow, effective stress increases, and compaction can continue.

sediment pile as a complex function of stratigraphy, tectonic deformation, and strain rate will result in a range of porosity loss trends for an initially homogeneous sedimentary unit at different spatial locations in the basin.

Furthermore, the present day porosity–depth curve includes many points, each of which may have experienced a different effective stress history, and thus the porosity loss curve cannot represent the effective stress history of any single point (Fig. 17). Thus it is not possible to completely eliminate errors in decompaction, even if a well calibrated porosity–depth curve is available. To accurately reconstruct the subsidence history and porosity of a given unit, it is necessary to know the effective stress history (and hence, if overpressuring occurred at any stage, the fluid flow characteristics of the system).

Development of overpressures is a very particular example of the time-dependence of compaction. However, consideration of the physical processes involved in compaction suggests that time-dependency may be a more general phenomenon. Compaction/porosity reduction involves a number of processes that modify the grain framework (Fig. 18), including :

(a) grain re-arrangement and breakage;
(b) elastic deformation;
(c) plastic deformation;
(d) dissolution of labile grains;
(e) pressure solution;
(f) chemical cementation.

Of these processes, only elastic deformation is likely to be truly instantaneous.

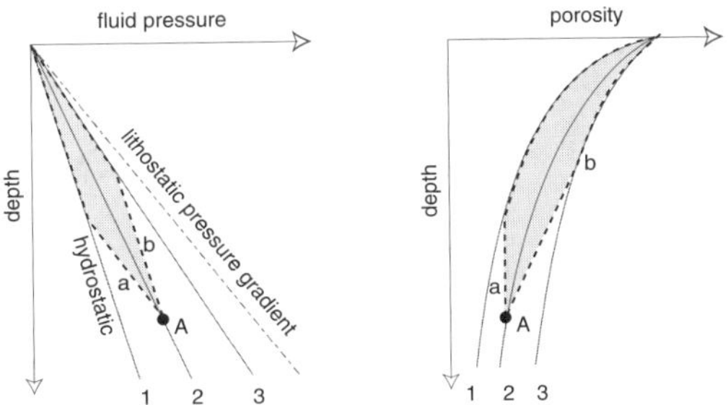

Fig. 17. Non-uniqueness of porosity–depth curves. A whole range of possible effective stress histories exists for any point on the porosity–depth curve. The point A could have first compacted normally, and then not at all (for example due to late overpressuring) or it could have been undercompacted initially (due to rapid burial) and then compacted later, once the overpressures had dissipated. Thus it is unlikely that the point A was reached simply by following the present day porosity loss curve.

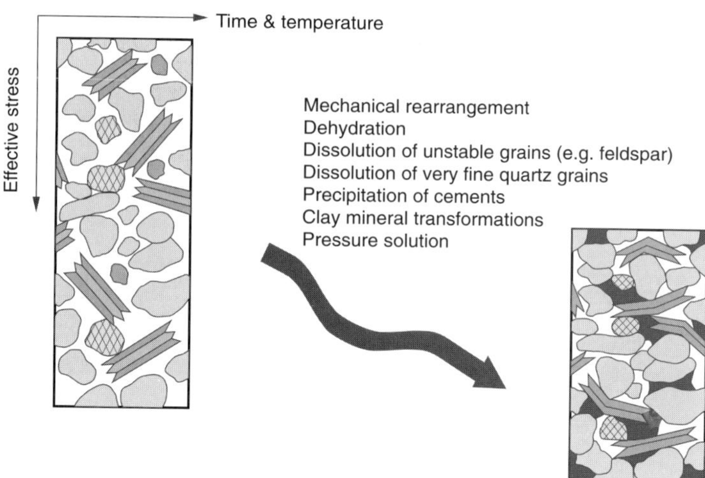

Fig. 18. Compaction in sandstones.

At low stress, compaction occurs by the sliding and rotation of grains. Time-dependence may be introduced via the friction between grain surfaces as suggested by de Waal (1986). In attempting to explain reservoir compaction as a result of hydrocarbon production, de Waal (1986) developed a theory of Dieterich (1978), who suggested that the frictional coefficient between grains depended on the average lifetime of grain contact points, called 'asperities' (Bowden and Taybor 1954; 1960). De Waal (1986) recognized that the lifetime of such asperities would depend on the strain rate. When the strain rate is low it takes a long time to reach the critical strain for breakage of an individual asperity, and new asperities have ample time to develop, resulting in a relatively high contact area between grains. At high strain rates, asperities exist for shorter periods resulting in a lower frictional coefficient between grains (de Waal 1986). Once loading ceases, compaction must continue until the frictional force between grains is sufficient to withstand the overburden load; i.e. until the grains have reached an equilibrium configuration. However, it is not known how this effect operates on a geological time scale.

Grain breakage may also exhibit a time effect, in that build-up of a critical strain is required before failure can occur. This critical strain may vary from grain to grain, depending on the stress distribution within the individual grains and the load carried by the surrounding grains. Breakage or slippage in the surrounding grains may cause the critical stress to be exceeded.

Plastic deformation which results in physical rearrangement of the crystal lattice by migration of dislocations, translational gliding, etc. may also be a mechanism of compaction. In the case of quartz grains it will only occur at high stresses and temperatures, but other components of sandstones, particularly mudstone clasts and intergranular clay, may behave plastically at lower stresses. These components will deform from a load-bearing to non-load-bearing state once the yield stress has been reached resulting in a period of deformation at constant load.

Progressive framework weakening due to grain dissolution may be a very important process in compaction of sandstones. Sandstones can contain an appreciable quantity of load-bearing labile grains such as feldspars, carbonates and volcanic fragments. During burial, increasing temperatures may trigger decomposition of such unstable grains (Giles & de Boer 1990) to form pore-filling clay minerals and cements (Fig. 18). For example, in the Ness sands of the Brent Group more than 20 % B.V. of feldspars are present in the shallowest samples, whereas samples buried to more than 4 km depth contain only a few percent feldspar (Giles *et al.* 1992). The removal of load-bearing components and their replacement with non load-bearing components means that the average stress carried by the load-bearing grains must increase, thus promoting further compaction.

The increase in stress associated with burial, together with the increased solubility of many minerals (including quartz) at higher stresses results in the process of pressure solution. Grain-to-grain contact stresses are generally much higher than those in the surrounding pore fluid. The layer of water molecules bound to the grain surface therefore experiences much higher pressures and allows higher solubility at grain-to-grain contacts than elsewhere on the grain. A concentration gradient is set up which allows diffusion of material from the grain contacts to the bulk pore fluid. Preferential dissolution at the grain-to-grain contacts causes grain interpenetration. For example in a quartz sandstone quartz will be dissolved at the grain contacts in response to increasing stress during burial. As the bulk pore fluid is generally already saturated or slightly oversaturated with silica, due to equilibrium with the quartz grains surrounding the pore, the quartz dissolved at the grain contacts will be precipitated in the pores surrounding the interpenetrating grains. The resulting increase in grain-to-grain contact area causes reduction of grain-to-grain stresses until equilibrium is attained. Mathematical and experimental descriptions of pressure solution and the related process of grain necking, have been developed, for instance by Weyl (1959), Kingery (1959), de Boer (1975; 1977*a–c*) and more recently Schutjens (1991*a,b*), Hickman & Evans (1992, grain necking) and Stephenson *et al.* (1992). The latter developed an ingenious theory of compaction by grain interpenetration based around energy — and volume — balance considerations. All of these theories predict some form of time-dependence for pressure solution, and hence time-dependency of porosity loss as a function of effective stress. In this case porosity loss and compaction are thus linked but are not independent.

Experimental evidence also indicates a time-dependency to compaction. Numerous triaxial experiments on unconsolidated sands (de Boer 1975; de Waal 1986) have shown that compaction continues following loading of a sample and subsequently holding the load constant. This phenomenon, commonly referred to as ‘creep’, appears to occur at all laboratory loading rates. Furthermore, compaction does not appear to stop on a laboratory time scale, although its rate falls with time. Thus even after 6 years (de Waal 1986) vertical strain continues. Continued, though slow, compaction over long time scales at constant stress may explain why triaxial compaction experiments on loose sand show porosities much higher than those found in unconsolidated sands in the subsurface at the same effective stress (Figs 7 & 19). Examination of samples which have undergone compaction experiments show a predominance of mechanical rearrangement and grain breakage, with little evidence of pressure solution except at high temperatures. Thus creep in this instance may be equated with purely mechanical processes.

Although it would appear from Fig. 3 that many data are available on porosity–depth behaviour, it still remains difficult to find data which can be unambiguously compared; i.e. sands which have the same composition, come from areas of similar geothermal gradient etc. Figure 19 shows porosity–effective stress trends which have been approximately converted from the corresponding depth trends for three quartz rich sandstone trends, and one representative experimental data set. The sandstone trends were selected from areas where

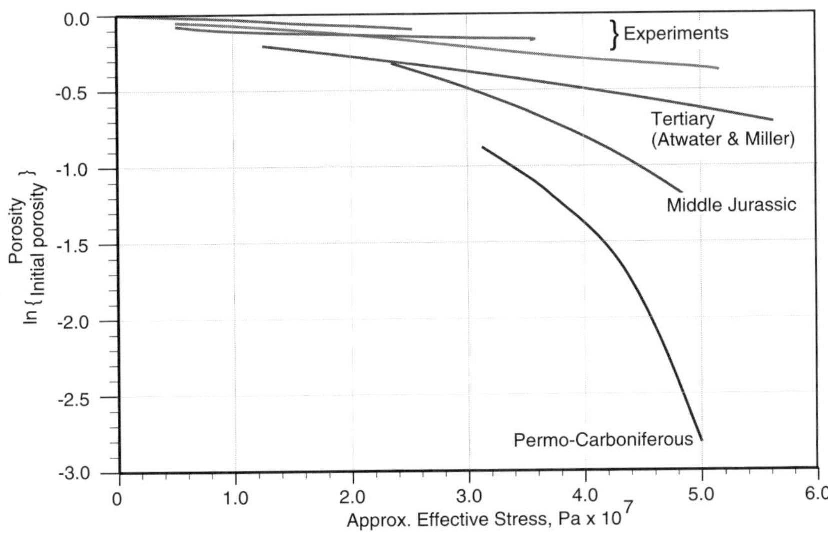

Fig. 19. Time dependency of compaction. Plotting porosity data from sandstones of similar composition, from areas with similar geothermal gradients (30 °C km^{-1}), and where none of the samples had undergone uplift (so that they are presently at their maximum effective stress) against effective stress shows that the oldest samples exhibit the lowest porosity (greatest porosity loss).

the rocks are presently at or very close to their maximum depth of burial and from areas of near normal geothermal gradients. These trends show that the porosity at a given effective stress decreases as the age of the rocks increases, with the experimental data (age = 0 Ma) showing the least porosity loss. It would appear therefore that time plays an important role in porosity loss and compaction.

Conceptual models

In attempting to put together a compaction model which includes time effects, the most obvious choice is provided by visco–elastic (Lockett 1972; Hunter 1983) or visco–elasto–plastic (Schneider *et al.* 1993) models. These provide mathematical descriptions that enable certain classes of behaviour to be approximated. The most general form of a visco–elastic model requires that the strain (ε) should be dependent on the imposed stress (σ), the stress rate ($\dot{\sigma}$), and higher order stress rate terms. Such models are most easily thought of as being composed of a number of elastic and viscous elements; an elastic element obeys Hooke's law and a viscous element shows a strain rate which is proportional to the applied stress. Viscous elements are included as a way of introducing time-dependence. Visco–elastic models have the advantage that physical analogue models can be constructed by placing springs and dash-pots (viscous elements) in series or parallel. Each viscous or plastic element introduced into the model brings with it a characteristic time constant. The combination of all the viscous/plastic elements included in the model controls the overall time-dependence of the modelled strain. The constants in the modelled solution can be calibrated from experimental or well log data. Although these models have proved immensely useful in metallurgy and plastics, they cannot provide any fundamental understanding of why a rock is behaving in a particular way. Furthermore, as we have no clear idea of the long term compaction behaviour of rocks, it is difficult to know the complexity of model to use.

A more fundamental, but much more difficult task is to build a model based on a physical description of the interaction of relevant processes. Given that the bulk volume is the sum of the solid and fluid volumes (eqn 2), the normalized volumetric strain rate (i.e. strain rate per unit volume) is:

$$V_b' = \frac{1}{V_b}\left(\frac{\partial V_b}{\partial t}\right) = \frac{1}{1-\phi}\left(\frac{\partial \phi}{\partial t}\right) + \frac{1}{V_s}\left(\frac{\partial V_s}{\partial t}\right) \quad . \quad (24)$$

By assuming that that compaction occurs at constant cross-sectional area this equation can be combined with eqn 15 to enable porosity loss curves to be converted to true compaction curves in depth. This requires a knowledge of the rate of porosity loss, which could be based on any of the aforementioned models. For instance, in a hydrostatically pressured basin simple porosity–depth curves might be used, together with information on the timing of dehydration reactions.

If the volumetric bulk strain rate can be described in terms of solid volume and porosity (eqn 24), the bulk strain rate (normalized with respect to bulk volume) can then be assumed to result from the sum of the strain rates (ω_i) associated with a number of processes (i = 1,I...I), which are weighted according to the fraction of minerals (m_j) taking part.

$$-\frac{1}{1-\phi}\left(\frac{\partial \phi}{\partial t}\right) = \frac{1}{V_s}\left(\frac{\partial V_s}{\partial t}\right) + \sum_{j=1}^{J}\sum_{i=1}^{I} m_j\omega_i \quad . \quad (25)$$

In principle, such a simplistic model will overestimate the strain rate by not allowing for the coupling between processes. For instance, porosity reduction and cementation would be expected to change the elastic constants of the rocks. In practice, many of the coefficients of the model will depend on porosity, thus introducing a feedback loop between processes. In the example shown in Fig. 20, the elastic term has been constrained to approximate that required to recover about 0.02 (i.e. 2 % porosity) on unloading, all mechanical processes have been assumed to be described by a visco–elastic creep function from Cristescu (1983) and by a pressure solution expression derived from de Boer (1975). Given these functions, eqn 25 was integrated numerically along constant loading paths (see Giles 1996 for more details). As shown in Fig. 20 the results are encouraging in that the simple model approximates well the natural data. Use of this approach would enable simple compaction models to be constructed which embody all the petrographic data on the processes taking place in the sandstone.

Conclusions

The use of default compaction curves can introduce significant errors into thermal history and pore-fluid pressure calculations, particularly where little well data are available to calibrate the model.

The use of locally calibrated porosity–depth curves will reduce calculation errors, but not eliminate them, because the observed present day porosity–depth curve represents a concatenation of samples with different effective stress histories, and hence it does not describe the precise route by which a given sample achieved its present day

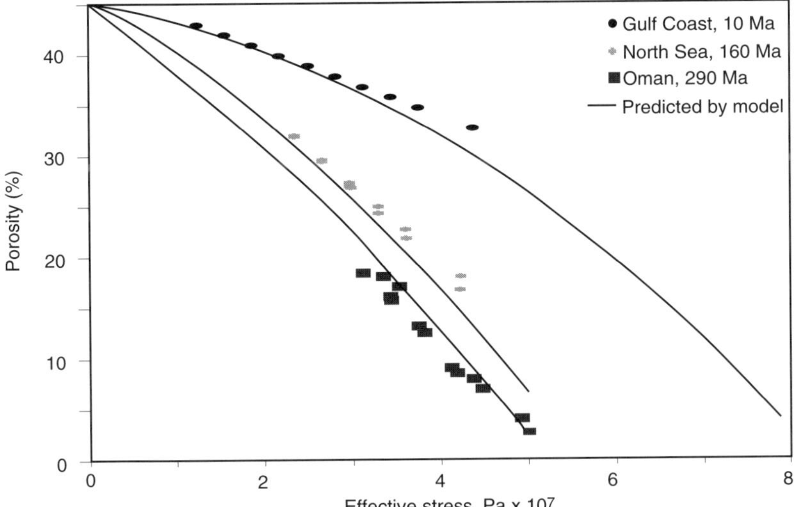

Fig. 20. Predicted versus observed porosity at different rates of loading. Porosity was predicted using eqn 25 with the same set of parameters to give three predicted porosity curves for different ages. Natural data were converted to approximate effective stresses dependent on steady burial through time. There is good agreement between the model and observations.

porosity. High-side and low-side scenarios should ideally be run to check the scale of the uncertainties.

Porosity–effective stress relationships are to be preferred over porosity–depth relationships. Such equations are more physically realistic, able to handle porosity hysteresis and therefore uplift, and inherently take overpressures into account.

The assumption that porosity loss is equivalent to compaction is fundamentally flawed. Such a hypothesis requires that solid volume is constant. Dehydration reactions within shales and evaporites, and cementation/dissolution reactions in sandstones all violate this assumption. Local calibration of porosity–depth/effective stress functions are needed.

Porosity loss and, by extension, compaction cannot be considered to be time-independent processes. Experimental and natural data all point to a clear time-dependence where porosity falls and compaction increases with time at constant effective stress.

Current compaction models do not provide good descriptions of either compaction or porosity loss. Thus new constitutive equations to describe compaction are required which are firmly based on the underlying physical processes contributing to compaction.

We appreciate helpfully critical reviews by Dr D. Waples and Dr G. Couples.

Appendix

Derivation of porosity loss — effective stress law (Giles 1996)

If porosity is assumed to be a function of the confining pressure (P_c) and the pore pressure (P_f), it follows that:

$$\mathrm{d}\phi = \left(\frac{\partial \phi}{\partial P_c}\right)_{P_f} \left\{ \mathrm{d}P_c - \left[\left(\frac{\partial \phi}{\partial P_f}\right)_{P_c} \Big/ \left(\frac{\partial \phi}{\partial P_c}\right)_{P_f} \right] \right\} \quad \text{(AP1)}$$

where:

$$\alpha' = \left\{ \left(\frac{\partial \phi}{\partial P_f}\right)_{P_c} \Big/ \left(\frac{\partial \phi}{\partial P_c}\right)_{P_f} \right\}. \quad \text{(AP2)}$$

Note that the laws relating bulk volume to effective stress, and porosity to effective stress are not the same (compare equation AP2 with equation 13). Hence: $\alpha \neq \alpha'$. The relationships between differential effective stress laws have been extensively studied by Berryman (1992) who has been able to derive relationships between the various poro-elastic coefficients.

If it is assumed that the confining and pore pressures are independent and that the poro-elastic coefficient for porosity changes is a constant, the differential effective stress may be defined as:

$$\mathrm{d}\sigma_{\mathrm{eff}} = \mathrm{d}P_c - |\,\mathrm{d}P_f \quad \text{(AP3)}$$

and a compressibility modulus for the rate of change of porosity with maximum confining pressure can be defined as:

$$\frac{1}{k_\phi} = -\frac{1}{\phi}\left(\frac{\partial \phi}{\partial P_c}\right)_{P_f} . \quad \text{(AP4)}$$

Experimental and natural evidence shows that both mechanical and chemical porosity loss are irreversible functions of effective stress. Consequently, the above expression may be modified so that it holds only if the effective stress is a maximum.

$$\frac{1}{k_\phi} = -\frac{1}{\phi}\left(\frac{\partial \phi}{\partial P_c}\right)_{P_f \sigma_{eff} = \max} \quad \text{(AP5)}$$

Substituting these relationships into equation AP3 gives:

$$d\phi = -\frac{\phi}{k_\phi} d\sigma_{eff = \max} \quad \text{(AP6)}$$

Assuming that both the compressibility modulus and the poro-elastic coefficient are constants, integration of this expression between the limits of $\phi = \phi_o$ at $\sigma_{eff} = 0$, and $\phi = \phi$ at σ_{eff}, gives:

$$\phi = \phi_o e^{-\frac{\sigma_{eff}}{k_\phi}} \quad \text{(AP7)}$$

where the effective stress is given by $\sigma_{eff} = P_c - \alpha P_f$. This equation can be rewritten in terms of a hydrostatic fluid pressure (P_{hy}) and an excess fluid pressure (overpressure) (P_{ex}): $\sigma_{eff} = P_c - \alpha(P_{hy} + P_{ex})$.

When σ_{eff} falls below $\sigma_{eff\ max}$, for example after unloading by erosion, the recovery in porosity is less than predicted by equation AP7. Possibly, the small amount of porosity recovered on unloading represents that portion of the strain which is truly elastic. A compressibility modulus for unloading can then be defined as:

$$\frac{1}{k'} = -\frac{1}{\phi_{min}}\left(\frac{\partial \phi}{\partial P_c}\right)_{P_f,\ \sigma_{eff} < \sigma_{eff\ max}} . \quad \text{(AP8)}$$

Substituting equation AP8 into AP3 gives:

$$d\phi = -\frac{\phi_{min}}{k'} d\sigma_{eff\ max} . \quad \text{(AP9)}$$

Integrating between ϕ_{min} at $\sigma_{eff\ max}$ and ϕ at σ_{eff}, gives:

$$\phi = \phi_{min} e^{\frac{1}{k'}(\sigma_{eff\ max} - \sigma_{eff}} . \quad \text{(AP10)}$$

Substituting for the minimum porosity in equation AP10 using equation AP7 gives:

$$\phi = \phi_o e^{-\frac{1}{k}\sigma_{eff\ max}} e^{\frac{1}{k'}(\sigma_{eff\ max} - \sigma_{eff})} . \quad \text{(AP11)}$$

Note $k \neq k'$.

Also note that if $\sigma_{eff} = \sigma_{eff\ max}$ the second exponential term in equation AP11 reduces to 1, so that AP11 reduces to AP7. Hence equation AP11 may be used to model both the loading and unloading phases of (de)compaction.

References

ALLAN, U. S. 1989. Model for hydrocarbon migration and entrapment within faulted structures. *Bulletin of the American Association of Petroleum Geologists*, **73**, 803–811.

ATHY, L. F. 1930. Density, porosity, and compaction of sedimentary rocks. *Bulletin of the American Association of Petroleum Geologists*, **14**, 1–24.

ATKINS, J. E. & MCBRIDE, E. F. 1992. Porosity and packing of Holocene river, dune, and beach sands. *Bulletin of the American Association of Petroleum Geologists*, **76**, 339–355.

ATWATER, G. I. & MILLER, E. E. 1965. The effect of decrease in porosity with depth on future development of oil and gas reserves in south Louisiana. *Bulletin American Association of Petroleum Geologists*, **49**, 334

BALDWIN, B. & BUTLER, C. O. 1985. Compaction curves. *Bulletin of the American Association of Petroleum Geologists*, **69**, 622–626.

BEARD, D. C. & WEYL, P. K. 1973. Influence of texture on porosity and permeability of unconsolidated sand. *Bulletin of the American Association of Petroleum Geologists*, **57**, 345–369.

BERRYMAN, J. G. 1992. Effective stress for transport properties of inhomogeneous porous rock. *Journal of Geophysical Research*, **97**, 17 409–17 424.

BOSWELL, P. G. H. 1961. *Muddy Sediments*. Cambridge University Press, Cambridge.

BOWDEN, F. P. & TAYBOR, D. 1954. *The friction and lubrication of solids: Part1*, Oxford University Press, London.

—— & —— 1960. *The friction and lubrication of solids: Part 2*, Oxford University Press, London.

BREDEHOEFT, J. D., DZEVANSHIR, R. D. & BELITZ, K. R. 1988. Lateral fluid flow in a compacting sand–shale sequence: South Caspian Basin. *Bulletin of the American Association of Petroleum Geologists*, **72**, 416–424.

BURST, J. F. 1969. Diagenesis of Gulf Coast clayey sediments and its possible relation to petroleum migration. *Bulletin of the American Association of Petroleum Geologists*, **53**, 73–93.

CRITESCU, N. 1983. *Rock Rheology*. Kluwer Academic Publishers, Dordrecht.

DALMUSS, K. F. 1958. Mechanics of basin evolution and its relation to the habitat of oil in the basin. *In:*

WEEKS, L. G. (ed.) *Habitat Of Oil*. American Association of Petroleum Geologists, Memoir **36**, 2071–2124.

DE BOER, R. B. 1975. *Influence of pore solutions on rock strength*. PhD thesis, University of Utrecht.

—— 1977*a*. Pressure solution: theory and experiments. *Tectonophysics*, **39**, 287–301.

—— 1977*b*. On the thermodynamics of pressure solution interaction between chemical and mechanical forces. *Geochimica et Cosmochimica Acta*, **41**, 249–256.

—— 1977*c*. Pressure solution experiments on quartz sand. *Geochimica et Cosmochimica Acta*, **41**, 257–264.

DE WAAL, J. A. 1986. *On the rate type compaction behaviour of sandstone reservoir rock*. PhD thesis, Technische Hogeschool Delft.

DICKINSON, G. 1953. Geological aspects of abnormal reservoir pressures in Gulf Coast Louisiana. *Bulletin of the American Association of Petroleum Geologists*, **37**, 410–432.

DIETERICH, J. H. 1978. Time dependent friction and the mechanics of stick–slip. *Pure and Applied Geophysics*, **116**, 790–806.

DZEVANSHIR, R. D., BURYAKOVSKIY, L. A. & CHILINGARIAN, G. V. 1986. Simple quantitative evaluation of porosity of argillaceous sediments at various depths of burial. *Sedimentary Geology*, **46**, 169–173.

EBERL, D. D. & HOWER, J. 1976. Kinetics of illite formation. *Bulletin of the Geological Society of America*, **87**, 1326–1330.

ENOS, P. & SAWATSKY, L. H. 1981. Pore networks in Holocene carbonate sediments. *Journal of Sedimentary Petrology*, **51**, 961–985.

FALVEY, D. A. & MIDDLETON. M. F. 1981. Passive continental margins. Evidence for a pre-break-up deep crustal metamorphic subsidence mechanism. *In: Proceedings 26th International Geological Congress, Geology of Continental Margins Symposium, Colloquium*, 103–114.

—— & DEIGHTON, I. 1982. Recent advances in burial and thermal geohistory analysis. J*ournal of the Australian Petroleum Exploration Association*, **22**, 65–81.

FOSTER, J. B. & WHALEN, H. E. 1966. Estimation of formation pressures from electrical surveys-offshore. Louisiana. *Journal of Petroleum Technology*, 165–171.

GALLAGHER, K. 1989. An examination of some uncertainties associated with estimates of sedimentation rates and tectonic subsidence. *Basin Research,* **2,** 97–114.

GALLOWAY, W. E. 1974. Deposition and the diagenetic alteration of sandstone in northeast Pacific arc-related basins. Implications for graywacke genesis. *Bulletin of the Geological Society of America*, **85**, 379–390.

GILES, M. R. 1996. *Diagenesis — a quantitative perspective*. Kluwer Academic Publishers, Dordrecht.

—— & DE BOER, R. B. 1990. Origin and significance of redistributional secondary porosity. *Marine & Petroleum Geology*, **7**, 378–397.

——, STEVENSON, S., MARTIN, S. V., CANNON, S. J. C., HAMILTON, P. J., MARSHALL, J. D. & SAMWAYS, G. M. 1992. *In:* MORTON, A. C., HASZELDINE, R. S., GILES M. R. & BROWN, S. (eds) *The Reservoir Properties of the Brent Group: A Regional Perspective*. Geological Society, London, Special Publications, **61**, 289–327.

GREENSMITH, J. 1978. *The Petrology of the Sedimentary Rocks*. Allen & Unwin, Hemel Hempstead.

GRIFFITHS, J. C. 1964. Statistical approach to the study of potential oil reservoir sandstones. *In:* PARKS, G. A. (ed.) *Computers in the mineral industry*. Stanford University Publications in Geological Sciences, 637–668.

HANOR, J. S. 1987. Evidence for kilometre-scale, thermohaline overturn of porewaters in the Louisiana Gulf Coast. *Nature*, **327**, 501–504.

HEDBERG, H. D. 1936. Gravitational compaction of clays and shales. *American Journal of Science*, **231**, 241–287.

HERMANRUD, C. 1993. Basin modelling techniques — an overview. *In:* DORÉ, A. G. *et al. Basin Modelling: Advances and applications*. Norwegian Petroleum Society Special Publications, **3**,1–34.

HICKMAN, S. H. & EVANS, B. 1992. Growth of grain contacts in halite by solution-transfer: Implications for diagenesis, lithification, and strength recovery. *In:* EVANS, B. & WONG, T. (eds) *Fault Mechanics and Transport Properties of Rocks*. Academic Press, London, 253–280.

HOSOI, H. 1963. First migration of petroleum in Akita and Yamagata Prefectures. *Economic Geology*, **49**, 43–55.

HUNTER, S. C. 1983. *Mechanics of continuous media*. Ellis Horwood Ltd.

JOWETT, E. C., CATHLES, L. M. & DAVIES, B. 1993. Predicting depths of gypsum dehydration in evaporitic basins. *Bulletin of the American Association of Petroleum Geologists*, **77**, 402–413.

KAMB, W. B. 1961. The thermodynamic theory of non-hydrostatically stressed solids. *Journal of Geophysical Research*, **66**, 259–271.

KINGERY, W. D. 1959. Sintering in the presence of a liquid phase. *In:* KINGERY, W. D. (ed.) *Kinetics of high temperature processes*. Technical Press, MIT and John Wiley & Sons, New York, 187–194.

KNIPE, R. J. 1992. The influence of fault zone processes and diagenesis on fluid flow. *In:* HORBURY, A. D. & ROBINSON, A. G. (eds) *Diagenesis and Basin Development*. American Association of Petroleum Geologists, Studies in Geology, **36**, 135–154.

KRUMBEIN, W. C. & MONK, G. D. 1942. Permeability as a function of the size parameters of unconsolidated sand. Petroleum Technology, AIME publication 1942, **5**, 1–11.

LANG, W. H. J. 1978. The determination of prior depth of burial, uplift and erosion using interval transit time. SPWLA, 19th Annual Log Symposium Transactions, paper B.

LAURENT, J., BOUTECA, M. & SARDA, J. P. 1990. *Pore Pressure Influence in the Poro-elastic Behaviour of Rocks:Experiemental Studies and Results*. Society of Petroleum Engineers Paper **20922**, 387–392.

——, ——, —— & BARY, D. 1993. *Pore-Pressure Influence in the Poro-elastic Behaviour of*

Rocks:Experimental Studies and Results. Society of Petroleum Engineers, Richardson, Texas, 117–122.

LAWRENCE, D. T., DOYLE, M. & AIGNER, T. 1990. Stratigraphic simulation of sedimentary basins: Concepts and calibration. *Bulletin of the American Association of Petroleum Geologists*, **74**, 273–295.

LOCKETT, F. J. 1972. *Non-linear Visco–elastic Solids.* Academic Press, New York.

MAGARA, K. 1968. Compaction and migration of fluids in Miocene mudstone, Nagaoka Plain, Japan. *Bulletin of the American Association of Petroleum Geologists*, **52**, 2466–2501.

MAXWELL, J. C. 1960. Experiments on the compaction and cementataion of sand. *In*: GRIGGS, D. T. & HANDIN, J. (eds) *Rock deformation*, Geological Society of America Memoir 79, Boulder, Colorado, 105–312.

—— 1964. Influence of depth, temperature and geologic age on porosity in quartzose sandstone. *Bulletin of the American Association of Petroleum Geologists*, **48**, 697–709.

MCBRIDE, E. F. 1989. Quartz cement in sandstones: a review. *Earth Science Reviews,* **26,** 69–112.

MEADE, R. H. 1966. Factors influencing the early stages of compaction of clays and sands-review. *Journal Sedimentary Petrololgy*, **36**, 1085–1101.

NAGTEGAAL, P. J. C. 1978. Sandstone instability as a function of burial diagenesis. *Journal of the Geological Society, London*, **135**, 101–105.

PATERSON, M. S. 1973. Non-hydrostatic thermodynamics and its geologic applications. R*eviews in Geophysics & Space Physics*, **11**, 355–359.

PROSHLYAKOV, B. K. 1960. Reservoir properties of rocks as functions of their depth and lithology. *Geol. Neft. Gaza.*, **12**, 24–29.

PRYOR, W. A. 1973. Permeability–porosity patterns and variations in some Holocene sand bodies. *Bulletin of the American Association of Petroleum Geologists*, **57**, 162–189.

RAYMER, L. L., HUNT E. R. & GARDNER, J. J. 1980. An improved sonic transit time to porosity transform. SPWLA 21st Annual Logging Symposium, Paper P, Tulsa, Oklahoma.

ROYDEN, L. & KEEN, C. E. 1980. Rifting processes and thermal evolution of the continental margin of eastern Canada determined from subsidence curves. *Earth & Planetary Science Letters*, **51**, 343–361.

RUBEY, W. W. & HUBBERT, M. K. 1959. Role of fluid pressure in mechanics of overthrust faulting. II. *Bulletin of the Geological Society of America*, **70**, 167–206.

SCHEIDEGGER, A. E. 1974. *The physics of flow through porous media (3rd ed.).* University of Toronto Press, Toronto.

SCHENK, C. J. 1983. Textural and structural characteristics of some experimentally formed aeolian strata. *Developments in Sedimentology*, **38**, 41–49.

SCHERER, M. 1987. Parameters Influencing Porosity in sandstones: A model for sandstone porosity prediction. *Bulletin of the American Association of Petroleum Geologists*, **71**, 485–491.

SCHNEIDER, F., BURRUS, J. & WOLF, S. 1993. Modelling overpressures by effective-stress/porosity relationships in low permeable rocks: empirical artifice or physcial reality? *In*: DORÉ, A. G. *et al.* (eds) *Basin Modelling: Advances and Applications.* Norwegian Petroleum Society Special Publications, **3**, 333–341.

SCHUTJENS, P. M. T. M. 1991*a*. Experimental compaction of quartz sand at low effective stress and temperature conditons. *Journal of the Geological Society, London*, **148**, 527–539.

—— 1991*b*. *Intergranular pressure solution in wet halite and quartz sands; an experimental investigation.* PhD thesis, State University of Utrecht, NL.

SCLATER, J. G. & CHRISTIE, P. A. F. 1980. Continental stretching: An explanation of the post Mid-Cretaceous subsidence of the Central North Sea Basin. *Journal of Geophysical Research*, **85**, 3711–3739.

SERRA, O. 1984. *Fundamentals of well-log interpretation 1. Acquisition of logging data.* Elsevier, UK.

SKEMPTON, A. W. 1970. The consolidation of clays by gravitational compaction. *Quarterly Journal of the Geological Society of London*, **125**, 373–411.

STEPHENSON, L. P., PLUMLEY, W. J. & PALCIAUSKAS, V. V. 1992. A model for sandstone compaction by grain interpenetration. *Journal of Sedimentary Petrology*, **62**, 11–22.

SYLTA, Ø. 1993. New techniques and their applications in the analysis of secondary migration. *In:* DORÉ, A. G. *et al.* (eds) *Basin Modelling: Advances and Applications.* Norwegian Petroleum Society Special Publication, **3**, 385–398.

TERZAGHI, K. & PECK, R. B. 1948. *Soil Mechanics In Engineering Practice.* Wiley, New York.

TICKELL, F. G. & HIATT, W. N. 1938. Effect of angularity of grains on porosity and permeability of unconsolidated sands. *Bulletin of the American Association of Petroleum Geologists*, **22**, 1272–1274.

VAN HINTE, J. E. 1978. Geohistory analysis — applications of micropaleontology in exploration geology. *Bulletin of the American Association of Petroleum Geologists*, **72**, 758–764.

VASSEUR, G. & VELDE, B. 1993. A kinetic interpretation of the smectite-to-illite transformation. *In:* DORÉ, A. G. *et al.* (eds) *Basin Modelling: Advances and applications.* Norwegian Petroleum Society Special Publications, **3**, 173–184.

WATTS, A. B. & RYAN, W. B. F. 1976. Flexure of the lithosphere and continental margin basins. *Tectonophysics,* **36**, 25–44.

WATTS, N. L. 1987. Theoretical aspects of cap-rock and fault seals for single and two-phase hydrocarbon columns. *Journal of Marine & Petroleum Geology*, **4**, 274–307.

WEEDMAN, S. D., GUBER, A. L. & ENGELDER, T. 1992. Pore pressure variation within the Tuscaloosa trend: Morganza and Moore-Sams fields, Louisiana Gulf Coast. *Journal of Geophysical Research*, **64**, 2001–2025.

WELLER, F. A. 1959. Compaction of sediments. *Bulletin of the American Association of Petroleum Geologists*, **43**, 273–310

WEYL, P. K. 1959. Pressure solution and the forces of crystallistion—A phenomenological theory. *Journal of Geophysical Research*, **64,** 2001–2025.

WOODSIDE, W. & MESSMER, J. H. 1961. Thermal conductivity of porous media (Parts I and II). *Journal of Applied Physics*, **32,** 1688–1707.

WOOLLER, D. A., SMITH, A. G. & WHITE, N. 1992. Measuring lithospheric stretching on Tethyan passive margins. *Journal of the Geological Society, London*, **149**, 517–532.

WYLLIE, M. R. J., GREGORY, A. R. & GARDNER, G. H. F. 1956. Elastic wave velocities in heterogeneous and porous media. *Geophysics*, **21**, 41–70.

ZIMMERMAN, R. W. 1991. *Compressibility of sandstones*, Developments in Petroleum Science, **29**, Elsevier, Amsterdam.

Simulation of oil expulsion by 1-D and 2-D basin modelling — saturation threshold and relative permeabilities of source rocks

A. OKUI[1,3], R. M. SIEBERT[2] & H. MATSUBAYASHI[4]

[1] *Technology Research Center, Japan National Oil Corporation, 1-2-2, Hamada, Mihama-ku, Chiba 261, JAPAN*

[2] *Conoco Inc., P. O. Box 1267, Ponca City, OK 74603, USA*

[3] *Present Address: Idemitsu Oil Development Co., Ltd., 2-3-4, Okubo, Shinjuku-ku, Tokyo 169, JAPAN*

[4] *Present Address: Japan National Oil Corporation, 2-2-2, Uchisaiwai-cho, Chiyoda-ku, Tokyo 100, JAPAN*

Abstract: In order for oil to accumulate in economic quantities, it first has to be generated and expelled from source rocks in sufficient quantities. In spite of long term efforts, the mechanism of oil expulsion from the source rocks is not completely understood. For modelling expulsion, the adoption of pressure-driven multiple-phase fluid flow governed by Darcy's law is widely accepted. However, relative permeabilities for fine-grained source rocks, which is an essential parameter for this model, are very difficult to specify. The conventional reservoir-rock curve is obliged to be used for the modelling. Simplification of the relative permeability model is generally used for one-dimensional basin modelling. The 1-D model also requires substantial optimization of the key parameter, saturation threshold.

Laboratory measurement of relative permeabilities for fine-grained rocks is very difficult, therefore, we carefully interpreted the analysed laboratory data on various sandstones to establish a relationship between relative permeability curves and pore geometry parameters, with a view to extrapolate the relationships developed in sandstones to fine-grained source rocks. It was found that the relative permeability curves, or irreducible water saturation are controlled by two factors ; grain size and clay content. Since the total surface area of the pore system becomes greater as the grain size decreases, we consider surface water adsorbed on the grain surface as a part of irreducible water. Since clay contains micro-porosity which cannot be displaced by oil due to high capillary pressure, we regard this to also play a role. Prediction of relative permeability curves for fine-grained rocks by both processes results in the curve with high irreducible water saturation.

The new relative permeability curves were tested by both one-dimensional and two-dimensional basin modelling. Test results indicate that the new curves can reproduce expulsion efficiency, locations of accumulations and leaking through cap rock, which is consistent with actual observations.

Basin modelling is giving new insights to oil and gas exploration, since it can evaluate the history of sedimenatry basins quantitatively and integrate many processes occurring in basins. Many geological and geochemical processes are too slow and complex for human beings to integrate quantitatively, but the evolution of computer techniques enables simulation and visualization of these processes. Explorationists generally form several hypotheses or scenarios by the evaluation of present day situations. Artificial experiments by basin modelling can be compared with these hypotheses. Since each module forming the whole basin modelling package is developed by the physical and chemical knowledge, basin modelling can constrain and realize these hypotheses and so reduce exploration risk.

Simulation techniques have been widely used among engineers. Quality, accuracy and reality of the simulations depend not only on the model itself, but also on the input parameters and numerical scheme. For the construction of the model, physical and chemical understanding of the process is essential. This is generally achieved by

OKUI, A., SIEBERT, R. M. & MATSUBAYASHI, H. 1998. Simulation of oil expulsion by 1-D and 2-D basin modelling — saturation threshold and relative permeabilities of source rocks. *In*: DÜPPENBECKER, S. J. & ILIFFE, J. E. (eds) *Basin Modelling: Practice and Progress*. Geological Society, London, Special Publications, **141**, 45–72.

laboratory experiments and the observation of actual data. However, most processes are time-dependent in geological scale, therefore, the compensation of time by another factor such as the temperature for pyrolysis experiments is necessary in the experiments (Tissot 1987; Burnham & Braun 1990; Takeda *et al.* 1990; Ungerer 1990). The key to building an accurate model is the extrapolation of the facts observed in human time-scale to geological time-scale.

The determination of the values for input parameters is also important. Generally, these parameters are coefficients, boundary conditions and initial conditions for the equations. Geologically and geochemically, the input parameters are rock properties such as porosity, permeability and thermal conductivity, fluid properties such as density and viscosity, and source rock properties such as activation energy and frequency factor. Some of the rock and fluid properties are expressed as functions of pressure and temperature. Since pressure and temperature vary with time and space in sedimentary basins, the rock and fluid properties change with time and space through simulation. Fortunately, some of the rock, fluid and source-rock properties are measurable in the laboratory. However, a problem is deciding whether the sample or measured value is representative of the entire basin and its history, since the measurement is generally made on a small piece of the sample collected at present day. A scale effect always exists. Another serious problem with input parameters is that some of them cannot be measured in the human time-scale, even if they are essential to the modelling. In this case, the value which is guessed by the users may be used. The accuracy of values can be increased by optimization processes, which matches simulated results with related observations. Theoretical or empirical models for input parameters may also be used. Input parameters are not always seriously considered in basin modelling, but are a key to successful simulations.

Expulsion or primary migration is the phenomena defined as the movement of oil and gas out of their source rocks (organic-rich, fine-grained rocks) to permeable carrier beds. Pepper & Corvi (1995) distinguished expulsion from primary migration and redefined expulsion as the 'release of oil and gas from kerogen to the inorganic pore network in source rocks', and primary migration as 'the movement through the inorganic pore network in source rocks out to carrier beds'. Pepper & Corvi's (1995) redefinition corresponds to the term called 'adsorption' in the previous work (Pepper 1991; Okui & Waples 1992; Sandvik *et al.* 1992). But in this paper, expulsion and primary migration are considered as almost the same term, as in the traditional sense.

Expulsion is one of the remaining targets in basin modelling research, since the mechanism or process is not fully understood. Expulsion is still the subject for debate among scientists, since the process is dynamic and takes a long geological time. Even if much oil is generated in the source rock, low expulsion efficiency may result in only small amounts of oil being accumulated in traps. With additional heating, this residual oil in the source rock will be cracked to gas and may be expelled as the gas phase.

One of the hurdles to overcome to construct an accurate expulsion model is the difficulty of laboratory experiments (JNOC 1987; 1992). Since the expulsion process is slow, the compensation of time by temperature and pressure is necessary to complete the process. However, the properties of fluids such as density and viscosity are temperature-dependent, so that the flow condition recorded at different temperatures from the subsurface cannot represent those occurring in nature. Lafargue *et al.* (1990; 1994) use a sophisticated procedure for the expulsion experiment to overcome the high-temperature problem. Their experiment is conducted in two stages; the generation of oil is achieved at around 300 °C as the first stage and after the sample cools down to around 100 °C, a pressure gradient is applied to conduct expulsion experiments. However, Lafargue *et al.* (1994) have concluded that the experiments alone are not sufficient to elucidate expulsion completely.

In this paper, we would like to discuss the simulation of oil expulsion in one-dimensional and two-dimensional basin modelling, emphasizing the key input parameters ; saturation threshold and relative permeabilities of source rocks, which are very sensitive to the simulated results. Knowing the problems with the laboratory experiments as mentioned above, the approach we took in this paper is more deductive and practical.

Current knowledge of expulsion

Oil and gas expulsion has been a matter of debate among scientists for a long time. There are several aspects about the debate; mechanism, fluid phase, driving force, and pathway. These discussions have been based mainly on theoretical consideration, well data and laboratory experiment. Before mentioning proposed models, the observations in basins and experiments are summarized as follows.

The parameter which can be easily and directly monitored is the concentration of oil or petroleum in source rocks. Measurements can be made by considering the amount of solvent extracts or S1 value from Rock-Eval pyrolysis. Even in an immature state, small amounts of oil or petroleum are recognized in source rock as initial bitumen, which is thought to be the inheritance from original living matter or early transformer in sediments (Tissot *et al.* 1971). The amount of oil (mg Extract g^{-1} TOC) in source rocks remains low value during a certain depth or temperature, but then appears to increase rapidly beyond a certain threshold in conjunction with the degradation of kerogen (Tissot *et al.* 1971; 1974). This observation was considered as evidence for kerogen as the source material for oil and gas. The amount of extract peaks at 100 mg Extract g^{-1} TOC for hydrocarbons, and 180 mg Extract g^{-1} TOC for hydrocarbons+resins +asphaltenes at 2500 m in the Paris basin (Tissot *et al.* 1971). Meissner (1978) reported a progressive increase of electrical resistivity with depth for the Bakken source rock in the Williston basin. This is interpreted as an increase of pore saturation with oil by generation with temperature.

Moreover, it was observed that the amount of oil in the source rock from the Mahakam Delta in Indonesia increased with maturity, but then depleted above a certain value of threshold (Tissot 1987). The increase in oil amount followed by depletion with depth in source rocks was also reported from the Niigata basin, Japan (Sekiguchi *et al.* 1984). The depletion is interpreted as the expulsion of oil out of the source rock. A time lag therefore exists between the onset of oil generation and expulsion, and the increase of oil concentration in source rock required to expel oil. This observation is also reported from laboratory experiment (Takeda *et al.* 1990; Suzuki *et al.* 1991; Lafargue *et al.* 1994). Compaction pyrolysis experiments (Takeda *et al.* 1990) revealed that the active expulsion was initiated when the concentration of oil reached a certain value and that this value varies with kerogen type. However, it should be also noticed that once the oil expulsion is started, the process does not stop even if the concentration of oil decreases below the threshold value (Sekiguchi *et al.* 1984; Tissot 1987; Suzuki *et al.* 1991).

It was believed that generated oil exists in the pore system of source rocks (Durand 1983; Ungerer 1990). Recently, the role of adsorption of kerogen has been emphasized for retaining oil in the source rock (Pepper 1991; Okui & Waples 1992; Sandvik *et al.* 1992; Pepper & Corvi 1995). It is known that organic matter or kerogen is oil-wettable and indicates fluorescence under U.V. light in some cases. Oil may be trapped in micropores in kerogen or adsorbed on the surface by the polarity of heavy compounds.

Mass balance calculations enabled scientists to access to the calculation of expulsion efficiency (Cooles *et al.* 1986; Mackenzie *et al.* 1987; Rullkotter *et al.* 1988; Pepper 1991; Pepper & Corvi 1995). The calculation is simply a ratio of expelled oil to generated oil, but the initial state of kerogen is difficult to evaluate directly from geochemical analysis on matured and expelled source rock. Therefore, it was assumed that the absolute amount of inert kerogen (Cooles *et al.* 1986) or mineral matrix (Rullkotter *et al.* 1988) remained constant as an unchanged reference. These calculation indicated that expulsion efficiency for various source rocks from the world increases as the potential increases. It was shown that the expulsion efficiency from the rich source rock of which initial S2 value is greater than 5 mg HC g^{-1} Rock is very efficient above 60 %. Pepper (1991) proposed the modification of the Cooles' diagram from initial S2 value to initial kerogen composition (Hydrogen Index) as the horizontal axis. The expulsion efficiency from organic-rich coals are relatively low and gave a scatter to the original diagram. Pepper (1991) concluded that kerogen composition controls the expulsion efficiency rather than the richness. Mackenzie *et al.* (1987) investigated cores of Kimmeridge Clay from the North Sea using Cooles' approach, focusing more on the microscopic scale. It was shown that the expulsion efficiency rises within mudstone–sandstone contact in two cores (30 % to 80 % and 80 % to 92 % in 10 m, respectively), which is interpreted to be a local effect caused by capillary pressure.

The mobility of oil and gas through a source rock has been also investigated. Absolute permeability is the rock parameter used to express the mobility of fluid assuming Darcy's type fluid flow. Magara (1978) reported the measurements of porosity and permeability on the cores from Japan and Canada. Dutta (1987) reported the measurements from the US Gulf Coast. Sandvik & Mercer (1990) measured oil permeability after brine saturation through source rocks. The problem remains whether these measurements represent the values as they occur in nature, since some were measured without overburden, and some were measured by air. Satisfactory measurements indicate that the permeability ranges between 10^{-4} and 10^{-5} Md at a porosity of about 10 %, where source rocks are matured. Okui *et al.* (1994*a*) discussed how mineralogy may alter the behaviour of permeability of the source rock. They found that the permeability of siliceous source rock decreases more rapidly than clayey source rock due to different diagenetic reactions.

Gross compositional fractionation during the expulsion from a source rock was observed both in cores from wells (Leythaeuser *et al.* 1988) and in laboratory experiments (Lafargue *et al.* 1990). In both cases, it was shown that saturated hydrocarbons were preferentially expelled over aromatic hydrocarbons and heavy compounds, which is in accordance with observations on the compositional differences between reservoired oil and shale bitumen. This effect may be attributed to the retention of polar compounds on mineral and/or kerogen surfaces. However, Leythaeuser *et al.* (1988) found that molecular compositional fractionation did not exist, which has been interpreted as the expulsion in bulk oil flow.

In order to account for these facts, several models have been proposed since the 1960s. At first, a model in which oil is expelled in water solution was proposed (Powers 1967; Burst 1969). In this model, the total amount of expelled oil depends on the amount of water expelled from sediments and the solubility of oil in water. With the establishment of oil generation from kerogen at a certain depth or temperature, it was realized that the most of water had already been expelled and little remained when oil was generated. Furthermore, McAuliffe (1980) showed that the solubility of oil is not large enough to make commercial accumulations and dramatically decreases as carbon number increases. This decrease of solubility with increase of carbon number is not consistent with the composition of accumulated oil, which is also inconsistent with this model.

The model of molecular diffusion was proposed by Leythaeuser *et al.* (1982), Kross (1988) and Kross *et al.* (1993). However, it was shown that the molecular diffusion is also an inefficient expulsion method except for gas (Kross 1988). Molecular diffusion favors dispersion rather than concentration, so that it is a leakage process through cap rocks, especially for gas.

A widely accepted model in the petroleum industry now is a pressure-driven bulk flow as a separate oil phase (Durand 1983, 1988; Ungerer 1990), which is in accordance with the previously mentioned observations (Mackenzie *et al.* 1987; Leythaeuser *et al.* 1988). Driving forces can be any excess pressure to hydrostatic pressure created by sediment loading, thermal expansion of fluids and hydrocarbon generation. In addition, capillary pressure and buoyancy play a minor role. Capillary pressure works only at the interface of separated fluids. The direction of capillary pressure depends on the geometry of the rock pore system. Therefore, capillary pressure works both as a driving force and resistance (seal) for the flow. For the expulsion, capillary pressure contributes locally at the boundary of the source rock and carrier rock (Mackenzie *et al.* 1987; Leythaeuser *et al.* 1988).

Rock pore systems and fractures can be the pathway for expulsion. It is interesting to note that the mechanism of microfracturing (du Rouchet 1981; Hunt 1990) is an expansion of the bulk flow model. It was proposed that if pore pressure in the source rock exceeds the strength of the source rock, hydraulic fractures are created, which can release the fluids (oil, gas and water) out of the source rock. This phenomena may be expressed as an enhancement of permeability for source rock in Darcy's equation. Bitumen or oil-filled fractures have been reported (du Rouchet 1981; Talukdar *et al.* 1987), which would be direct evidence for the expulsion through microfractures. But this process is still a matter of debate.

Expulsion model in basin modelling

Only when the process of expulsion is fully understood, can an efficient model be produced. As discussed in the previous section, a pressure-driven bulk flow as a separate oil phase is widely accepted as an expulsion mechanism to explain the observations in basins and experiments, even though the expulsion process has not been fully understood. In two-dimensional basin modelling, Darcy's law with the relative permeability concept was tested and is generally used to model expulsion as a pressure-driven bulk flow of a separate oil phase. (Welte & Yalcin 1988; Ungerer *et al.* 1990; Hermanrud 1993; Okui *et al.* 1996).

Darcy's law is an empirical relationship between specific discharge, hydraulic head and hydraulic conductivity, proposed in 19th century, which can be theoretically derived from the Navier–Stokes equation under some assumptions. Hubbert (1940) introduced the term, hydraulic potential, which accounts for the energy by fluid pressure and elevation, and re-wrote the Darcy's equation. Darcy's law has been widely used and well established in reservoir engineering (Muskat 1949; Amyx *et al.* 1960; Cosse 1993; Fig. 1). This law can be applicable to secondary migration of oil and gas, since carrier rocks are equivalent to reservoir rocks in a basin scale. However, the application of Darcy's law to fine-grained rocks has not been confirmed. Different physical and chemical characteristics of fine-grained rocks may result in different flow behaviour from reservoir rocks.

Relative permeability is a parameter measured in the laboratory which expresses the behavior of multi-phase fluid flow through rocks (Caudle *et al.* 1951; Amyx *et al.* 1960). A rock plug saturated with one fluid is placed in the equipment and then another fluid is injected under a ceratin pressure. Pressure drop, production rate of both fluids, and saturations in the sample are monitored. An accurate but time-consuming method is to inject two fluids into the rock in a certain ratio and then measure the production rates when the ratio of

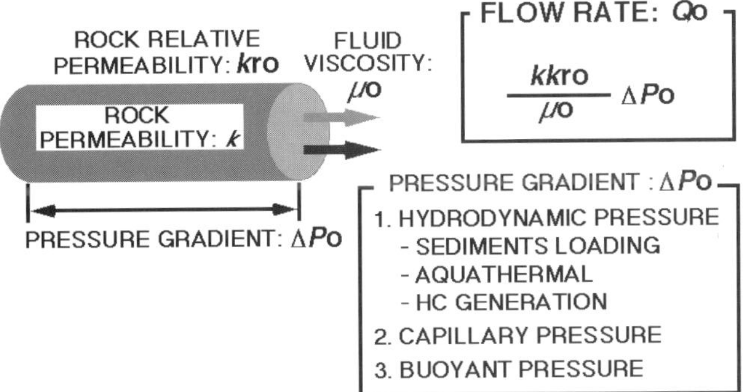

Fig. 1. Darcy's law adapted for multi-phase fluid flow with relative permeability concept.

produced fluids become equal to injected fluids, by which time the saturations of fluids are expected to be steady-state. Hence, relative permeability is considered to be an empirical relationship or record of these displacement experiments under a certain condition . The results are expressed as a diagram which reflects the mobility of each fluid phase as a function of saturation (Fig. 2). Generally, the mobility of one fluid increases as the saturation increases, but the relationship is not simple enough to be linear. Furthermore, the end points of each relative-permeability curve normally do not intersect with the axes at 100 % saturation. The point where the relative permeability to oil is equal to zero is defined as 'residual oil saturation (Sor)', and the point where the relative permeability to water is equal to zero is 'irreducible water saturation (Swirr)'. A hysteresis is also observed, indicating that the behaviour of the curves is different in cases where saturation of one fluid is increased from that where the saturation is decreased (Fig. 2). For an oil–water system, the case of increasing oil (non-wetting phase) saturation is defined as 'drainage', and decreasing oil (non-wetting phase) saturation is 'Imbibition'. The wettability of the rock also effects the curves.

Laboratory-derived relative permeability curves are sometimes anomalous, which raises questions regarding sampling procedures, core preservation and the methods of analysis. However, they also

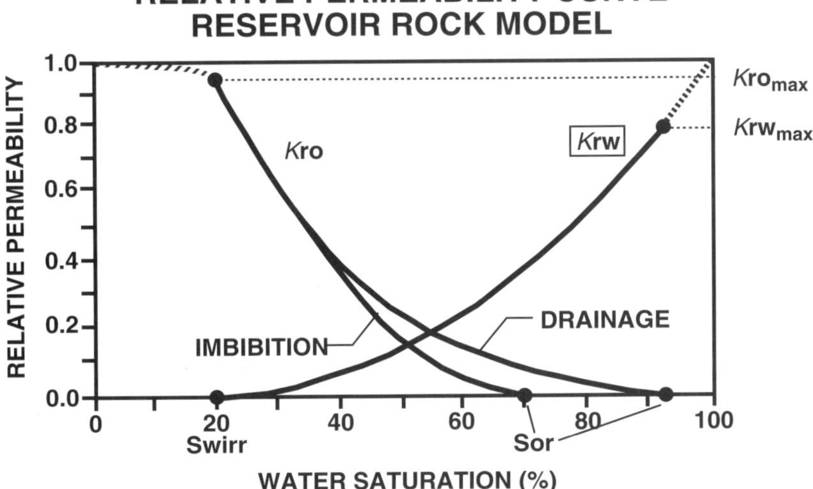

Fig. 2. Typical relative permeability curves for reservoir rocks. *k*ro : relative permeability to oil; *k*rw : relative permeability to water; Swirr : irreducible water saturation; Sor : residual oil saturation. The drainage is the curve derived from increasing the non-wetting phase saturation and the imbibition curve is derived from decreasing the non-wetting phase saturation.

suggest that other factors such as the pore geometry or the lithologic character of rocks may play an important role. The geological link between the behaviour of relative permeability curves and pore geometry of the rock has not been well established. Since laboratory measurements are very difficult, the relative permeability for fine-grained rocks is another pitfall in modelling oil and gas expulsion. Most basin modelling software is limited to using the analogous curves measured from reservoir rocks (Ungerer *et al.* 1990; Hermanrud 1993; Rudkiewicz & Behar 1994). However, since it is expected that the behaviour of fluid flow in organic-rich fine-grained source rocks is quite different from reservoir rocks, the relative permeability for fine-grained rocks is not likely to be analogous to reservoir sections.

Fluid flow in a basin occurs in three special dimensions. In one-dimensional basin modelling, only burial and thermal history are simulated and a simplified expulsion model is required. One of the widely-used expulsion models in one-dimensional basin modelling is an overflow or saturation threshold model (Ungerer 1993). Porosity is derived from the burial history and compaction behaviour, and the amount of oil generated is calculated from the thermal history and source rock potential to give the saturation in the source rock. The mass of oil generated should be converted to volume by the density of oil. If the calculated saturation exceeds a given threshold value, expulsion starts. All the excess saturation above this threshold value is assumed to be expelled in each time step.

One of the problems of this model is the value for saturation threshold, which cannot be directly derived from any geochemical parameter. A 20 % to 30 % saturation threshold has been recommended (Durand 1988), which might appear to be based on the residual oil saturation in the imbibition relative permeability curve. The threshold value may vary with geological and geochemical setting, such as kerogen type. Therefore, it is recommended that the threshold value should be optimized by measurable geochemical parameter such as S1 from Rock-Eval pyrolysis (Ungerer 1993).

Relative permeabilities for fine-grained source rocks

The adaptation of Darcy's law to multi-phase fluid flow with a relative permeability concept may improve the modelling of oil and gas expulsion. Pressure-driven bulk flow appears to be the principal expulsion mechanism (Mackenzie *et al.* 1987; Leythaeuser *et al.* 1988), but the application of a reservoir-type relative permeability curve is suspected to be incorrect.

Wettability

Wettability is another factor controlling multi-phase fluid flow and relative permeability curves besides pore geometry (Amyx *et al.* 1960; Anderson 1987). Wettability is a rock parameter which effects the behaviour of capillary pressure as well as interfacial tension. Capillary pressure works at the interface between two immiscible fluid phases. When a non-wetting phase enters a pore, capillary pressure works as a resistance which makes difficult to be injected into finer pores. On the other hand, capillary pressure works as a driving force to inject the wetting phase into pores. Therefore, this behaviour affects multi-phase fluid flow and hence the relative permeability curve.

Mechanisms of wettability are not fully understood, but it is believed that inorganic rock minerals exhibit water-wet characteristic and organic matters exhibit oil-wet character (Fig. 3; Amyx *et al.* 1960; Baker 1980; Pepper 1991). Rock minerals have maintained a film or layer of structured water at its surface by hydrogen bonding. Organic matter adsorbs oil at its surface by polar bonding. These molecular-scale physical and chemical characteristics appear to control the wettability of materials. Wettability of the whole rock may depend on the wettability of constituting materials and their contents. Generally, good source rocks contain several weight percent of organic matter, which corresponds to only about 10 % by volume due to the low density effect of organic matter. Therefore, most source rocks except rich coal seams are predominatly water-wet systems. However, the key factor is the distribution and connectivity of the organic matter. If the organic matter creates a pore network, the generated oils can be easily expelled through this oil-wet network. McAuliffe (1979) proposed that 5 % by weight organic matter content is the threshold above which the kerogens forms a network. Kuo *et al.* (1995) also found that 4 or 6 % by weight is necessary, which suggests that this mechanism only works for very rich source rock. Therefore, we assume that the system in the most of source rocks are water-wet as a whole and eliminate the data of oil-wetted reservoirs for this study.

Drainage or imbibition

The saturating direction also affects multi-phase fluid flow and relative permeability (Fig. 3; Amyx *et al.* 1960; Raimondi & Torcaso 1964). In a water-wet system, the process of increasing the non-

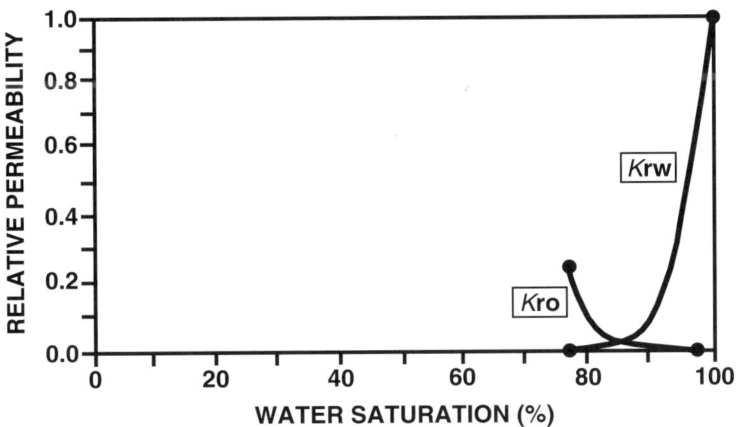

Fig. 3. Relative permeability curves for fine-grained rocks with absolute permeability = 10^{-4} mD produced by Okui & Waples (1993). Reprinted with permission of Elsevier.

wetting phase (oil) saturation is defined as 'drainage', and the process of decreasing non-wetting (oil) saturation is 'imbibition'. Generally, the imbibition process causes the non-wetting phase (oil) to lose its mobility at a higher value of saturation (corresponding to residual oil saturation) than the drainage process. In other words, the movement of oil by the drainage process can occur at lower saturations than imbibition. This may be because small oil globules can flow with water through larger pores, but the connectivity of the continuous oil phase is easily lost when depleting oil.

The process of oil and gas expulsion may be dynamic and unstable. If the rate of oil generation exceeds the rate of expulsion, the saturation in the source rock increases which corresponds to the drainage process. In the opposite case, the saturation decreases which corresponds to the imbibition process. Actual expulsion may start with the increase of oil and gas saturation by rapid generation (drainage process), and after peak generation, the saturation may be decreased by rapid expulsion triggered by high saturation (imbibition process). Since we have more interest in the onset of expulsion, we assume that the drainage process represents the saturation history for oil and gas expulsion in this study.

Previous work

The understanding of relative permeability or multi-phase fluid flow has been mainly studied in permeable reservoir rocks (Amyx *et al.* 1960; Morgan & Gordon 1970; Morrow 1970, 1971; Wardlaw & Cassan 1979; Wardlaw 1980; Moraes 1991; Melas & Friedman 1992; Coskun *et al.* 1993; Okui & Waples 1993), since laboratory measurements on low-permeable fine-grained rocks present challenges. The measurement of liquid relative permeability is limited to silt size, whose absolute permeability is about 1 to 10 Md. Reservoir engineers have always been interested in the end points of the relative permeability curves, since recovery of oil is related to residual oil saturation and reserve is related to irreducible water saturation.

Wardlaw & Cassan (1979) investigated 27 sandstone cores from various depths, ages and geographical locations. The results of relative permeability tests were related to petrophysical data such as porosity and permeability, as well as to petrographic observations and direct observations of pore structure made from resin casts. 27 variables for 27 sandstone were statistically analysed and correlation coefficients were obtained for each pair of variables. It was shown that high recovery efficiency is significantly related to high porosity, small pore-to-throat size ratio, small mean particle size and low percentage of carbonate, but, significantly, not to absolute permeability. Wardlaw (1980) conducted two laboratory experiments on artificial samples made of glass, and emphasized that the pore-to-throat size ratio is of first-order importance in affecting residual oil (non-wetting phase) saturation. It was demonstrated that large pore-to-throat size ratio resulted in high residual saturation. Moraes (1991) also calculated the pore-to-throat size ratio of lacustrine deltaic and turbiditic sandstones in Brazil. The pore size was measured on thin section and the throat size was derived from a mercury injection test. Moraes (1991) concluded that the pore-to-throat size ratio is probably a major control on residual oil saturation. Primary porosity in these sandstones mostly occur as 'islands' of small, isolated, commonly triangular pores apparently connected by small pore throats. The superposition of different diagenetic processes and products significantly modified the original pore structures.

Studies on irreducible water saturation have also been conducted (Morrow 1970, 1971). Morrow (1970) used a drainage column with random packing of equal spheres such as glass beads. An air–water system was introduced and the irreducible wetting-phase saturation was measured as it varied with variable fluid properties (density, surface tension, viscosity, viscoelasticity), wettability, solid properties (grain size and its distribution, grain shape, heterogeneity), and consolidation. The results indicated that high irreducible water saturation is obtained in the case of heterogeneous packing, which is formed by distributing clusters of fine beads in a matrix of coarser beads. Morrow (1970) reasoned that coarse beads can drain water, but hydraulic conductivity to fine bead clusters was lost before the drainage pressure became sufficient for draining water from fine beads clusters. Morrow (1971) conducted another series of experiments using unconsolidated reservoir cores. The cores were crushed, mixed and packed in a cell. These packed samples were resaturated and irreducible water saturation measured. The irreducible water saturation after mixing indicated much lower values than the original one, which was interpreted to show that the original heterogeneity of packing was eliminated by the mixing process. Moraes (1991) revealed that high irreducible water saturation is related to high residual oil saturation due to high pore-to-throat size ratio. Melas & Friedman (1992) investigated the Jurassic Smackover carbonates in Alabama and Florida, and found a relationship between throat size distribution and irreducible water saturation. The rock exhibited a high irreducible water saturation indicating a wider and broader throat-size distribution.

Morgan & Gordon (1970) investigated rock samples by microscopic examination on thin section, and emphasized the effect of surface area on the relative permeability curves. Morgan & Gordon (1970) concluded that rocks with small pores have a larger surface area and so a larger irreducible water saturation that leaves little room for the flow of fluids. Okui & Waples (1993) tried to establish the link between pore geometry and relative permeabilities using the dataset in Morgan & Gordon (1970). Okui & Waples (1993) found that the grain size is one of the parameters which effects relative permeabilities. The extrapolation of a quantitative relationship found in this range of reservoir rocks to fine-grained source rocks enabled us to successfully produce a relative permeability for fine-grained rocks, which suggests much higher irreducible water saturations than reservoir rocks (Fig. 3). Maximum relative permeability to oil at irreducible water saturation was also decreased in fine-grained rocks. The simulation of expulsion by applying this new set of curves successfully reproduced the expulsion behaviour observed in basins (Waples & Okui 1992; Okui & Waples 1993).

Theoretical consideration

Theoretical consideration of relative permeability is rarely found (Colonna *et al.* 1972). Only mathematical treatment such as fractal has been tried (Helba *et al.* 1992). Interpretation of actual data must be improved by theoretical consideration. Therefore, the behaviour of relative permeability curves were briefly discussed before working with actual data.

First, consider a simple pipe model (Fig. 4). We assume that there are only two straight pipes in a rock with different diameters which are saturated with water as a wetting phase and are suffered to displacement of water by oil (Fig. 4). If flowing pressure is below capillary pressure of both pores, oil cannot enter any pore, which has a water saturation of 100 %, relative permeability to water of 1 and relative permeability to oil of 0. At the moment when the flowing pressure exceeds the capillary pressure of the larger pore, oil is suddenly injected into the larger pore, at which time water saturation is decreased by the volume of larger pore. The relative permeabilities are proportional to the area of the base of each pipe. A flowing pressure larger than capillary pressure of the finer pipe can inject oil into all pores, which gives 0% water saturation. The relative permeability to water is 0 and the relative permeability to oil is 1.

In the next case, we look at three pipes with different diameters (Fig. 4). The procedure of displacement by oil is almost the same as in the previous case. When increasing flowing pressure, non-wetting oil is injected from larger pores to finer pores according to the flowing pressure. Water saturation steps down when flowing pressure reaches the capillary pressures corresponding to each pore size. The relative perme-

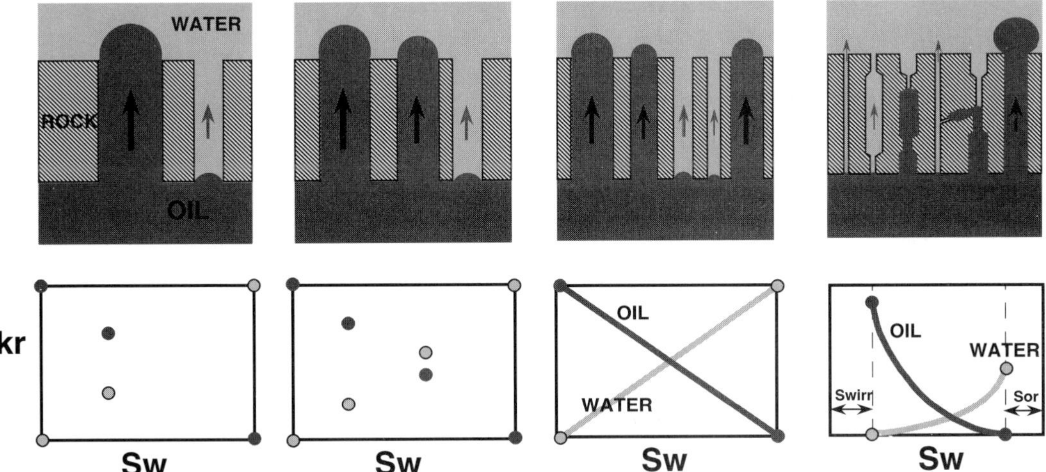

Fig. 4. Multiple-phase fluid flow in various pore structures. The process of displacement of water by oil is expressed as a relative permeability diagram. Surface water is not considered. The combination of pores and throats produces curvature and the shift of endpoints in natural pores.

abilities also change proportionally to the base area of each pipe as oil is injected.

If there are straight pipes with various sizes (Fig. 4), the displacement process with increasing flowing pressure proceeds almost continuously from finer pores to larger pores and will give a linear relationship. Since straight pipes are assumed, the volume of each pipe is proportional to the area of each pipe, which results in the liner relationship between saturation and relative permeability. The assumption of straight pipes and the ignorance of bond water at the surface of the pipes are the reasons why irreducible water saturation and residual oil saturation are nil.

Actual pore networks in rocks are not as simple as the pipe model (Fig. 4). Pores are not straight, do not have smooth surfaces, adsorb wetting fluid, are connected by narrow throats, are distributed in three-dimensions and so on. If oil enters pores connected by narrow throats and flowing pressure cannot overcome the capillary pressure of the narrow throat, that pore is capped by oil and cannot contribute permeability to any fluid. Therefore, relative permeability is decreased by the base area of that pore, even though some oil is trapped and contributes to the increase in the saturation. This process causes curvatures of the relative permeability curves. The water which cannot be displaced due to this capping remains as irreducible water saturation. It is suggested that the degree of complexity in natural pore networks provides the variation in relative permeability curves.

Methodology

Relative permeability curves appear to have a relationship to pore geometry. But pore geometry is a fuzzy term. It appears to be difficult to express pore geometry by only one measurable parameter. Pore geometry and the relative permeability curves could be expressed by multiple parameters. Since actual measurement of relative permeability for fine-grained rocks is almost impossible, the objective is to establish a relationship in reservoir rocks that can be extrapolated to fine-grained source rocks using the method of Okui & Waples (1993).

Database

We have compiled about 150 relative permeability datasets from over the world. There are many parameters to express relative the permeability curve and pore geometry. In order to obtain an approximate relationship, it is preferable to select appropriate parameters representing each category. Absolute air permeability was selected for pore geometry, since it is routinely measured and depends on pore or throat size distribution (Okui & Waples 1993). Irreducible water saturation is the favorable parameter for relative permeability, since the drainage process is assumed and maximum relative permeability to oil appears to be related to irreducible water saturation (Okui & Waples 1993).

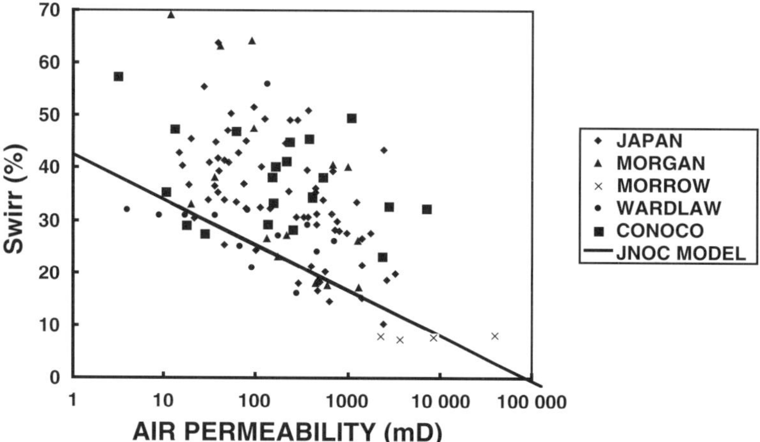

Fig. 5. Plot of Swirr versus log absolute permeability (air) for sandstones. JAPAN came from the Japanese reservoir rocks in the Pliocene. MORGAN, MORROW, and WARDLAW refers to Morgan & Gordon (1970), Morrow (1970), and Wardlaw & Cassan (1979), respectively. CONOCO was obtained from CONOCO core storage for this study.

The data for irreducible water saturation was plotted against air permeability for 150 collections (Fig. 5). JAPAN data came from actual measurements on Japanese Pliocene aged reservoir rocks. MORGAN, MORROW, and WARDLAW refers to Morgan & Gordon (1970), Morrow (1970), and Wardlaw & Cassan (1979), respectively. CONOCO data was obtained for this study. These samples were collected in CONOCO core storage and analysed by CONOCO laboratory. JNOC MODEL represents the relationship derived from the study by Okui & Waples (1993) for clay-free clean reservoir rocks and this relationship is interpreted as being caused by the change in grain size. The data indicate a scatter from 1 mD to 10 D permeability and from 5 % to 70 % irreducible water saturation, but the distribution appears to be triangular. Note that the plotted area almost overlaps with the initial MORGAN dataset and the JNOC MODEL traces the bottom of the distribution. This triangular shape suggests that another factor controls the distribution.

Interpretation of data

For additional interpretation, we will focus on 21 CONOCO samples (Table 1), since these samples were selected for this study and the complete dataset is available. Unsteady-state relative permeability measurements, absolute permeability measurement (air, brine, and brine with overburden), high-pressure mercury injection test (up to 10 000 psi (700 ksc)), and helium porosity measurements were conducted in CONOCO laboratory. Microscopic examination including grain size measurements and 400 point counting was also conducted.

Cores were systematically selected according to grain size and clay content. These variations are very important, to allow an efficient investigation of the character of the pore network. First, the irreducible water saturation of these 21 samples was plotted against air permeability grouped by grain size (Fig. 6a) and clay content (Fig. 6b), which were qualitatively defined by the microscopic examination. In Fig. 6a, each group (fine, medium and coarse grain size) is clearly separated, which is consistent with the Okui & Waples' (1993) study on the data of Morgan & Gordon (1970). However, the distribution of each group is elongated and its direction is intersected with the JNOC MODEL trend, which suggests that another factor also controls the distribution in each group. Also in Fig. 6b, each group (clean, moderate and clayey) is separated. The elongation of each group appears to be parallel to the JNOC MODEL trends, which infers that grain size controls the distribution in each group. As a first approximation, the irreducible water saturation appears to be controlled by both grain size and clay content based on this qualitative evaluation.

In order to verify this idea, quantitative parameters were derived and the relationships to the irreducible water saturation were evaluated. First, the irreducible water saturations were plotted against grain size, grain sorting and clay content (Fig. 7). However, these plots indicate that there is no clear relationship with only one parameter and

Table 1. *List of reservoir samples used for the interpretation of relative permeability curves in this study. All samples are supplied by CONOCO Inc.*

Sample name	Microscopic examination		He porosity	*k*air	Swirr
	grain size	clay content	%	mD	%
KVQ 3117	Fine	Clayey	17.95	3.13	57.3
MNG 11790.4	Fine	Clean	10.6	11.02	35.6
MNG 11789.1	Fine	Clean	11	13.62	47.5
MNG 11804.7	Fine	Clean	11.4	18.76	29.1
LEX 3474	Fine	Clean	12.7	40.68	27.5
KKZ 3272	Fine	Clayey	19.8	61.42	47.1
JZA 3220C	Fine	Clean	14.3	145.97	29.4
LWE 7417	Fine	Moderate	23.1	145.95	33.4
LAZ 9968	Fine	Moderate	21.6	155.49	38.2
LGW 3116A	Medium	Moderate	17.2	183.51	40.3
KEB 1840	Medium	Clayey	19.2	235.07	41.3
LWF 8104	Fine	Clean	24.9	273.18	28.3
KFQ 7584	Fine	Clayey	22	300.01	44.9
JVY 9210	Medium	Moderate	19.62	402.53	34.3
LAZ 10147	Medium	Moderate	22.3	415.58	45.5
LAZ 10112	Medium	Moderate	25.3	492.07	34.6
JUO 9	Fine	Moderate	21.7	602.75	38.2
LVY 10206	Coarse	Clayey	20.6	1244.45	49.6
MMD 2011	Medium	Clean	22	2887.2	23.2
LAZ 10118	Coarse	Clayey	25.1	2913.11	32.8
LVY	Coarse	Clean	22.8	7674.42	32.4

several parameters contribute to the irreducible water saturation. Therefore, we divided the irreducible water saturation into two portions; SwirrG which corresponds to the irreducible water saturation calculated by JNOC MODEL (minimum value of the distribution from all data at a certain permeability), and SwirrC which corresponds to actual measurement minus SwirrG (the excess of irreducible water saturation over SwirrG, Fig.6b). We expected that SwirrG would be controlled by grain size, and SwirrC by clay content, based on the qualitative evaluation discussed before.

The plot of SwirrG against average grain size derived from microscopic examination (Fig. 8a) indicated a correlation as expected. SwirrG increases as grain size decreases. However, the data indicated a scatter which may be due to grain sorting. In order to consider the sorting effect, we introduced a parameter 'effective pore area (m^2 g^{-1})'. Effective pore area was derived from a mercury injection test, in which the variation in injected volume with injected pressure was monitored. Injected pressure can supply the information about pore and throat size. From pore size and injected volume under a certain pressure, total surface area of mercury-injected pores can be estimated. The pressure drop during relative permeability measurements can be converted to equivalent mercury injected pressure by interfacial tensions at air–mercury and water–oil phase boundaries. This converted mercury injected pressure can supply the information about displaced pores by oil during relative permeability measurements. This procedure enables us to calculate total surface area displaced by oil during relative permeability measurement. The total surface area is better than the grain size since it includes the information on the distribution. So the effective pore areas increase as grain size decreases, and its good correlation with SwirrG confirms that SwirrG is strongly related to the surface area of pore system (Fig. 8b).

SwirrC was plotted against clay content, derived from point counting under a microscope (Fig. 9a). There is a scatter, but SwirrC roughly increases as clay content increases. Morrow's (1971) experiment allowed us to consider the SwirrC originating from microporosity in clay minerals. Now consider the porosity in each clay. Assuming the clay minerals were compacted as well as the whole rocks, we define the clay content multiplied by the porosity of whole rock as microporosity in the whole rocks. Minor contributions from dissolved porosity with narrow throats connected to main pores are also involved. The SwirrC appears to correlate with the ratio of microporosity to total porosity (volume occupied by microporosity in total porosity) with some scatter (Fig. 9b). The scatter may due to the error in the estimation of porosity in each clay. It is suggested that the water

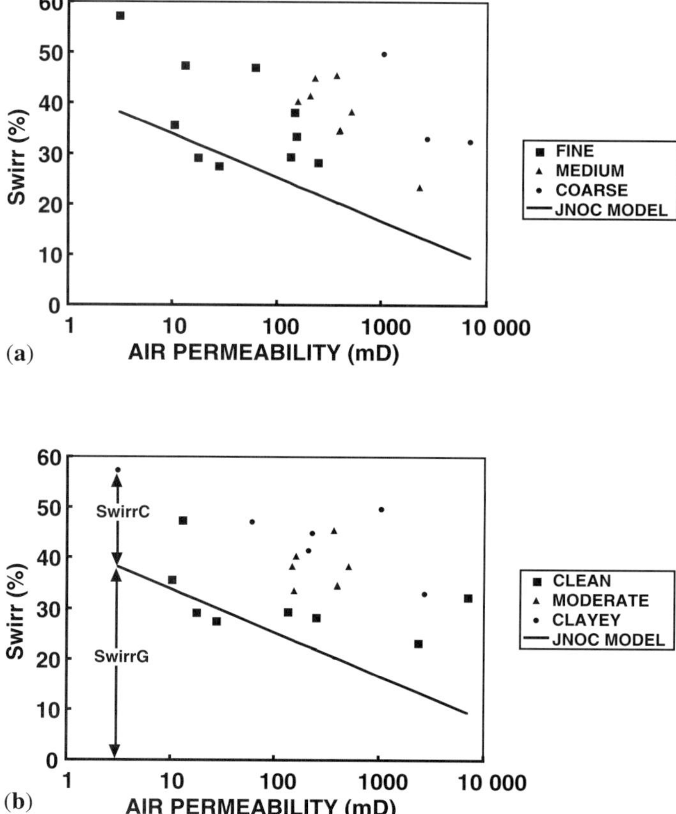

Fig. 6. Plot of Swirr versus log absolute permeability (air) for CONOCO samples. (**a**) grouped by grain size, (**b**) grouped by clay content. Each group is clearly separated, which may indicate that grain size and clay content control Swirr. Swirr is divided into SwirrG and SwirrC. SwirrG corresponds to a minimum value of Swirr distribution (Fig. 5) and SwirrC is Swirr minus SwirrG.

in microporosity cannot be replaced by oil due to the high capillary pressure of the narrow throat and remains as irreducible water saturation.

Model for fine-grained rocks

These interpretations conclude that irreducible water saturation in reservoir sandstones can be controlled by two factors: surface water adsorbed on a mineral's surface and microporosity in clay minerals. Surface water is also called bonded or structured water and exists as a molecular film surrounding minerals by a hydrogen bond due to a local electronic imbalance of the crystal, or local capillary pressure due to the roughness of mineral surface. This mechanism also causes the water-wet character of sandstones. Overlap of both effects (surface water and microporosity) causes the complexity of the behaviour of irreducible water saturation and hence relative permeability curves (Fig. 10). The clue to constructing relative permeability curves for fine-grained rocks is the understanding of the curve in reservoir sandstones as discussed above. The curve for fine-grained rocks can be derived by the extrapolation from sandstones using an adequate scale parameter. However, we have to keep in mind the differences between sandstones and source rocks (shales), which are, for example, organic matter, mineralogy, grain shape, and grain size.

Source rocks generally contain more organic matter than reservoir rocks, which affects wettability. As mentioned before, the behaviour of multi-phase fluid flow is suddenly altered, if this organic matter forms a network. The threshold of organic richness to form the network is supposed to be above 5 % (McAuliffe 1979; Kuo *et al.* 1995). Therefore, in most source rocks, organic matter

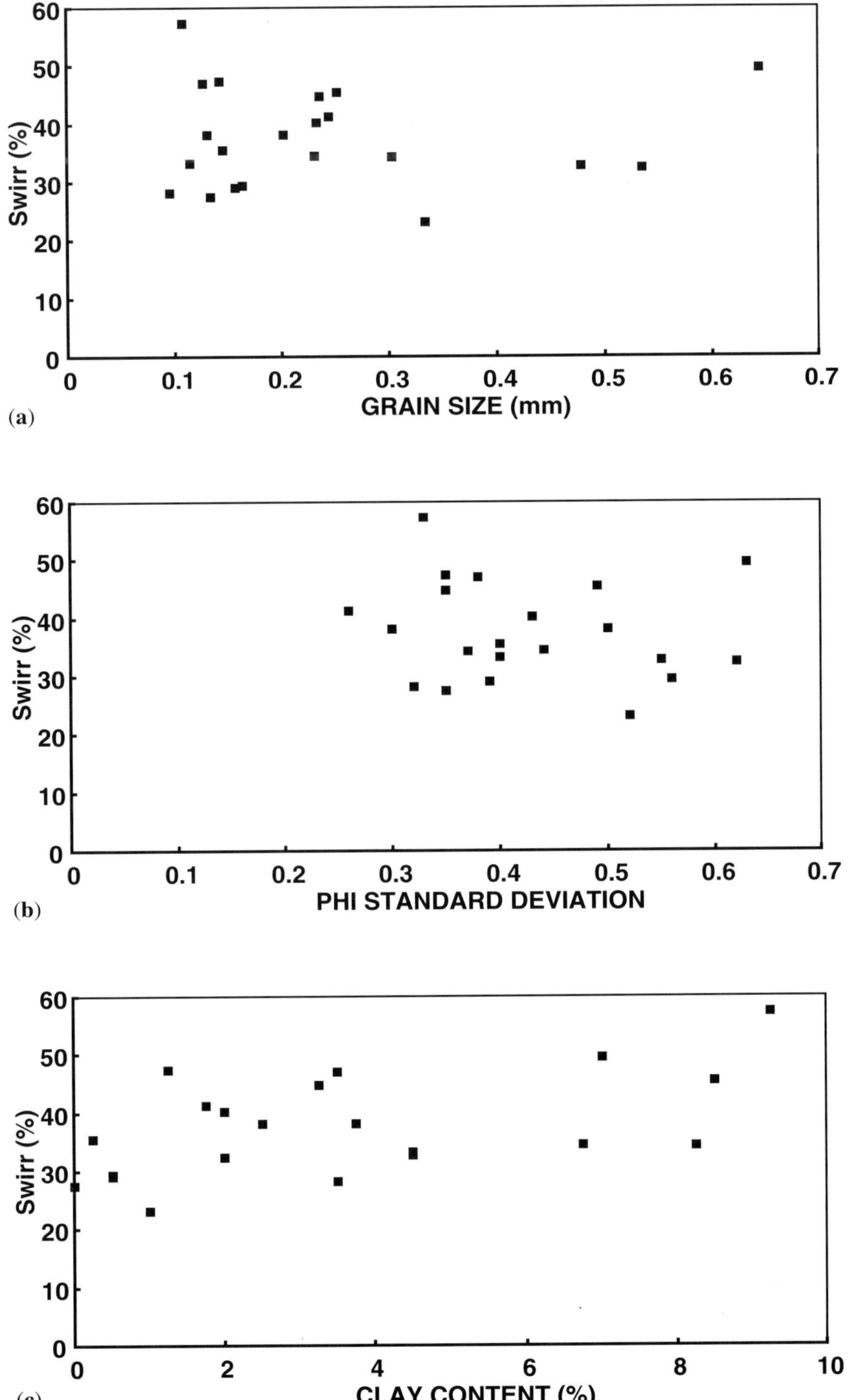

Fig. 7. Plot of Swirr versus (**a**) grain size, (**b**) grain sorting, and (**c**) clay content. All parameters were derived from microscopic examination and 400 point counting.

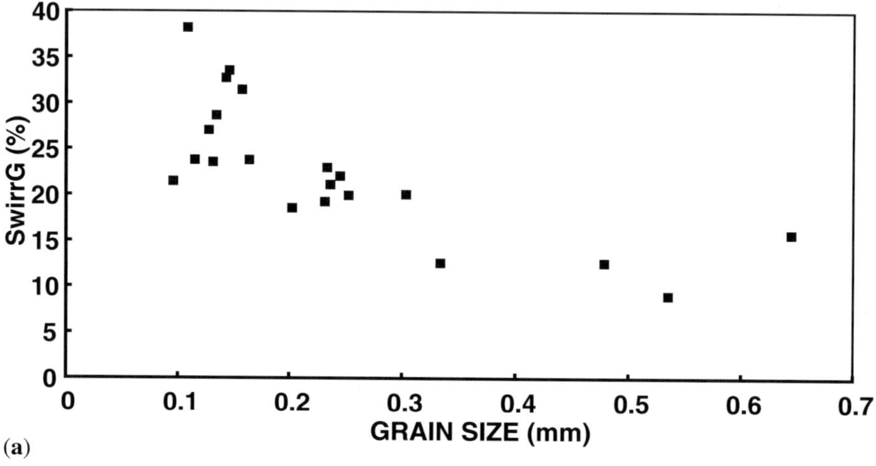

(a)

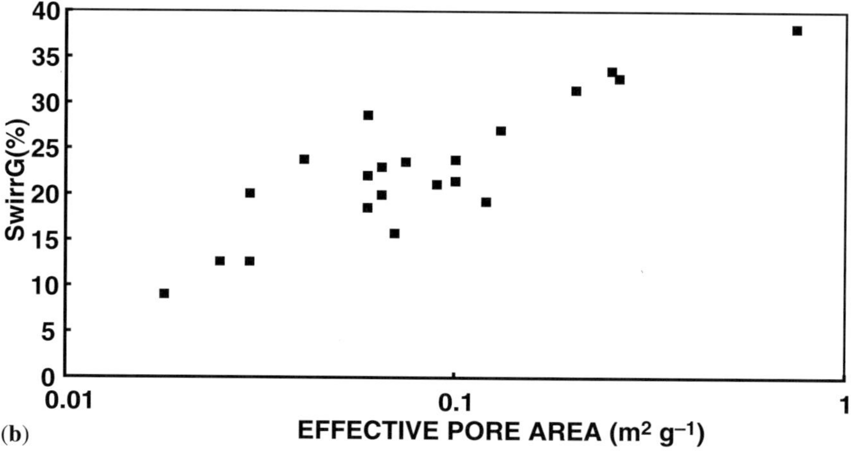

(b)

Fig. 8. Plot of SwirrG versus (**a**) grain size and (**b**) effective pore area. Effective pore area increases as grain size increases. Effective pore area is the total area of the pores occupied by oil during relative permeability measurements. SwirrG is related to the total surface area of the pores occupied by oil.

cannot form the network and oil (non-wetting phase) has to flow through a water-wet pore system, which may be similar to reservoir sandstones. Source rocks or shales are regarded as containing more clay minerals than sandstones. Clays are generally phyllo silicate minerals charged at surface by a local electronic imbalance. Quartz and feldspar minerals which are the main components of sandstones have more rhombohedral and tetragonal shapes. This difference in grain shape may bring about the different pore system in source rocks. Additionally, different crystal structures and hence different strength of hydrogen bonds may bring about the different thickness of surface water. Sandstones, especially shaley sandstones may have a bimodal distribution of grain size; one mode corresponds to the quartz grain size and the other corresponds to clay sizes, which also results in bimodal pore and throat size distribution. When displacing water in the pores created by quartz and feldspar by oil, the clay pores cannot be injected and it remains as irreducible water. The range of grain sizes in source rocks (shales) may be narrower than in bimodal shaley sandstone.

Recognizing these problems, we simply used the extrapolation by grain size and clay content as the scale parameters, since the interpretation of actual

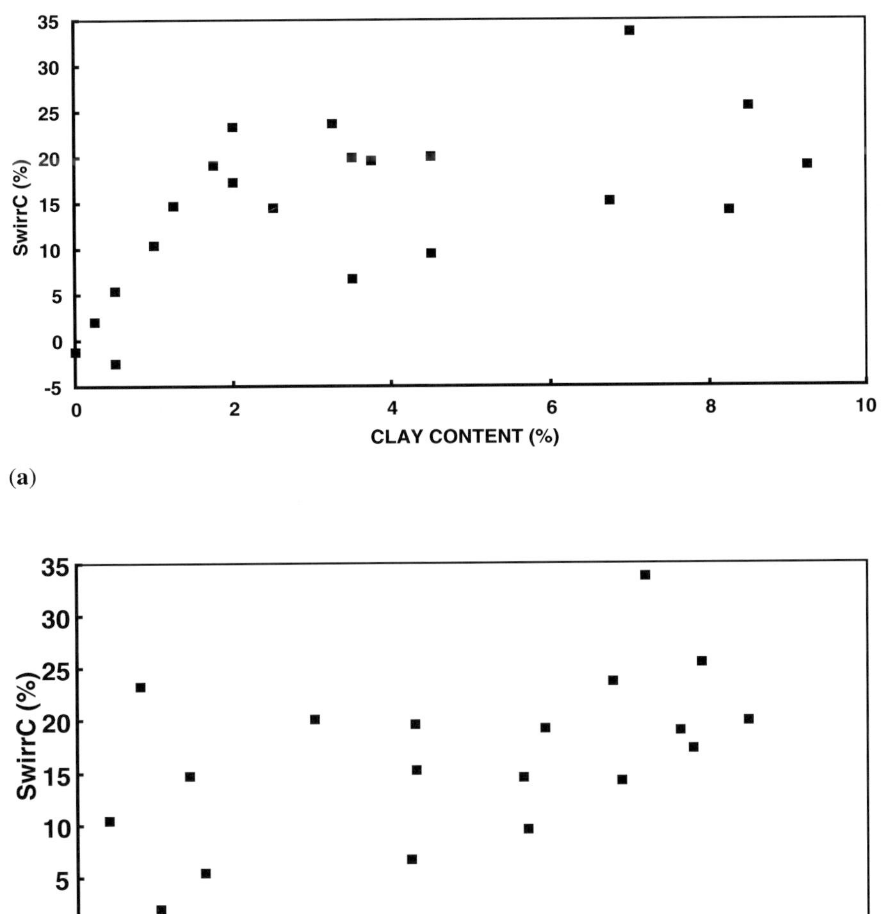

Fig. 9. Plots of SwirrC versus (**a**) clay content and (**b**) micro-porosity in total porosity. Micro-porosity is defined as (clay content) × (porosity + dissolved porosity). SwirrC is correlated with the volume occupied by micro-porosity in total porosity.

data indicated that these factors appear to control irreducible water saturation. Either decreasing grain size or replacing quartz and feldspar by clay mineral is the process by which fine-grained source rocks (shales) are approached. As discussed above, both processes increase irreducible water saturation, which suggest that source rocks should have high irreducible water saturation. Therefore, we used grain-size trends for the extrapolation and absolute permeability for the scaling parameter, as Okui & Waples (1993) did. As grain size decreases, pore size decreases, but assuming constant thickness of surface water, the ratio of the volume occupied by surface water in pores increases. The empirical relationship for this process is as follows :

$$\text{Swirr} = 42.5 - 8.62 \times \log_{10}k \quad (1)$$

$$\text{kro@Swirr} = 0.567 \times \exp(0.218 \times \log_{10}k) \quad (2)$$

where Swirr is the irreducible water saturation in %, kro@Swirr is the maximum relative permeability of oil at Swirr, and k is the absolute permeability in mD. Eqs 1 and 2 cause the relative permeability change shown in Fig. 11. In the case of very low absolute permeability, irreducible water saturation exceeds 100 % by this equation,

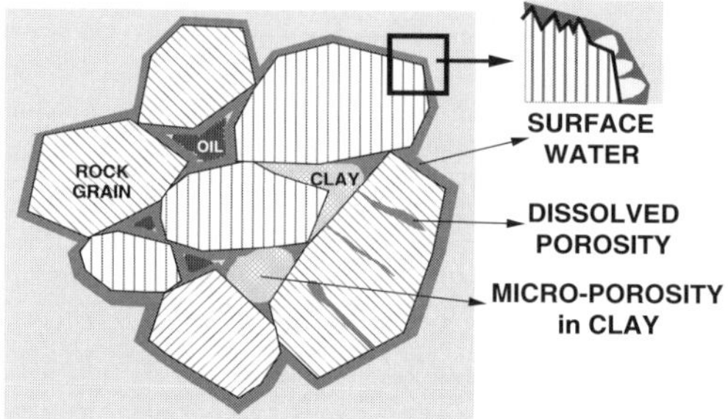

Fig. 10. Schematic diagram indicating nature of irreducible water in reservoir rocks. Bond water is adsorbed on the mineral surface and the micro-porosity in clay contributes to the irreducible water. Dissolved porosity with a narrow throat also plays a minor role.

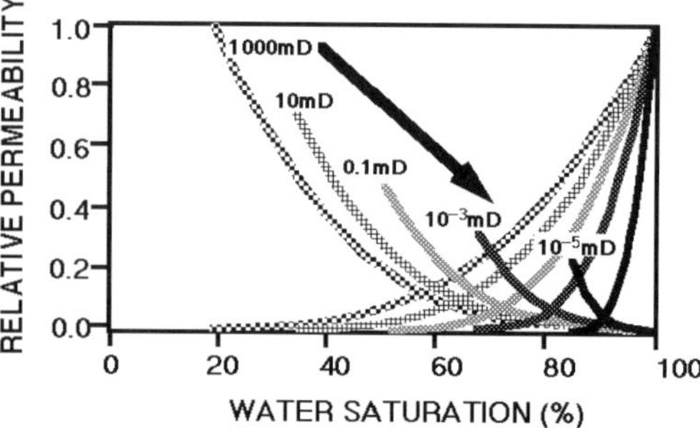

Fig. 11. Evolution of relative permeability curves as a function of absolute permeability derived from this study. Decreases in absolute permeability is the result of throat or pore size decreases.

which may suggest that all pore volume is occupied by irreducible or surface water, or that the equation should be modified in this range.

Basically, eqns 1 and 2 express the change of relative permeability curves by the decrease of pore and throat size (absolute permeability) caused by grain size decrease. The decrease of pore and throat size is also achieved by compaction, therefore, this model can also be applied to the compaction of source rocks (Fig. 12). Compaction causes fluid expulsion from pores, and assuming constant thickness of surface water, expelled water is composed of free water, and hence the ratio of surface water (irreducible water) in pore space increases.

Test of new curve by 1-D basin modelling

New relative permeability curves for fine-grained rocks were tested to evaluate their validity. One-dimensional basin modelling to calculate oil expulsion efficiency was conducted and the efficiency was compared with the estimates derived from actual geochemical analyses. Oil expulsion efficiency from actual data is based on the mass balance method of BP's scientists (Cooles *et al.* 1986; Pepper 1991; Pepper & Corvi 1995).

The model used for the calculation is the simple one-dimensional or zero-dimensional model referred to in Okui & Waples (1993). This model simulates the generation and expulsion of oil and gas for a certain source rock, assuming a constant subsidence rate and geothermal gradient through time. Porosity reduction is assumed to be exponential (Athy 1930) and absolute permeability is given as a function of porosity by the Kozeny–Carman equation (Ungerer *et al.* 1990). Hydrocarbon generation is calculated by the first-order parallel reaction kinetics model (Tissot *et al.* 1987; Waples *et al.* 1992). The total volume of fluid expelled is assumed to be equal to the porosity reduction in each time step. The volume of each

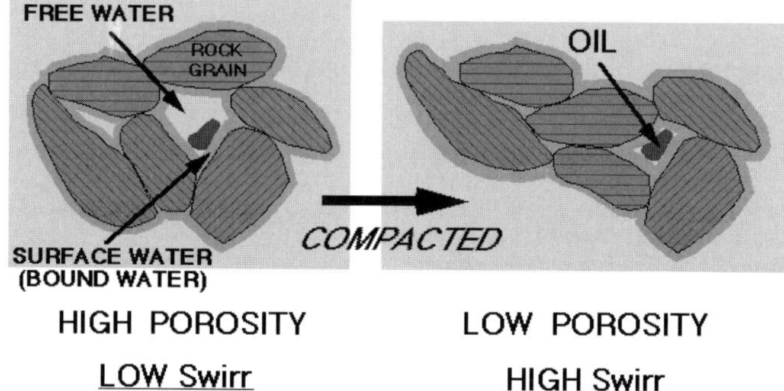

Fig. 12. Relative permeability model given as a function of compaction. Compaction of source rock also results in pore size decrease and hence our model can be applied.

fluid (water, oil and gas) is proportional to the relative permeability and pressure gradient of each fluid, but inversely proportional to the viscosity of each fluid. Relative permeability is given as a function of absolute permeability as discussed above. The pressure gradient of the non-wetting phase (oil and gas) is assumed to be 1.5 times greater that of water due to capillary pressure and buoyancy. Viscosity is given as a function of temperature.

For modelling oil expulsion efficiency, subsidence rate and geothermal gradient are assumed constant as 80 m Ma^{-1} and 3.5 °C per 100 m, respectively. Compaction is assumed to be normal (porosity exponential decreases with depth). Kinetic parameters are from Lawrence Livermore's National Laboratory measurements on type II kerogen (Waples *et al.* 1992). Initial bitumen content is assumed as 50 mg HC g^{-1} TOC. This framework gave the following history; oil generation started at depth of 2700 m, temperature of 105°C and porosity of 15%, peaked at 3500 m, 135°C and 10 %, and ended at 4300 m, 160°C and 7 %.

Oil expulsion efficiency is defined as BP's PEE_L (Cooles *et al.* 1986; Pepper 1991; Pepper & Corvi 1995), which is the ratio of the mass of oil expelled to the mass of oil generated. Assuming type II kerogen, oil expulsion efficiency is calculated by varying source rock potentials. The results are drawn on the diagram in Pepper (1991; Fig. 13). PEE_L is plotted against Hydrogen Index as a source rock potential in this diagram. For the calculation, the empirical relationship of organic matter richness (TOC) with kerogen composition (Hydrogen Index) derived from our database was applied, which is:

$$TOC = 0.000114 \times HI^{1.7} \quad (3)$$

where TOC is total organic carbon content in % and HI is Hydrogen Index in mg HC g^{-1} TOC. This relationship may depend on several factors such as the origin of the organic matter and sedimentary environment. Again we assumed type II marine planktonic kerogen for the calculations. Oil expulsion efficiencies were calculated from the relative permeability curves for reservoir rocks and fine-grained source rocks. Conventional reservoir-type curves have a constant irreducible water saturation of 20 % and residual oil saturation of 5 %, which corresponds to a drainage curve. New source rock curves as described were used and the residual oil saturation decreases from 5 % to almost 0 % as compaction proceeds. The adsorption capacity of kerogen was also varied from 0, 100 to 200 mg OIL g^{-1} TOC.

Figure 13 indicates that oil expulsion efficiency increases as source rock potential increases. The expulsion efficiency of the reservoir rock curves is lower than the new source-rock curves, especially for moderate to poor source rocks. Most of the measured data on actual source rocks (Pepper 1991) indicate a good match with the predictions using the new source rock curves, especially for moderate to poor source rocks. This suggests that the new source rock curves can better reproduce the behaviour of oil expulsion efficiency observed in actual source rocks than the reservoir rock curves. Lower efficiency for four of the rich source rocks cannot be addressed without access to any

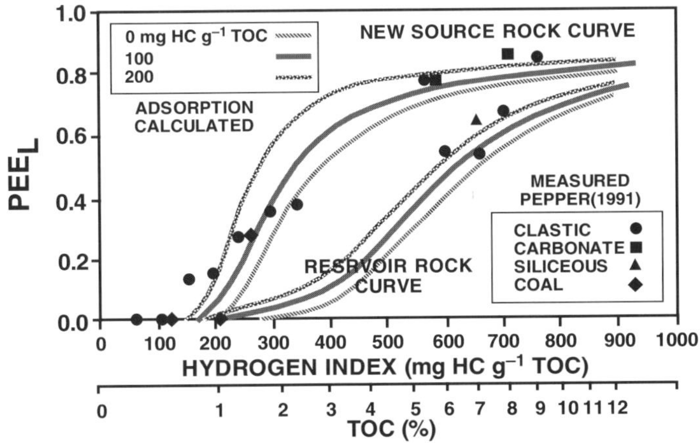

Fig. 13. Oil expulsion efficiency calculated by one-dimensional basin modelling (Okui & Waples 1993), applying a new source rock curve and reservoir rock curve as relative permeability. The results are drawn on Pepper's (1991) diagram. Good matching with the measurements on actual source rocks is obtained for the new curve.

detailed geological information about BP's data (Fig. 13). The geology for the simulations was simplified. For example, if overpressuring was to develop and porosity reduction ceases, the high porosity prevents an oil saturation increase and expulsion is hindered.

A series of simulations using the new source rock curves suggests that the expulsion efficiency reaches almost 80 %, if Hydrogen Index exceeds 600 mg HC g^{-1} TOC, even though the adsorption capacity varies. The expulsion efficiency rapidly increases from 0 to 60 % in the range of 200 to 400 mg HC g^{-1} TOC (Fig. 13). In this range, the expulsion efficiency is also sensitive to the adsorption capacity. For the source rock of 300 mg HC g^{-1} TOC, the expulsion efficiency changes from 30 %, 40 % and 60 %, as the adsorption capacity decreases from 200, 100 to 0 mg HC g^{-1} TOC, respectively (Fig. 13).

Tests of new curve by 2-D basin modelling

In order to test the validity of new relative permeability curves for fine-grained source rocks, two-dimensional basin modelling coupled with fluid flow (pressure reconstruction) was also conducted. The simulations assumed a conventional reservoir rock curve and a new source rock curve as the relative permeability for fine-grained source rock were conducted and compared. The computer model we used is SIGMA-2D, which is developed by the Technology Research Center of the Japan National Oil Corporation (Okui *et al.* 1994*a*,*b*, 1995, 1996). The comparison was done in the Niigata basin, Japan and the North Sea, Norway.

SIGMA-2D

SIGMA-2D (2-Dimensional Simulator for Integration of Generation, Migration and Accumulation) is a finite difference code developed by the Technology Research Center of Japan National Oil Corporation. SIGMA-2D can simulate the generation, migration and accumulation of oil and gas (three-phase fluid flow) in a two-dimensional cross-section. The modelling is divided into three categories; Geological, Generation and Migration. The Geological modelling is responsible for the reconstruction of burial and compaction of sediments, tectonic and hydraulic fracturing, fluid flow and heat flow. The compaction is calculated based on effective stress laws and fluid flow conditions, which is governed by Darcy's law. Pressure increase is achieved either by sediments loading, fluid expansion or hydrocarbon generation. Tectonic fracturing is calculated by simplified strain analysis and hydraulic fracturing is predicted based on pore pressure distribution. Conductive and convective heat flow creates the temperature distribution in the section.

The generation modelling is responsible for the calculation of maturation of organic matter (vitrinite reflectance and sterane epimerization) and generation of oil and gas. A 1st-order kinetic reaction model is applied for these calculation and multiple parallel reactions are especially used for the generation.

The migration modelling is responsible for the calculation of expulsion, secondary migration, PVT (Pressure, Volume, Temperature) conditions (dissolution of fluid) and sealing (accumulation). The expulsion and secondary migration is calculated based on Darcy's law using the relative permeability concept. Maximum dissolved capacity of gas into oil is calculated based on pressure and temperature for certain oil and gas types. Free gas which corresponds to the excess of this capacity can migrate as a separate phase. The migrating oil and gas can be trapped, based on the capillary pressure concept.

Four governing equations with the terms related to the above phenomena are solved simultaneously in each time step at each grid point in the section. These equations are the mass conservation equation for water, oil and gas, respectively, and the energy (heat) conservation equation. Rock and fluid properties have to be given as input parameters. The rock properties related to fluid flow such as permeability, relative permeability and capillary pressure are primarily given as a function of porosity for each lithology. The properties related to heat flow such as thermal conductivity and heat capacity are given as a function of porosity, temperature and pressure.

SIGMA-2D has been applied to over 30 basins throughout the world. They are mainly the basins in Japan and Southeast Asia, but the applications to the North Sea and Middle East are also successful.

Niigata basin, Japan

The Niigata basin is a back-arc rift basin which formed in Miocene times (Fig. 14; Nakayama 1987). After the rifting in Early Miocene times, this basin was stable, and siliceous source rock beds were deposited under a deep marine environment during the Miocene. In Early Pliocene times, terrestrial sediments transported as turbidites filled up the basin and deposited the primary reservoir sediments. In Middle Pliocene times, the tectonic setting was changed to become compressive, which created structures. During the same period, kitchen areas subsided deeply and the source rock was matured. Biomarker data indicates that most of the oils accumulating in traps have a

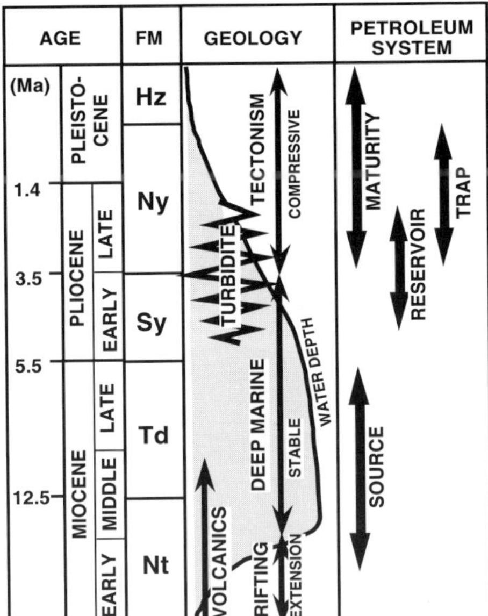

Fig. 14. Petroleum system in the Niigata basin. Niigata is a Tertiary back-arc basin in Japan. The primary source rock is in the Miocene and the primary reservoir is in the Pliocene. Structuring and maturity of source rock occurred after the middle Pliocene.

good correlation with the source rock in the Middle Miocene section (Nt and lower Td). One-dimensional maturity modelling indicates that the source rock beneath accumulations are immature which suggest that oil generated from the Middle Miocene source rock must have occurred in the deeply subsided trough area, and then migrated vertically and horizontally to the Pliocene reservoir rock in the last 3 to 4 million years. Overpressuring is observed below the Miocene caused by rapid sedimentation after Middle Pliocene times.

A section across an offshore area in the Niigata basin was modelled (Fig. 15; Okui *et al.* 1994*a*, 1995). The thermal history was optimized by measured temperature and vitrinite reflectance data from the wells. Rift-type heat flow history with present heat flow of about 1.1 HFU was derived. Rock properties, mainly porosity and absolute permeability of shales were calibrated to measured pressure and porosity data. Geochemical data demonstrate that the best source rock interval has about 2 % TOC and contains type II kerogen of 400 mg HC g^{-1} TOC Hydrogen Index. The source rocks in Niigata basin are not as rich as typical type II source rocks. We assumed that the Nt, lower Td and upper Td have source rock potential (Nt, TOC = 1.5 %, HI = 250 mg HC g^{-1} TOC; LTd, TOC = 2 %, HI = 400 mg HC g^{-1} TOC; UTd, TOC = 1.2 %, HI = 250 mg HC g^{-1} TOC). Kinetic parameters were acquired by pyrolysis data on actual source rock samples. Activation energy has a peak at 50 kcal mol^{-1} with a frequency factor of 6.28×10^{13} sec^{-1}. Conventional reservoir-type curves have a constant irreducible water saturation of 20 % and residual oil saturation of 5 %, which corresponds to a drainage curve. The new source rock curve is as described before.

Simulated results by a new source rock curve (Fig. 15a) indicates that oils have been generated and expelled since Middle Pliocene times and have migrated vertically through a tectonic fracture zone. These oils have then migrated horizontally after reaching one of the turbidite fans distributing widely in the basin. The trap marked AGN was charged by oils very recently. The tectonic fracturing was predicted by the procedure developed by Larson *et al.* (1993). The simplified strain analysis was based on the change of the distance between nodal blocks and predicted that tectonic fracturing had occurred in the slope area after Middle Pliocene times when the tectonic setting changed from extensional to compressive. The behaviour of deformation for each rock is defined as ductility in this model. Since the ductility of a rock increases as effective stress increases (Handin *et al.* 1963), overpressuring which decreases effective stress, enhanced the creation of fracture zones. Without the fracturing, oils cannot be expelled out to carrier beds since source rock potentials is not high and generation is very late (the time available for expulsion and migration is very short).

In contrast, the simulation by the conventional reservoir rock curve cannot reproduce oil expulsion, migration and accumulation in the Niigata basin, even when tectonic fracture zones are simulated (Fig. 15b). Instead, source rocks were matured and reached the oil saturation of about 30 %. The maximum oil saturation of source rocks in the case of the new source-rock curve is about 7 %. The oil saturation required for active expulsion has not been reached in the case of reservoir type curve due to the leanness of source rock and higher porosity caused by overpressuring. This comparison also implies that active expulsion does not occur around residual oil saturation (the saturation where oil starts movement in the drainage process).

North Sea, Norway

The North Sea is one of the most famous oil producing provinces in the world. The North Sea is a rift basin evolved in conjunction with opening of the Atlantic Ocean. The geology and geochemistry of the North Sea is well described (Cornford 1984;

NIIGATA BASIN : JAPAN
NEW SOURCE-ROCK TYPE CURVE
(CONTOUR : Z FRACTURE PERMEABILITY(md))
0 (Ma)
FORMATION
OIL SATURATION
Hz/Ny
LNy1
LNy2
USy
LSy
UTd
LTd
Nt
AGE(Ma)
DISTANCE (km)
DEPTH (km)
0.1 m/year
0.1 m/year
(%)
please set or modify some items and finally click 'ok'

(a)

NIIGATA BASIN : JAPAN
RESERVOIR-ROCK TYPE CURVE
(CONTOUR : Z FRACTURE PERMEABILITY(md))
0 (Ma)
FORMATION
OIL SATURATION
Hz/Ny
LNy1
LNy2
USy
LSy
UTd
LTd
Nt
AGE(Ma)
DISTANCE (km)
DEPTH (km)
0.1 m/year
0.1 m/year
(%)
please set or modify some items and finally click 'ok'

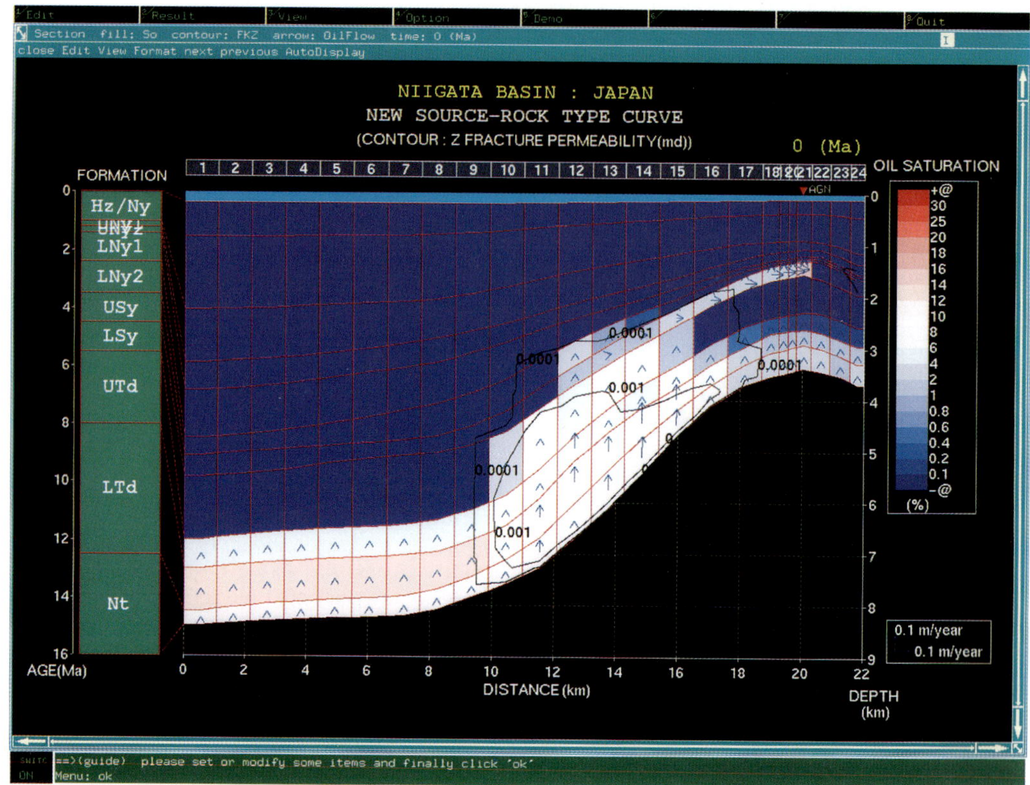

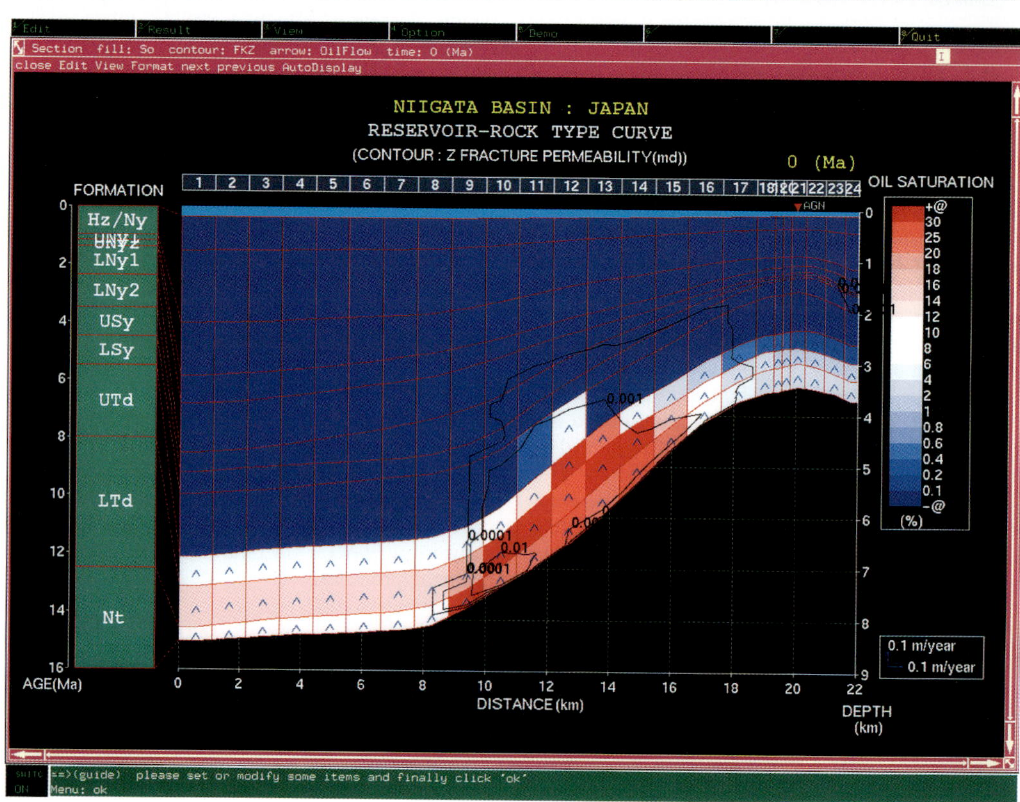

(b)

Thomas *et al.* 1985; Pegrum & Spencer 1990). There were several rift events discussed in the North Sea, but the most important one was during the Jurassic (Fig. 16). This rift event resulted in the change of sedimentary environment from fluvial in Triassic to Early Jurassic, deltaic in Middle Jurassic, restricted marine in Late Jurassic and open marine in Cretaceous and later. One of the most important reservoir rocks is the Brent sandstone in the Middle Jurassic, in which many giant oil fields have been discovered. The primary source rock is the Kimmeridgian Clay, which is called the Draupne source rock in the Norwegian sector. However, recent biomarker works (Chung *et al.* 1992; Gormly 1994) have indicated that several oil families exist even in the Northern North Sea and the contribution from Heather source rock and Brent coal is discussed.

A section crossing the Northern North Sea merged from several seismic lines was simulated (Fig. 17). All four structures in the section were drilled. Oil and gas were discovered in the Brent sandstone of two structures in the centre of the section. The structure located in the left is also producing oil from the Brent sandstone. However, the Brent sandstone in the structure of the right was dry, but a thin oil column was discovered in the Palaeocene sandstones. A thermal history was optimized from measured temperature and vitrinite reflectance data from the wells. A rift-type heat flow history with present heat flow from 1.0 to 1.3 HFU (varying with location in the section) was derived. Rock properties, mainly porosity and absolute permeability of shales were calibrated to measured pressure and porosity data. The observed overpressuring was reproduced. The Draupne and Heather Formation were assumed to be the source rocks and applied IFP's kinetics (Tissot *et al.* 1987; Waples *et al.* 1992). The Brent coal was not involved as a source rock in this study. The Draupne source rock was assumed to have 4 % TOC and 100 % type II kerogen (HI = 500 mg HC g^{-1} TOC), and the Heather source rock was assumed to have 3 % TOC and mixing of 50 % type II and 50 % type III kerogen (HI = 350 mg HC g^{-1} TOC). The conventional reservoir-type relative permeability curves have a constant irreducible water saturation of 20 % and residual oil saturation of 5 %, which corresponds to the drainage curve. Once again the new source-rock curve is as described earlier.

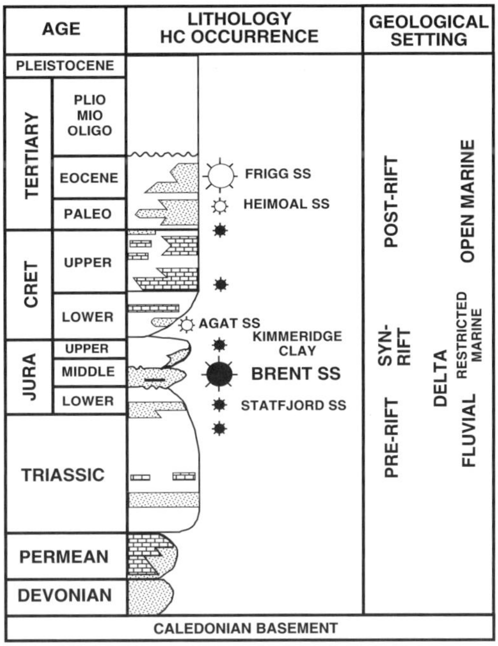

Fig. 16. Geological framework in the Northern North Sea. The North Sea was formed as a Jurassic rift system. The primary source rocks is in the Upper Jurassic (the Draupne and Heather formation) and the primary reservoir is in the Middle Jurassic (the Brent sandstone).

The results of simulations (Fig. 17a) indicate that oils were generated from Late Cretaceous times (80 Ma) in the Draupne and Heather source rock. These oils were first expelled and down-migrated into the Brent carrier beds in the centre of the Viking Graben, and then has migrated horizontally along the Brent sandstone network from the Graben to the traps. The Brent sandstone network transferred the fluid and pressure from the deeply subsided Graben to the traps, which caused bleeding off of the overpressure from the Brent Sandstone in the centre of the Graben. This process created the pressure gradient from the Draupne source rock to the Brent carrier bed, which has been maintained due to rapid sedimentation from Cretaceous times and maintained down-migration from the Draupne to the Brent. The first oil had already arrived at the nearest structure at the end of the Cretaceous (63 Ma).

Fig. 15. Comparison of two-dimensional basin modelling (JNOC's SIGMA-2D) in the Niigata basin, applying a new source rock curve (**a**) and reservoir rock curve (**b**) as relative permeability. The new source rock curve can reproduce expulsion, migration and accumulation better than the reservoir curve. Colour code indicates oil saturation, counter line indicates fracture permeability in vertical direction and arrow indicates oil flow velocity.

NORTHERN NORTH SEA : NORWAY
NEW SOURCE-ROCK TYPE CURVE
(CONTOUR : TEMPERATURE(C))
0 (Ma)
FORMATION
PLIO
MIO
OLIG
EOC
ROG
JORS
KYRRE
TRYG
ROD
DRAU
HEAT
BRT
DUN
AGE(Ma)
DISTANCE (km)
DEPTH (km)
OIL SATURATION
(%)
OIL FLOW
0.01 m/Y
0.01 m/Y

(a)

NORTHERN NORTH SEA : NORWAY
RESERVOIR-TYPE CURVE
(CONTOUR : TEMPERATURE(C))
0 (Ma)
FORMATION
PLIO
MIO
OLIG
EOC
ROG
JORS
KYRRE
TRYG
ROD
DRAU
HEAT
BRT
DUN
AGE(Ma)
DISTANCE (km)
DEPTH (km)
OIL SATURATION
(%)
OIL FLOW
0.01 m/Y
0.01 m/Y

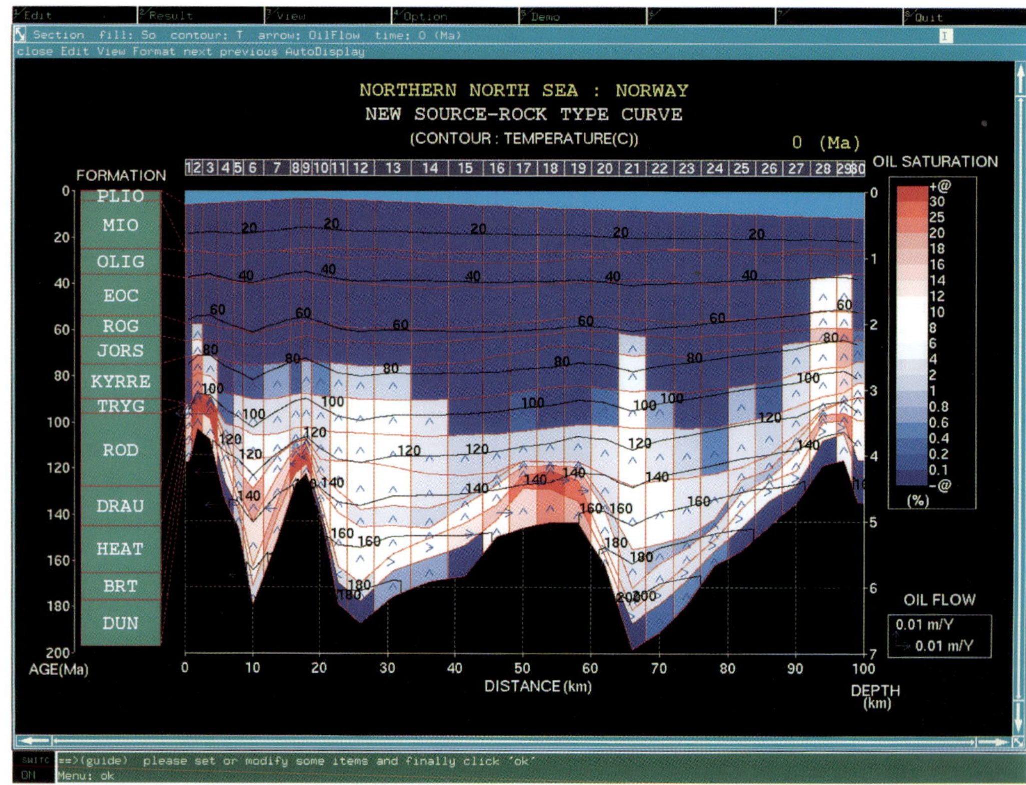

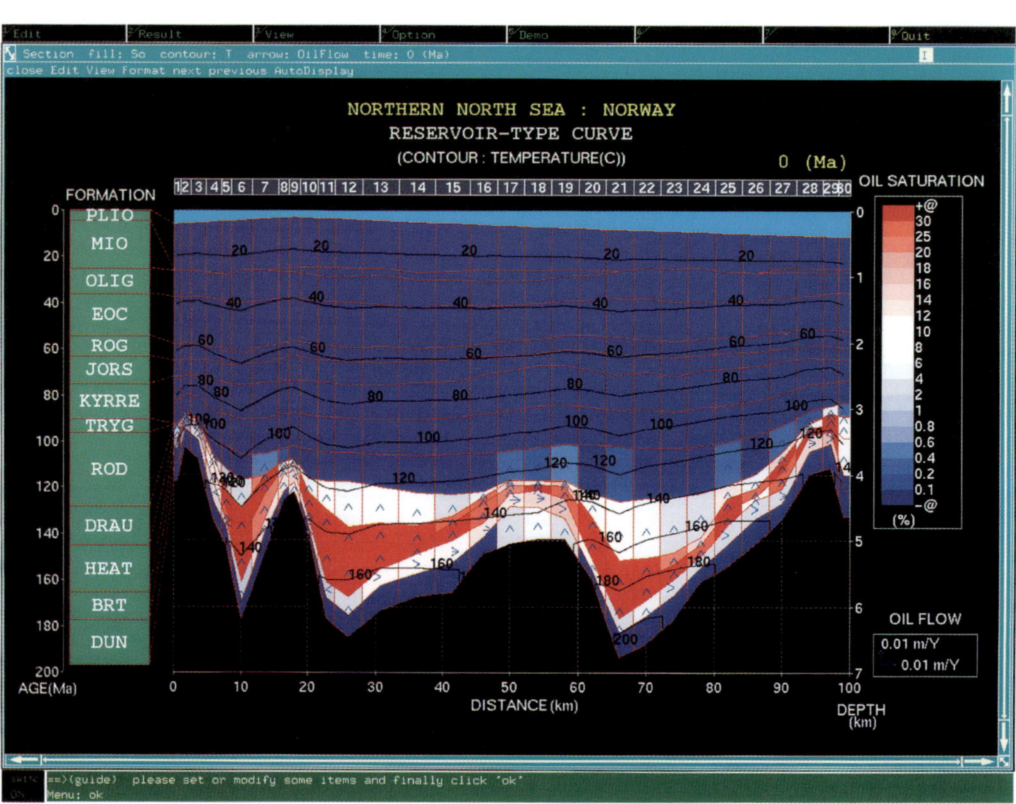

(b)

The structure located on the right is small, close to the deeply-subsided Viking Graben, and its Brent reservoir is located at a relatively shallow depth. This geological setting allowed oil to leak out to Cretaceous and Cenozoic formations. This structure received oils earlier and in a shallower position than others. The oil column grew rapidly due to its geometry. Seal capacity (capillary pressure) is given as a function of porosity and permeability in SIGMA-2D. If a seal exists in a shallow position in the sedimentary column, the seal capacity is small. A thicker oil column may be held in deeper positions. In the structure on the right, the critical oil column was easily reached and leaking started. In this case, the seal is also the source rock and the generation of oil may help the increase of oil saturation in the seal. It was simulated that leaked oils reach the Palaeocene strata at present, which is consistent with actual exploration results. The maximum oil saturation in the Draupne and Heather source rocks is 28 % and 15 %, respectively. Calculated velocity for oil expulsion is 5×10^{-5} m a^{-1} in a vertical direction and 7×10^{-6} m a^{-1} in a horizontal direction, which is consistent with England *et al.* (1987).

Simulations using a reservoir-type curve (Fig. 17b) cannot reproduce the active oil expulsion and the leaking through the seal. The reason for this is the same as the simulation in the Niigata basin in that the oil saturation required for active oil expulsion is higher in this case and hence expulsion was delayed. The amount of expelled oil is small, and these oils did not arrive at the structure to produce a column that would leak. The maximum oil saturation in the Draupne and Heather source rock is 48 % and 35 %, respectively.

Discussions: relative permeability and saturation threshold

Applications to one-dimensional and two-dimensional basin modelling supported the idea that the relative permeability model for fine-grained source rocks developed in this study can simulate the behaviour of oil expulsion better than the conventional reservoir rock curve. The modellings also indicated that oil saturation in source rocks during expulsion is controlled by a dynamic system. If the rate of oil generation is greater than the rate of expulsion, the oil saturation increases. On the other hand, if the rate of expulsion is greater than generation, the saturation decreases. The rate of generation depends on thermal history (heating rate), source rock kinetics and richness, and the rate of expulsion depends on the permeability relative to oil which is a function of oil saturation.

It was found that active oil expulsion does not occur around the residual oil saturation of realtive permeability curves, which is the saturation point where oil starts its movement. This saturation has been considered to be the saturation threshold for oil expulsion. When oil saturation exceeds residual oil saturation, oil starts flowing, but the amount of oil in total flowing fluid is little. Therefore, oil expulsion is not active and the oil saturation increases, since the rate of oil generation is greater than the rate of oil expulsion. The permeability relative to oil becomes greater than the permeability relative to water, when almost half of the free water space is occupied by oil, which corresponds to (100-Swirr)/2 (Fig. 18). Above this value, the permeability relative to oil is greater than to water, and hence active expulsion can be expected. However, even when the oil saturation exceeds this value, the saturation may increase if the rate of generation is high. In the case of rich source rocks, the rate of generation is high, so that the saturation will further increase close to the irreducible water saturation. Around the irreducible water saturation, almost all free pore space is occupied by oil, and therefore, very active expulsion must occur. When the oil expulsion becomes active, the oil saturation is kept almost constant as a maximum value since the rate of expulsion is almost equal to the rate of generation. This implies that the minimum oil saturation required for active expulsion is around (100-Swirr)/2, but maximum oil saturation during expulsion depends on the rate of generation. Richer source rock can reach higher maximum oil saturation. The maximum oil saturation can be regarded as the equilibrium point where the rate of generation is equal to the rate of expulsion and corresponds to the saturation threshold in one-dimensional basin modelling.

The maximum oil saturation during expulsion appears to be related to irreducible water saturation which may be a function of compaction (permeability or porosity) as suggested by the new relative

Fig. 17. Comparison of two-dimensional basin modelling (JNOC's SIGMA-2D) in the Northern North Sea, applying a new source rock curve (a) and reservoir rock curve (b) as relative permeability. The new source rock curve appears to reproduce expulsion, migration, accumulation and leaking better than the reservoir curve. Colour code indicates oil saturation, counter line indicates temperature and arrow indicates oil flow velocity.

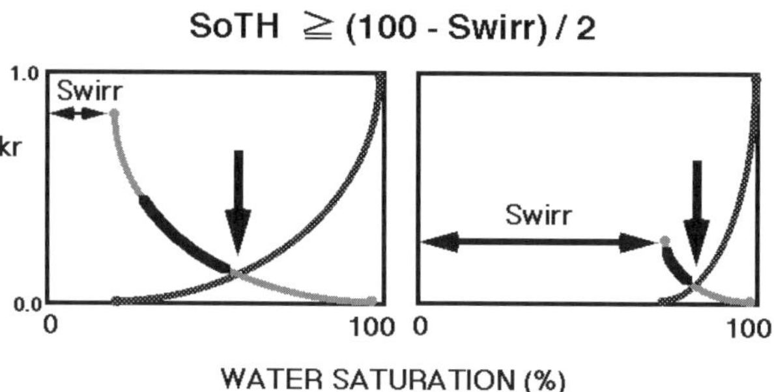

Fig. 18. Definition of saturation threshold by relative permeability curve. Active expulsion occurs when the oil saturation exceeds (100-Swirr)/2, where *k*ro becomes greater than *k*rw. Maximum oil saturation during expulsion is the point where generation rate and expulsion rate are equalized.

permeability model. Therefore, there is a possibility that the maximum oil saturations are given as a function of compaction. Figure 19 demonstrates this relationship. Free water saturation (100-Swirr) and (100-Swirr)/2 are drawn as a function of porosity. Since the irreducible water saturation is given as a function of absolute permeability in the new realtive permeability model, the Kozeny–Carman equation is assumed for the conversion from permeability to porosity. The (100-Swirr) and (100-Swirr)/2 decrease with porosity reduction, which implies that compacted source rock can expel oil with lower saturation. Expected maximum oil saturation at 10 % porosity is between 10 and 20 %.

In order to test this idea, about 300 Rock-Eval data were collected with porosity values from the Niigata basin, Japan (Fig. 20). The porosity is mainly derived from electrical logging (density log, sonic log and neutron-porosity log). The oil saturation of each rock was calculated from the S1 of rock-eval and porosity data using following equation:

$$So = C \times (S1/1000) \times (\rho_{rock}/\rho_o) \times (\phi/(1-\phi)) \times 100 \quad (4)$$

where So is oil saturation in %, S1 is from Rock-Eval in mg HC g^{-1} TOC, ρ_{rock} is density of source rock in g cm^{-1} (2.55 is used in this study), ρ_o is density of oil in g cm^{-1} (0.85 is used in this study), and ϕ is porosity in fraction. C is a factor to convert measured S1 value to *in-situ* oil mass. The S1 does not cover all oils due to the vapourization of light oil. We gave 2 for the C in this study (Cooles *et al.* 1986). The (100-Swirr)/2 value is calculated as a

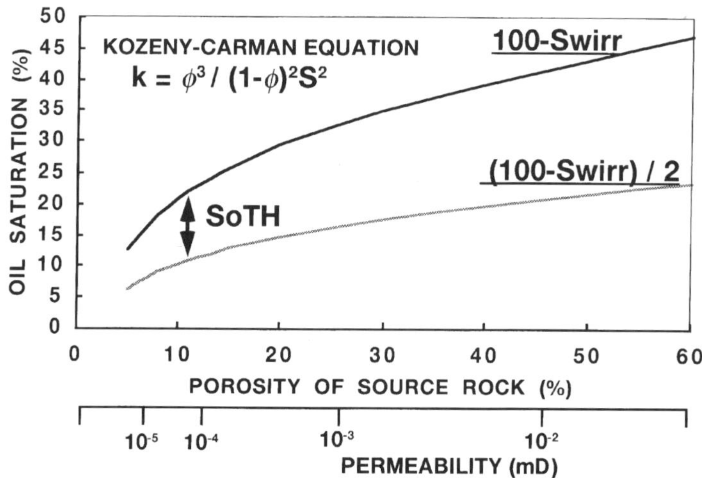

Fig. 19. Evolution of Swirr and (100-Swirr)/2 as a function of compaction (porosity) suggested by our model. Maximum oil saturation (saturation threshold) should be between (100-Swirr)/2 and (100-Swirr).

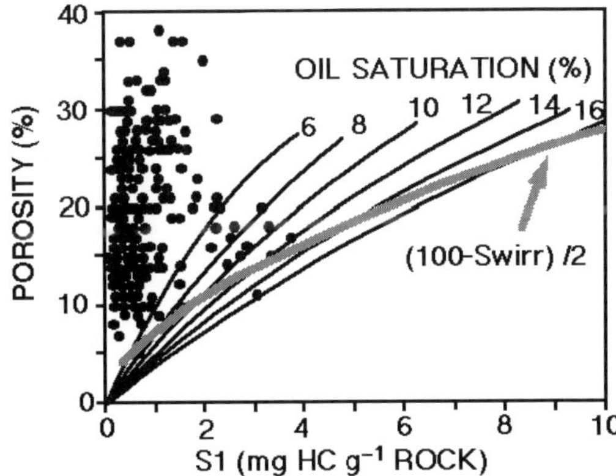

Fig. 20. Oil saturation in samples from the Niigata basin. Oil saturation is calculated by porosity from E-log and S1 from Rock-Eval pyrolysis. The (100-Swirr)/2 value is calculated as a function of porosity. Measured data do not exceed (100-Swirr)/2, which suggests (100-Swirr)/2 as the maximum oil saturation.

function of porosity and drawn on Fig. 20 as a thick line. The (100-Swirr)/2 decreases as porosity decreases and appears to trace the lower limit of the distribution of oil saturation, which suggests that the maximum oil saturation at various degrees of compaction in the Niigata basin is almost equal to (100-Swirr)/2.

Conclusions

Long-term research efforts by many scientists have revealed that oil expulsion is best expressed as pressure-driven bulk flow. The expulsion is generally modelled by Darcy's law with the relative permeability concept. Relative permeability and saturation threshold are the key parameters for modelling.

The relative permeability curves for reservoir rocks appears to be controlled by surface water and micro-porosity. Knowing the system in reservoir rocks, we constructed the relative permeability curves for fine-grained source rocks extrapolated by permeability as a scaling factor. These new curves have much higher irreducible water saturations up to 80 % at a porosity of 10 %. This model can be also applied to the compaction of source rocks.

One-dimensional and two-dimensional basin modelling using the new source-rock curves can reproduce expulsion and its efficiency, location of accumulations and leaking through cap rocks better than conventional reservoir-rock curves. Oil expulsion efficiency increases as source rock potential increases and reaches 80 % for rich source rock.

Active oil expulsion does not occur around residual oil saturation. Oil expulsion is activated when half of the free water is replaced by oil (at So = (100-Swirr)/2), where relative permeability to oil becomes greater than relative permeability to water. Maximum oil saturation during expulsion, which is regarded as saturation threshold, should be between (100-Swirr)/2 and (100-Swirr) according to source rock richness. For rich source rocks, the maximum saturation reaches higher levels due to rapid generation. Since irreducible water saturation depends on the degree of compaction, maximum oil saturation should be given as a function of porosity as well as source rock richness.

We would like to thank JNOC and CONOCO for the permission to publish this paper. We are also greatful to our colleagues in JNOC and CONOCO who helped us for measurement, coding, and supply of data. Useful comments by J. Iliffe, D. W. Waples, A. S. Pepper, L. C. Kuo, K. Nakayama and T. Tokunaga are greatfully acknowledged. Part of this work was carried out during the collaborative research between JNOC and CONOCO.

References

AMYX, J. W., BASS, D. M. & WHITTING, R. L. 1960. *Petroleum reservoir engineering*, McGraw-Hill, New York.

ANDERSON, W. G. 1987. Wettability literature survey — Part 5: The effects of wettability on relative permeability. *Journal of Petroleum Technology*, **39**, 1453–1468.

ATHY, L. F. 1930. Density, porosity and compaction of sedimentary rocks. *American Association of Petroleum Geologists Bulletin*, **14**, 1–24.

BAKER, C. 1980. Primary migration : The importance of water–mineral–organic matter interactions in the source rocks. *In*: ROBERTS, W. H. & CORDELL, R. J. (eds) *Problems of petroleum migration*, American Association of Petroleum Geologists Studies in Geology, **10**, 19–31.

BURNHAM, A. K. & BRAUN, R. L. 1990. Development of a detailed model of petroleumn formation, destruction, and expulsion from lacustrine and marine source rocks. *Organic Geochemistry*, **16**, 27–39.

BURST, J. F. 1969. Diagenesis of Gulf coast clayey sediments and its possible relation to petroleum migration. *American Association of Petroleum Geologists Bulletin*, **53**, 73–93.

CAUDLE, B. H., SLOBOD, R. L. & BROWNSCOMBE, E. R. 1951. Further developments in the laboratory determination of relative permeability. *Petroleum Transactions, AIME*, **192**, 145–150.

CHUNG, H. M., WINGERT, W. S. & CLAYPOOL, G. E. 1992. Geochemistry of oils in the Northern Viking Graben. *In*: HALBOUTY, M. T. (ed.) *Giant Oil and*

Gas Fields of the Decade 1978–1988, American Association of Petroleum Geologists Memoir, **54**, 277–296.

COLONNA, J., BRISSAUD, F. & MILLET, J. L. 1972. Evolution of capillary and relative permeability hytsteresis. *Society of Petroleum Engineers Journal*, **12**, 28–38.

COOLES, G. P., MACKENZIE, A. S. & QUIGLEY, T. M. 1986. Calculation of petroleum masses generated and expelled from source rocks. *Organic Geochemistry*, **10**, 235–245.

CORNFORD, C. 1984. Source rocks and hydrocarbons of the North Sea. *In*: GLENNIE, K. W. (ed.) *Introduction to the Petroleum Geology of the North Sea*. Blackwell, Oxford, 171–204.

COSKUN, S. B., WARDLAW, N. C. & HARVERSLEW, B. 1993. Effects of composition, texture and diagenesis on porosity, permeability and oil recovery in a sandstione reservoir. *Journal of Petroleum Science and Engineering*, **8**, 279–292.

COSSE, R. 1993. *Basics of reservoir engineering*. Editions Technip, Paris.

DURAND, B. 1983. Present trends in organic geochemistry in reserach on migration of hydrocarbons. *Advances in Organic Geochemistry 1981*. Wiley & Sons, 117–128.

—— 1988. Understanding of HC migration in sedimentary basins (present state of knowledge). *Organic Geochemistry*, **13**, 445–459.

DUTTA, N. 1987. Fluid flow in low permeable porous media. *In*: DOLIGEZ, B. (ed.) *Migration of Hydrocarbons in Sedeinetary basins*. Editions Technip, Paris, 567–588.

ENGLAND, W. A., MACKENZIE, A. S., MANN, D. M. & QUIFLEY, T. M. 1987. The movement and entrapment of petroleum fluids in the subsurface. *Journal of Geological Society , London*, **144**, 327–347.

GORMLY, J. R., BUCK, S. P. & CHUNG, H. M. 1994. Oil-source correlation in the North Viking Graben. *Organic Geochemistry*, **22**, 403–413.

HANDIN, J., HAGER, R. V., Jr, FRIEDMAN, M. & FEATHER, J. N. 1963. Experimental Deformation of sedimentary rocks under confining pressure : pore pressure tests. *American Association of Petroleum Geologists Bulletin*, **47**, 1–50.

HELBA, A. A., SALML, M., SCRLVEN, L. E. & DAVIS, H. T. 1992. Percolation theory of two-phase relative permeability. *SPE Reservoir Engineering*, February, 123–132.

HERMANRUD, C. 1993. Basin modelling techniques — an overview. *In*: DORE, A. G. *et al.* (eds) *Basin Modelling : Advances and Applications*, Elsevier, 1–34.

HUBBERT, M. K. 1940. The theory of ground-water motion. *Journal of Geology*, **48**, 785–944.

HUNT, J. M. 1990. Generation and migration of petroleum from abnormally pressured fluid compartments. *American Association of Petroleum Geologists Bulletin*, **74**, 1–12.

JNOC 1987. Measurement of hydrocarbon generation pressure (in Japanese). *Annual Report of Technology Research Center, JNOC*, 50–53.

—— 1992. Relationship between oil expulsion and oil saturation in source rocks (in Japanese). *Annual Report of Technology Research Center, JNOC*, 66–70.

KROSS, B. M. 1988. Experimental investigation of the molecular migration of C_1–C_6 hydrocarbons: Kinetics of hydrocarbon release from source rocks. *Organic Geochemistry*, **13**, 513-523.

——, HANEBECK, D. & LEYTHAEUSER, D. 1993. Experimental investigation of the molecular migration of light hydrocarbons in source rocks at elevated temperature. *In*: DORE, A. G. *et al.* (eds) *Basin Modelling : Advances and Applications*, Elsevier, 277–291.

KUO, L. C., HARDY, H. H. & OWILI-EGER, A. S. C. 1995. Quantitative modeling of organic matter connectivity in source rocks using fractal geostatistical analysis. *Organic Geochemistry*, **23**, 29–42.

LAFARGUE, E., ESPIRTALIE, J., JACOBSEN, T. & EGGEN, S. 1990. Experimental simulation of hydrocarbon expulsion. *Organic Geochemistry*, **16**, 121–131.

——, ——, BROKS, T. M & NYLAND, B. 1994. Experimental simulation of primary migration. *Organic Geochemistry*, **22**, 575–586.

LARSON, K. W., WAPLES, D. W., FU, H. & KODAMA, K. 1993. Predicting tectonic fractures and fluid flow through farctures in basin modelling. *In*: DORE, A. G. *et al.* (eds) *Basin Modelling : Advances and Applications*, Elsevier, 373–383.

LEYTHAEUSER, D., SCHAEFER, R. G. & YUKLER, A. 1982. Role of diffusion in primary migration of hydrocarbons. *American Association of Petroleum Geologists Bulletin*, **66**, 408–429.

——, —— & RADKE, M. 1988. Geochemical effects of primary migration of petroleum in Kimmeridge source rcoks from the Brae field area, North Sea. I: Gross composition of C_{15+}-spiluble organic matter and molecular composition of C_{15+}-saturated hydrocarbons. *Geochimica et Cosmochimica Acta*, **52**, 701–713.

MACKENZIE, A. S., PRICE, I. & LEYTHAEUSER, D. 1987. The expulsion of petroleum from Kimmeridge clay source-rocks in the area of the Brae Oilfield, UK continental shelf. *In*: BROOKS, J. & GLENNIE, K. (eds) *Petroleum Geology of North West Europe*, Graham & Trotman, London, 865–877.

MAGARA, K. 1978. *Compaction and Fluid Migration, Practical Petroleum Geology*. Elsevier, Amsterdam.

MCAULIFFE, C. D. 1980. Oil ans gas migration: chemical and physical constraints. *In*: ROBERTS, W. H. & CORDELL, R. J. (eds) *Problems of petroleum migration*, American Association of Petroleum Geologists Studies in Geology, **10**, 89–107.

MEISSNER, F. F. 1978. Petroleum geology of the Bakken Formation Williston Basin, North Dakota and Montana. *The Economic Geology of the Williston Basin*, Montana Geological Survey, 207–227.

MELAS, F. F. & FRIEDMAN, G. M. 1992. Petrophysical charcteristics of the Jurassic Smackover Formation, Jay Field, Conecuh Embayment, Alabama and Florida. *American Association of Petroleum Geologists Bulletin*, **76**, 81–100.

MORAES, M. A. S. 1991. Diagenesis and microscopic heterogenity of lacustrine deltaic and turbiditic sandstone reserviors (Lower Cretaceous), Potiguar

Basin, Brazil. *American Association of Petroleum Geologists Bulletin*, **75**, 1758–1771.

MORGAN, J. T. & GORDON, D. T. 1970. Influence of pore geometry on water–oil relative permeability. *Journal of Petroleum Technology*, **10**, 1199–1208.

MORROW, N. R. 1970. Irreducible wetting-phase saturations in porous media. *Chemical Engineering Science*, **25**, 1799–1815.

—— 1971. Small-scale packing heterogenities in porous sedimentary rocks. *American Association of Petroleum Geologists Bulletin*, **55**, 514–522.

MUSKATT, M. 1949. *Physical principles of oil production*. McGraw-Hill, New York.

NAKAYAMA, K. 1987. Hydrocarbon-expulsion model and its application to Niigata area, Japan. *American Association of Petroleum Geologists Bulletin*, **71**, 810–821.

OKUI, A. & WAPLES, D. W. 1992. The influence of oil expulsion efficiency on the type of hydrocarbons accumulating in traps. *9th Offshore South East Asia Conference Preprints*, OSEA 92151.

—— & —— 1993. Relative permeability and hydrocarbon expulsion from source rocks. *In*: DORE, A. G. *et al.* (eds) *Basin Modelling : Advances and Applications*, Elsevier, 293–301.

——, HARA, M. & MATSUBAYASHI, H. 1994*a*. Overpressuring in a Japanese Tertiary basin simulated by two-dimensional basin model "SIGMA-2D" (abstract). *AAPG Hedberg Research Conference "Abnormal Pressure in Hydrocarbon Environments" Abstracts.*

——, —— & —— 1994*b*. The analysis of secondary migration by two-dimensional basin model "SIGMA-2D" (abstract). *1994 AAPG Annual Convention Official Program*, 227–228.

——, —— & —— 1995. Three-dimendional assessment of oil migration in a a Japanese basin by two-disemsional basin mdoel "SIGMA-2D" (abstract). *1995 AAPG Annual Convention Official Program*, 72A–73A.

——, ——, FU, H. & TAKAYAMA, K. 1996. SIGMA-2D : A simulator for the integration of generation, migration, and accumulation of oil and gas. *Proceedings of VIIIth International Symposium on the Observation of the Continental Crust Through Drilling*, 365–368.

PEGRUM, R. M. & SPENCER, A. M. 1990. Hydrocarbon plays in the Northern North Sea. *In*: ENGLAND, W. A. & FLEET, A. J. (eds) *Classic Petroleum Provinces*. Geological Society, London, Special Publications, **50**, 441–470.

PEPPER, A. S. 1991. Estimating the petroleum expulsion behaviour of source rocks: a novel quantitative approach. *In*: ENGLAND, W. A. & FLEET, A. J. (eds) *Petrolum Migration*. Geological Society, London, Special Publications, **59**, 9–31.

—— & CORVI, P. J. 1995. Simple kinetic models of petroleum formation. Part III: Modelling an open system. *Marine and Petroleum Geology*, **12**, 417–452.

POWERS, M. C. 1967. Fluid-release mechanisms in compacting marine mudrocks and their importance in oil exploration. *American Association of Petroleum Geologists Bulletin*, **51**, 1240–1254.

RAIMONDI, P. & TORCASO, M. A. 1964. Distribution of the oil phase obtained upon imbibition of water. *Petroleum Transactions, AIME*, **231**, Part II, 49–65.

DU ROUCHET, J. 1981. Stress field, A key to oil migration. *American Association of Petroleum Geologists Bulletin*, **64**, 74–85.

RUDKIEWICZ, J. L. & BEHAR, F. 1994. Influence of kerogen type and TOC content on multiphase primary migration. *Organic Geochemistry*, **21**, 121–133.

RULLKOTTER, J., LEYTHAEUSER, D., HORSFIELD, B., LITTKE, R., MANN, U. *ET AL.* 1988. Organic matter maturation under the influence of a deep intrusive heat source: A natural experiment for quantification of hydrocarbon generation and expulsion from a petroleum source rcok (Toarrcian shale, Nothern Germany). *Organic Geochemistry*. **13**, 847–856.

SANDVIK, E. I. & MERCER, J. M. 1990. Primary migration by bulk hydrocarbon flow. *Organic Geochemistry*, **16**, 83–89.

——, YOUNG, W. A. & CURRY, D. J. 1992. Expulsion from hydrocarbon sources: the role of organic absorption. *Organic Geochemistry*, **19**, 77–87.

SEKIGUCHI, K., OMOKAWA, M., HIRAI, A. & MIYAMOTO, Y. 1984. Geochemical study of oil and gas accumulation in 'Green Tuff' reservoir in Ngaoka to Kashiwazaki Region (in Japanese with English abstract). *Journal of the Japanese Association for Petroleum Technology*, **49**, 56–64.

SUZUKI, M., TAKEDA, N. & MACHIHARA, T. 1991. Study on petroleum generation by compaction petrolysis, Part 2: The amount of expelled and residual pyrolyzate with increasing maturity (abstract). *15th Meeting on the European Association of Organic Geochemists Poster Abstracts*, 3276–328.

TAKEDA, N., SATO, S. & MACHIHARA, T. 1990. Study on petroleum generation by compaction petrolysis — I. Construction of a novel pyrolysis system with compaction and expulsion of pyrolyzate from source rock. *Organic Geochemistry*, **16**, 143–153.

TALUKDAR, S., GALLANGO, O., VALLEJOS, C. & RUGGIERO, A. 1987. Observations on the primary migration of oil in the La Luna source rocks of the Maracaibo Basin, Venezuela. *In*: DOLIGEZ, B. (ed.) *Migration of Hydrocarbons in Sedimentary basins*. Editions Technip, Paris, 59–77.

TISSOTT, B. 1987. Migration of hydrocarbons in sedimentary basins: a geological, geochemical and historical perspective. *In*: DOLIGEZ, B. (ed.) *Migration of Hydrocarbons in Sedimentary basins*. Editions Technip, Paris, 1–19.

——, CALIFET-DEBYSER, Y., DERGO, G. & OUDIN, J. L. 1971. Origin and evolution of hydrocarbons in early Toarcian Shales, Paris Basin, France. *American Association of Petroleum Geologists Bulletin*, **55**, 2177–2193.

——, DURAND, B., ESPITALIE, J. & COMBAZ, A. 1974. Influenece of nature and diagenesis of organic matter in formation of petroleum. *American Association of Petroleum Geologists Bulletin*, **58**, 499–506.

——, PELET, R. & UNGERER, B. 1987. Thermal history of sedimentary basins, maturation indices and kinetics of oil and gas generation. *American Association of Petroleum Geologists Bulletin*, **71**, 1445–1466.

THOMAS, B. M., MOLLER-PENDERSEN, P., WHITAKER, M. F. & SHAW, N. D. 1985. Organic facies and hydrocarbon ditributions in the Norwegian North Sea. *Petroleum Geochemistry in Exploration of the Norwegian Shelf.* Norwegian Petroleum Society, 3–26.

UNGERER, P. 1990. State of the art of research in kinetic modelling of oil formation and expulsion. *Organic Geochemistry*, **16**, 1–25.

—— 1993. Modelling of petroleum generation and expulsion — an update to recent reviews. *In*: DORE, A. G. *et al.* (eds) *Basin Modelling : Advances and Applications*, Elsevier, 219–232.

——, BURRUS, J., DOLIGEZ, B., CHENET, P. Y. & BESSIS, F. 1990. Basin evaluation by integrated two-dimensional modeling of heat transfer, fluid flow, hydrocarbon generation, and migration. *American Association of Petroleum Geologists Bulletin*, **74**, 309–335.

WAPLES, D. W. & OKUI, A. 1992. A sensitivity analysis of oil expulsion (abstract). *1992 AAPG Annual Convention Official Program,* 137.

——, SUIZU, M. & KAMATA, H. 1992. The art of maturity modeling. Part 2: Alternative models and sensitivity analysis. *American Association of Petroleum Geologists Bulletin*, **76**, 47–66.

WARDLAW, N. C. 1980. The effect of pore structure on displacement efficiency in reservior rocks and in glass micromodels. *SPE Preprints*, SPE 8843.

—— & CASSAN, J. P. 1979. Oil recovery efficiency and the rock-pore properties of some sandstone reservoirs. *Bulletin of Canadian Petroleum Geology*, **27**, 117–138.

WELTE, D. W. & YALCIN, M. N. 1988. Basin Modelling — a new comprehensive method in petroleum geology. *Organic Geochemistry*, **13**, 141–151.

Some thoughts on porosity reduction — rock mechanics, overpressure and fluid flow

D. W. WAPLES[1] & G. D. COUPLES[2]

[1]*Consultant, 9299 William Cody Drive, Evergreen, CO 80439 USA*

[1]*Department of Geology and Applied Geology, University of Glasgow, Glasgow G12 8QQ, UK (Present address: Department of Petroleum Engineering, Heriot-Watt University, Edinburgh EH14 4AS, UK)*

Abstract: A currently popular paradigm, that porosity reduction occurs as a direct consequence of the effective stress acting on the rock framework grains, is mechanistically incorrect. The commonly observed covariance between porosity and effective stress does not reflect a cause-and-effect relationship. Instead, it arises because both low effective stress and slow porosity reduction are consequences of the inability of compacting rocks to expel their pore fluids quickly enough to maintain normal fluid pressures. The process of porosity loss is here divided into sequential steps, and we argue that the expulsion of pore fluids is the rate-determining step leading to overpressuring. Thus, Darcy's law assumes equal importance with the relationships describing the mechanical compaction of sediments. In this paper we describe how compaction can be treated as a Coulomb–plastic response that is functionally dependent on effective mean stress, deviatoric stress, and the state of compaction. In the next generation of basin models, a mechanistically correct approach is needed, combining both rock mechanics and hydrogeology.

Our purpose here is to investigate the process(es) of porosity reduction (i.e. mechanical compaction) in an effort to identify the step which is most important in controlling the rate of porosity loss in typical basin settings. This topic is of current interest because of the increasing importance of overpressured conditions in exploration and development, and because overpressure and under-compaction seem intimately related in many cases. Indeed, the present popular paradigm is that porosity can predict overpressure. This paper assesses the compaction phenomenon, providing a basis for evaluating whether this paradigm is well-founded, or is merely a coincidence that cannot be relied upon.

In order to permit us to identify the most important process(es) of compaction, we must divide the complex real physical system into major component steps. We begin by separating the evolution of subsiding and compacting rocks into conceptual steps. This approach allows us to show that, for cases representing normal basin subsidence, the primary control on compaction (porosity reduction) is the permeability. This result contrasts markedly with suggestions that porosity loss is directly related to the rate of sedimentation. We also consider the mechanical aspects of compaction, viewing this as a specific style of deformation, and relating it to a 'typical' basin setting. We suggest that compaction can be treated as a form of plastic yielding such that the stress state of compacting (yielding) rocks can be estimated from multi-parameter yield surfaces. Finally, we assess the practice of predicting overpressure from porosity measurements, and we conclude that this approach to overpressure prediction may be quite inappropriate, and hence may lead to dangerous over-confidence during drilling.

Steps in porosity reduction

We suggest that the compaction of rocks can be separated into four sequential steps:

(1) A load is applied to the system of sediment+pore fluid. In many cases, this load will be the addition of new sediment at the top of the stratigraphic column.
(2) The framework of grains deforms (yields), leading to a (slight) reduction in pore volume. This step includes simple grain re-arrangement and 'true' distortional strain.
(3) The contained pore fluid(s) experience a pressure increase due to the reduction in pore space. This step results in a net transfer of some of the applied load from the framework grains to the pore fluid.
(4) The now (slightly) overpressured pore fluid flows to sites with lower potential energy (for example, the surface), if this is possible.

WAPLES, D. W. & COUPLES, G. D. 1998. Some thoughts on porosity reduction — rock mechanics, overpressure and fluid flow. *In*: DÜPPENBECKER, S. J. & ILIFFE, J. E. (eds) *Basin Modelling: Practice and Progress*. Geological Society, London, Special Publications, **141**, 73–81.

In processes which occur as a series of steps, any one of the steps can, in principle, control the overall rate of the process. In chemical reactions, the slow step in a reaction sequence is called the 'rate-controlling step', the 'rate-limiting step', or the 'rate-determining step'. In the discussion which follows, we will borrow this terminology, and the slow step in the porosity-reduction sequence will be called the rate-determining step (RDS).

As appreciated by Magara (1978, p. 49), all four of the steps listed above should be considered *a priori* as possible rate-determining steps for the process of porosity reduction. However, identification of the rate-determining step is of little practical importance in cases where fluid pressures remain normal during compaction. In such cases, all four steps proceed approximately together, and any transient overpressure (as must be created to expel pore fluids) does not persist. Compaction in normally pressured settings is driven only as fast as the loading occurs, and so we may safely conclude that sedimentation is the rate-determining step (RDS) for such situations.

In circumstances where the pore fluid becomes overpressured (even by a small amount) for more than a brief time, the system becomes more interesting, both from a practical point of view, and a theoretical one. The existence of overpressure (or underpressure, but we do not consider this case here) means that the system has entered a non-linear mode, and that there is an energy gradient which can drive transient phenomena. In such situations, one of the four steps must be acting as the RDS for the system.

Below, we assess each of the steps in turn, asking whether or not it could be the primary RDS in overpressured systems. For this analysis, we choose a simple setting in which sedimentation is the only potential loading phenomenon. Such circumstances are probably widely applicable in the majority of overpressure environments (but see below). We evaluate four cases in which each of the four steps is, in turn, considered to be the 'slow' RDS.

Sedimentation as the RDS ?

Let us assume that the first step in the process, the application of load (by sedimentation), is the RDS in the sequence leading to overpressuring (Fig. 1). If this were the primary cause of the disequilibrium, then overpressuring would show a strong correlation with low sedimentation rates. Such an empirical correlation is not what is observed; in fact, the opposite correlation exists — high sedimentation rates and overpressure are commonly associated. Even if sedimentation were still somehow thought to be the RDS, any abnormal pressures created would dissipate rapidly (a consequence of the fourth step being fast). Thus, sedimentation cannot be the RDS in overpressured systems. As noted above, however, it may be the RDS in the compaction of normally pressured systems.

Grain response as the RDS ?

Can the grain response be the RDS in overpressured systems (Fig. 2)? This question requires us to consider the 'flow' of deforming rocks. When rocks yield, they change shape. (This is true except for pure 'hydrostatic' compression associated with the volumetric compaction that occurs under high isotropic stress.) The rate of this shape change, or the strain rate, can be related to the causative stress state by a rheologic model. As an example, a viscous model (e.g. $\dot{\varepsilon} = \eta\ \sigma$) reflects a simple, linear relationship between stress and strain rate. More complicated models can address non-linearities, but all such generally accepted models have the common characteristic that they predict a monotonic increase in strain rate as the deviatoric stress increases.

For a given rock material, such as a fine-grained sediment subsiding in a basin, let us imagine a loading that causes compaction, but no overpressure. Now let us imagine that the sedimentation rate is increased (a situation that is empirically associated with overpressure development). The extra load associated with the increased sediment column is expected to cause an increase in the stress. According to the simple viscous model, the strain rate of the rock will also increase, which we take to mean that the rate of grain response increases.

Since we are attempting to assess whether a *slow* grain response rate is the RDS, it seems that, for this to be true, we would be forced to infer that a slower-than-usual grain response is associated with

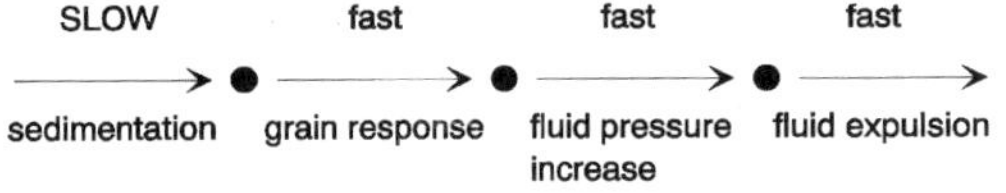

Fig. 1. Sequence of steps in overpressuring. Sedimentation shown as slow step.

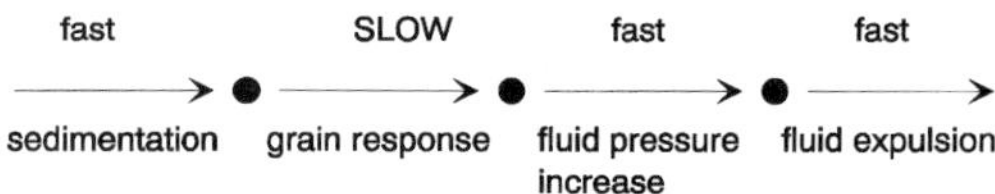

Fig. 2. Sequence of steps in overpressuring. Grain response shown as slow step.

a faster-than-usual sedimentation rate. This is the opposite to what is observed, and so we must conclude that the rate of grain response is not the RDS in typical overpressure settings.

Transfer of load to the fluid as the RDS ?

Figure 3 shows the transfer of load to the pore fluid as the RDS in overpressure development. However, it is physically impossible for this step in the compaction process to be the RDS because pressure in fluid is transmitted at the speed of sound (within pore spaces). Thus, fluid pressure within deformed pores will increase instantaneously (geologically speaking) when the grain framework experiences deformation (step 2). Therefore, step 3 is always fast and cannot be the RDS.

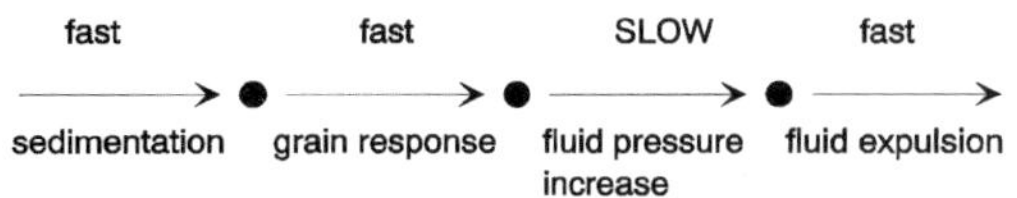

Fig. 3. Sequence of steps in overpressuring. Transfer of load to fluid shown as slow step.

Fluid expulsion as the RDS ?

At this point, we could conclude that step 4 must be the RDS since the preceding steps have been rejected as being incompatible with the evidence. However, that form of argument pre-supposes that our separation into steps is without any error. Because of the importance of determining the overall rate-limiting step in overpressure development, we here continue the analysis to see if step 4 is plausible as the RDS.

If fluid expulsion is rate-determining (Fig. 4), then only the hindered flow of pore fluids can result in overpressure, regardless of the rate of sedimentation, the rate of grain response, or the rate of transfer of load to the pore fluid. If we assume that fluid flow is governed by Darcy's law, then the factors which hinder that flow must be represented by the terms in Darcy's equation, as expressed for multiple fluid phases (written for the 1-D case to simplify the notation):

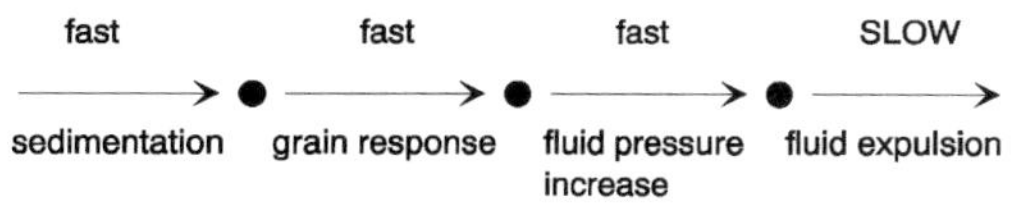

Fig. 4. Sequence of steps in overpressuring. Fluid expulsion shown as slow step.

$$q = \sum \frac{k \times k_{ri} \times A \times \phi_i}{\mu_i \times L}$$

(summed for all n fluid phases)

Here, q is the total flow rate of all n fluid phases per unit time across an area A, k is the absolute permeability, k_{ri} is the relative permeability to the ith fluid phase, φ_i is the drop in fluid potential along the path length L, and μ_i is the viscosity of the ith phase.

For a given block of rock+fluid, the factors that can hinder the flow of pore fluids are:

(a) low fluid-potential gradient;
(b) low absolute permeability;
(c) high fluid viscosity;
(d) or low relative permeability.

If the system is significantly overpressured, then there is by definition a significant fluid-potential gradient somewhere within the system. The only way that this gradient can fail to directly affect a given volume is when that volume is wholly within a hydraulically-connected 'cell' of overpressure. In this case, there is no local gradient, but there remains a gradient at the boundaries of the cell, and potential energy is not the limiting factor in motivating pore fluids to move to lower-pressure parts of the system.

The obvious candidate for causing slow fluid flow is low absolute permeability. Most of the attention given to the subject thus far has been directed towards this aspect of the rock+fluid system (Bredehoeft & Hanshaw 1968; Luo & Vasseur 1992; Deming 1994; He & Corrigan 1995; Neuzil 1995), but Benzring (1994) has shown, through experiments, that overpressuring can occur when *relative* permeabilities to all fluids are greatly reduced. Such a mechanism may be of particular importance in actively generating source rocks (Waples & Okui 1992). High-viscosity fluid should also contribute to overpressuring, as may be the case for the Monterey Formation of Southern California (with its high-sulfur, low-gravity oil; Waples & Okui 1992), but this model has not been explored in detail.

It should be noted that these independent variables in Darcy's equation need not necessarily have very low absolute values. They need only be small (or large, for viscosity) in comparison with the energy imparted to the system (for example, as a consequence of compaction of the framework grains). Some of the variables have natural variations covering one or two orders of magnitude, but permeability may vary by a factor of more than 10^{10}. Because of this extreme range, permeability can be fairly high (relatively speaking), and yet overpressure can develop if sedimentation rates are very high. Conversely, when sedimentation is slow, even a low-permeability rock may remain normally pressured.

Finally, it is worth noting that Terzaghi & Peck (1948, pp. 75–76) provide evidence that the expulsion of fluids is the RDS in compaction. They note a time effect on clay consolidation: the length of time required to achive a certain amount of compaction (under constant load) depends strongly on the layer thickness. The logical explanation for this time effect is that the rate of compaction is controlled by the rate of fluid flow, and that rate is strongly dependent on the length of the flow path (refer to Darcy's equation, above). These observations are inconsistent with any of steps 1–3 being the RDS of compactional deformations.

Thus we see that restricted fluid flow is a very plausible candidate for the RDS in overpressure development. Nevertheless, it must again be emphasized that flow restrictions are not a sufficient cause of overpressure: the system needs to have energy sources which exceed the rate of dissipation (cf. Neuzil 1995). In the discussion thus far, we have focused on sedimentation as the drive mechanism for compaction, but we broaden our considerations in the discussion. First, however, we need to elaborate on the mechanics underlying step 2.

Mechanics of compaction

Consider an 'element' of sediment+pore fluid that is subsiding within a basin (Fig. 5). In a typical basin, this element is surrounded by other elements which were deposited at the same time, and it is overlain by elements deposited subsequently. The weight of these overlying elements constitutes the primary cause of the compaction of 'our' element (ignoring tectonic distortions). In this simple setting, the rate of loading is directly associated with the rate of sedimentation (this is the same situation as assumed in the preceding section).

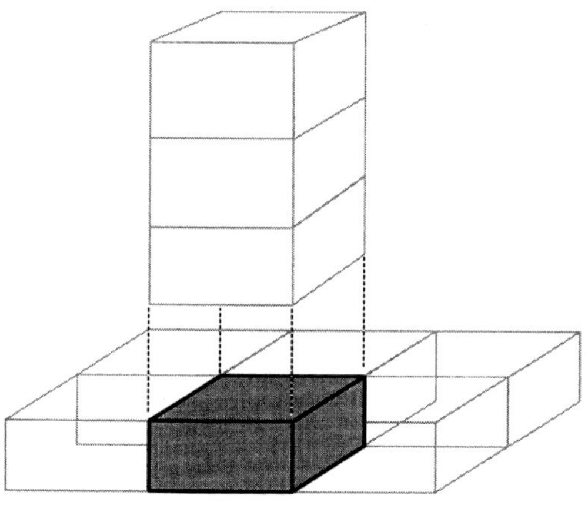

Fig. 5. Typical element of rock+fluid in a basin. Note surrounding elements, and column of overlying elements. In usual circumstances, vertical dimension of elements decreases with depth (compaction) as fluids are expelled.

As the element subsides in the basin and becomes buried, it compacts — that is, its vertical dimension becomes smaller. In most situations, it seems that such elements do not experience changes in their horizontal dimensions, and so the compaction is uniaxial. If we ignore chemical/ mineralogical changes, and especially if we ignore chemical mass-transport, then this vertical compaction is directly identified with porosity loss. The empirical reduction in porosity with depth of burial (Baldwin & Butler 1985; Sclater & Christie 1980) reflects the near-universality of this compactional evolution of sediments.

In mechanical parlance, the porosity loss experienced by our element during its burial is a compactant deformation. This deformation is permanent (non-reversible), and it can therefore be cast into terms derived from plasticity theory. Plasticity attempts to define a yield function (in stress space) which bounds the stable (non-yielding) states of stress which can be attained by a material. The yield function explicitly specifies those stress states where yielding occurs. The yielding of rocks, of course, shows a strong dependence on the mean stress (Handin & Hager 1957; Byerlee 1978; Stearns *et al.* 1981), and such behaviour (Fig. 6) is representative of a Coulomb–plastic material (Drucker & Prager 1959). Note that this representation of the plastic yield envelope is similar to that used for depicting the Mohr-Coulomb criterion (which is often displayed in τ-σ space; see Jones & Addis 1985, 1986; Jamison 1992), but that the plastic approach differs by having a 'cap' on the yield envelope.

Our element will be characterized by this form of yield function. In a subsiding basin, rocks are experiencing compaction, and we may therefore assume that their stress state is at yield. Because of the compactant, distortional (quasi-uniaxial strain) deformation, we may also assume that the stress state position (note that the stress state is represented by a point in this approach, as opposed to a circle in the usual Mohr–Coulomb depiction; Jamison 1992) is somewhere on the 'cap' of the yield envelope. An increment of deformation caused by this yielding produces both a volumetric strain component, and a distortional strain component. We can also assume that the mean stress axis is actually the effective mean stress (based on experimental work which shows that the concept of effective stress is valid; see, for example, Handin *et al.* 1963).

We also expect that typical basin rocks which

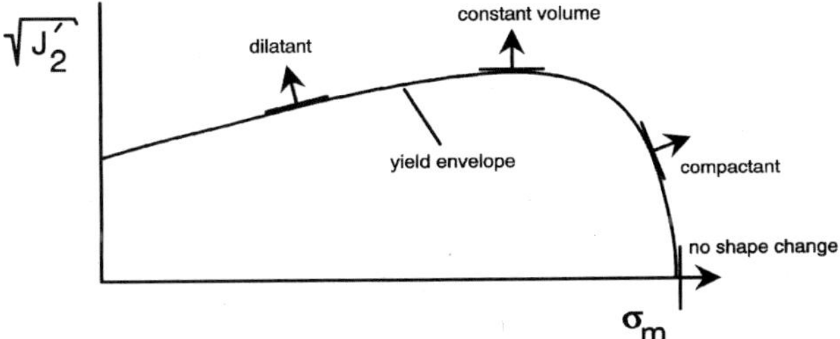

Fig. 6. Coulomb–plastic yield function with parameters σ_m and $J_2'^{1/2}$. Tangent lines and normal vectors indicate types of strain at yield, if stress axes are associated with auxiliary coordinates giving volume strain and distortional strain (see Hill 1950; Drucker & Prager 1952; Shield 1954; de Jong 1959; Jenike & Shield, 1959)

experience burial compaction generally become stronger. That is, burial and the resulting deformation alter the material's behaviour such that a state of stress which was previously capable of causing deformation becomes stable for the 'new', more compacted material. In short, the yield function must change as a result of the compactional deformation. In schematic fashion, this change may be viewed as a series of yield functions arrayed along a 'deformation' axis (Fig. 7). If these functions are continuously linked together (not shown), the deformation of the material can now be described via a yield surface. (Note: this treatment owes much to unpublished work by M. Fahy; see D'Onfro *et al.* 1994 for an illustration of its application; and compare Jones & Addis 1986.) For our purposes, let us associate the 'deformation' axis of this array of yield functions with 'compaction' (porosity loss).

Let us turn again to our element which is compacting (that is, undergoing compactant deformation) as it is buried in the hypothetical basin. Based on the approach outlined above, this element has a stress state that lies on the cap portion of the yield surface representing the mechanical behaviour of the material. Now let us assume that additional loading occurs as a result of further sedimentation. This loading is partitioned onto both the matrix and the pore fluid. Assuming that the rock matrix is still at yield (that is, the effective stress state lies on the yield surface), the resultant compactant deformation reduces the remaining pore space. The small increase in pore pressure caused by the distortion of the matrix (i.e. a pore volume reduction) provides energy to the fluid system, and the pore pressure is expected to be (slightly) elevated as a result. The slightly increased pore pressure causes the effective mean pressure to decrease, and the stress point moves slightly inside the yield surface (Fig. 8), slowing or stopping the deformation of the matrix. Resumption of compaction (porosity loss) is therefore dependent on relief of the overpresssure. If overpressure cannot be relieved, then the rocks fail to undergo further compaction, leading to 'under-compaction' or what is called compaction disequilibrium.

Discussion

Our view of overpressure development, in 'typical' basin subsidence situations, is that the rates of sequential processes are crucial considerations. We argue that it is the retarded expulsion of pore fluids that is the critical, limiting step in the overall process of creating overpressure, and, by analogy

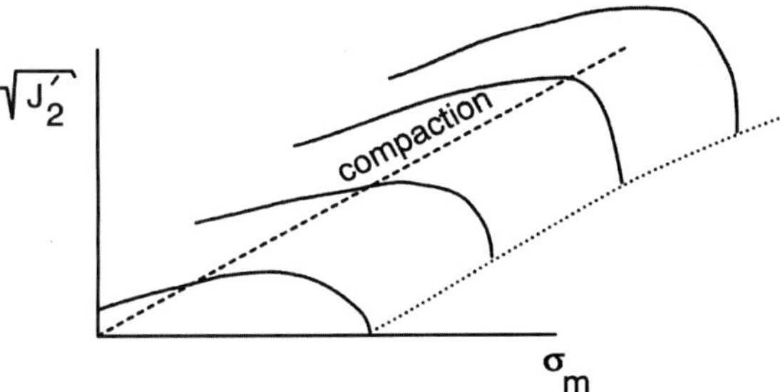

Fig. 7. Illustration of yield envelopes changing as compaction increases (see D'Onfro *et al.* 1994).

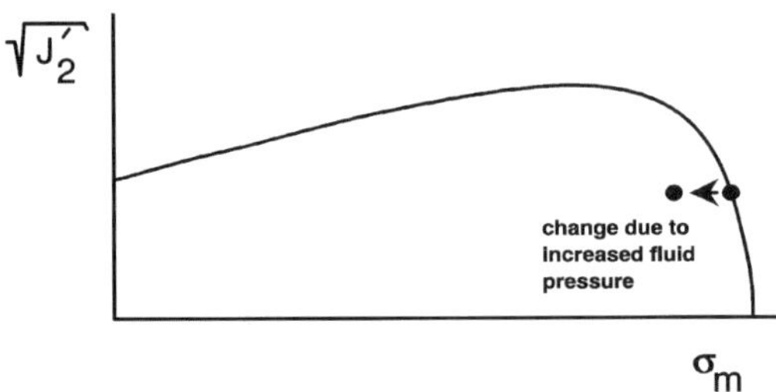

Fig. 8. Stress state at yield (compaction) altered by increase in pore pressure (decrease of effective stress). Poro-elastic considerations (see Engelder & Fischer 1994; Kümple 1991; Miller 1995) will simultaneously alter the porosity (not shown).

with reaction kinetics, we call this the rate-determining step. Low absolute permeability is probably the most common means by which fluid flow is retarded, but there may well be auxiliary factors such as poor relative permeabilities, or high fluid viscosities. In overpressured settings, fluid-potential gradients are usually not a factor which limits flow.

A number of experimental and theoretical works have shown that 'effective stress' should have a primary control on the deformation behaviour of porous materials. Our analysis supports this view of effective stress in overpressured settings, showing that effective stress is a valid mechanical concept when included into a complete analysis of the behaviour of the rock+fluid system. Thus, we are able to propose that the compactional behaviour of sediments undergoing burial can be suitably described using a Coulomb–plastic approach where one parameter of the yield function is the effective mean stress, and where another parameter is an invariant of the deviatoric stress tensor. A third parameter allows for explicit consideration of the evolution of the porosity. An elevated pore pressure allows sediments to retain a higher porosity than would otherwise be predicted for their state of burial (Wilkinson *et al.* 1997).

The results of our contribution, as summarized in the preceding two paragraphs, may not surprise many readers, and there may be some who will question why we troubled to 'discover' the obvious. The reason we have undertaken this analysis is so that we can attempt to sort out a problem in basin modelling which we perceive to be developing. This problem appears to be an example of a situation where cause and effect have become confused.

Some geologists performing basin analysis do not delve deeply into the mechanistic causes of overpressuring. General empirical observations like 'overpressuring occurs most commonly where sedimentation rates are high' and 'overpressuring occurs more frequently in fine-grained rocks than in sandstones' meet the needs of many geologists. However, basin modelling requires that compaction be calculated, and some developers of basin models have adopted a paradigm in which mechanical porosity reduction is controlled by the effective stress on the rock grains. In this simple, effective stress paradigm, when effective stresses are high, porosity is rapidly reduced. Conversely, when effective stresses are low, porosity is reduced more slowly. This model correctly predicts that when overpressuring occurs (causing decreased effective stress), porosity reduction is retarded.

Unfortunately, this relationship seems to have become extended into the form of a paradigm on whose origin one can only speculate. Terzaghi & Peck (1948, pp. 74–78) suggested that the nearly instantaneous transfer of loading-induced stress to the pore fluids increases the pore pressure and creates a fluid potential gradient that causes expulsion of the pore fluids. The loss of those fluids in turn permits grain responses that represent compaction. (Our view — as expressed above — is essentially the same.) These concepts were later slightly altered to indicate that '... for a given clay there exists for each value of porosity φ some maximum value of effective compressive stress which the clay can support without further compaction.' (Hubbert & Rubey 1959). This statement is functionally equivalent to saying that there is a minimum porosity associated with any given maximum effective stress. Extending this line of reasoning, Hottman & Johnson (1965) stated that '... the porosity φ at a given burial depth D is dependent on the fluid pressure p. If the fluid pressure is abnormally high (greater than hydrostatic), the porosity will be abnormally high for a given burial depth.'

The crux of the problem which we see developing lies in mistakenly equating a series of equivalencies with a mechanistic relationship. It is no error to believe that porosity covaries with effective stress (but see below), but it *is* an error to believe that this covariance necessarily implies cause and effect.

In fact, the increasingly popular paradigm, in which effective stress is a cause of porosity reduction, directly contradicts the predictions from Darcy's law. If the fluid potential gradient is high (overpressure), Darcy's law predicts greater fluid flow, and hence greater porosity reduction, than if the fluid potential gradient is low (e.g. normal pressure). The effective-stress paradigm, in contrast, predicts less fluid flow when effective stress is lower (overpressure), because the grain framework in such a case does not collapse. Thus, the effective stress paradigm, as it has come to be employed by many in basin analysis, is incompatible with the ideas reviewed in this paper. Instead, we argue that when overpressuring occurs, fluid flow is controlled by Darcy's law. Thus, the present effective stress paradigm is mechanistically incorrect for overpressured systems.

Why, then, does the effective stress paradigm make reasonable predictions of the amount of porosity reduction? Simply because both effective stress and porosity reduction have a common cause. Effective stress does not cause fluid to flow (only a fluid potential gradient can do that), but low effective stress is caused by overpressuring, and overpressuring is caused by a lack of good flow characteristics (step 4: low permeabilities, or high viscosities). Thus low effective stress is caused by a chain of relationships leading back to the principal cause: poor flow characteristics of the fluid–rock system. As permeabilities increase, or as viscosity decreases, fluid flow is enabled, and more porosity reduction can occur. (Or, if loading ceases or slows, excess fluid pressure will, with time, dissipate, and then compaction will proceed.) Thus, covariance between effective stress and porosity reduction is natural and expected for common circumstances.

However, the simple scenario which we have so far considered is not the only one in which overpressure occurs. The amount of porosity reduction in an abnormally pressured system is dependent on the behaviour of the overall fluid system. For example, there may be other events which impose high fluid pressures, and, if the response of the fluid is slow (due to any of the possibilities discussed under step 4), then any of these events may be considered as the root cause of overpressure. Presently, various workers are highlighting the role of hydrocarbon generation or conversion as sources of elevated fluid pressures (e.g. Bredehoeft *et al.* 1994; Luo *et al.* 1994; Yassir & Bell 1994; Darby *et al.* 1998). Aquathermal pressuring is another process which could contribute increments of overpressure to the fluid system irrespective of the compactional state of the rock framework (Barker 1972; Luo & Vasseur 1992, but compare the arguments of Miller & Luk 1993). In fact, there may be numerous simultaneous sources creating excess fluid pressure (Neuzil 1995; Swarbrick 1995; Swarbrick & Osborne 1996). The important common thread is that overpressure is significant only if the rate of dissipation is slow — that is, step 4 is the RDS for overpressured systems.

Another situation is recently gaining attention in overpressure analysis. Here we refer to those circumstances where high fluid pressures are generated off-site, but where that pressure is communicated to other parts of the system. Darby *et al.* (1997, 1998) describe such an example from the Central Graben of the North Sea. They argue that both compaction disequilibrium and hydrocarbon generation/cracking are occurring in deep kitchen areas of the graben, and that the high pressures created by these processes have been transmitted onto nearby structural highs. In these latter sites, pressure–depth and porosity–depth relationships are 'out of sync', and therefore cannot be used to make predictions of fluid pressures. Such situations alert us to the need to develop better methods of predicting conditions ahead of the drill bit.

Our purpose here has been to emphasize the intimate relationship between overpressure and fluid flow, and the concurrent relationship with rock mechanics. Further advances in basin modelling will necessitate that both mechanical and fluid systems be addressed, and we will also need to appreciate that one-dimensional simplifications can prove to be seriously erroneous. We anticipate that exciting new advances will be made in these areas leading to a new and better generation of basin models.

Conclusions

The concept of effective stress not only survives our analysis, it is even more strongly supported by evidence arising from overpressured situations. Compaction is critically dependent on fluid expulsion. Therefore, the calculation of porosity reduction necessarily requires a proper consideration of the fluid system coupled to the rock mechanics.

However, the popular effective stress paradigm is misleading because its presumed *cause* is actually an *effect*. Covariance of porosity and effective stress may occur in simple settings, but there are serious limitations to the use of this relationship.

In situations where processes other than compaction affect the fluid pressure, porosity–depth relationships, and porosity–overpressure relationships, are unreliable.

A new generation of basin models needs to be developed to incorporate the concepts outlined in this paper.

J. Vizgirda critically read an earlier draft and contributed many stimulating ideas to D. Waples that significantly improved the manuscript. G. Couples wishes to acknowledge a long-term debt to M. Fahy for fruitful discussions and tutelage in advanced rock mechanics. Z. He provided an extremely helpful review of an earlier version of the manuscript, but he should not be held accountable for any remaining errors. H. Lewis and an anonymous reviewer made several suggestions for improving the present version.

References

BALDWIN, B. & BUTLER, C. O. 1985. Compaction curves. *Bulletin of the American Association of Petroleum Geologists,* **69,** 622–626.

BARKER, C. 1972. Aquathermal pressuring: role of temperature in development of abnormal pressure zone. *Bulletin of the American Association of Petroleum Geologists,* **56,** 2068–2071.

BENZRING, W. 1994. The 'vapor-lock' pressure seal — a non-conformable seal creating potential buoyant forces within young clastic basins. *In*: LAW, B. E., ULMISHEK, G. & SLAVIN, V. (eds) *Abnormal Pressures in Hydrocarbon Environments.* American Association of Petroleum Geologists Hedberg Research Conference.

BREDEHOEFT, J. D. & HANSHAW, R. B. 1968. On the maintenance of anomalous fluid pressures: I. Thick sedimentary sequences. *Geological Society of America Bulletin,* **79,** 1097–1106.

——, WESLEY, J. B. & FOUCH, T. D. 1994. Simulations of the origin of fluid pressure, fracture generation, and the movement of fluids in the Uinta Basin, Utah. *Bulletin of the American Association of Petroleum Geologists,* **78,** 1729–1747.

BYERLEE, J. D. 1978. Friction of rock. *Pure and Applied Geophysics,* **116,** 615–626.

DARBY, D., HASZELDINE, R. S. & COUPLES, G. D. 1997. Pressure cells and pressure seals in the central North Sea. *Marine and Petroleum Geology*, **13**, 865–878.

——, —— & —— 1998. Central North Sea overpressure: insights into fluid flow from one- and two-dimensional basin modelling.*This volume.*

D'ONFRO, P., FAHY, M. F. & RIZER, W. 1994. Geomechanical model for fault sealing in sandstone reservoirs. *American Association of Petroleum Geologists Annual Convention Official Program,* **3,** 130–131.

de JONG, J. J. 1959. *Statics and dynamics of the failable zone of a granular material.* Waltman, Delft.

DEMING, D. 1994. Factors necessary to define a pressure seal. *Bulletin of the American Association of Petroleum Geologists,* **78,** 1005–1009.

DRUCKER, D. C. & PRAGER, W. 1952. Soil mechanics and plastic analysis or limit design. *Quarterly Applied Mathematics,* **10,** 157–165.

ENGELDER, T. & FISCHER, M. P. 1994. Influence of poro-elastic behavior on the magnitude of minimum horizontal stress, S_h, in overpressured parts of sedimentary basins. *Geology,* **22,** 949–952.

HANDIN, J. W & HAGER, R. V. 1957. Experimental deformation of rocks under confining pressure; tests at room temperature on dry samples. *Bulletin of the American Association of Petroleum Geologists,* **41,** 1–50.

——, ——, FRIEDMAN, M. & FEATHER, J. N. 1963. Experimental deformation of sedimentary rocks under confining pressure: Pore pressure tests. *Bulletin of the American Association of Petroleum Geologists,* **47,** 717–755.

HE, Z. & CORRIGAN, J. 1995. Factors necessary to define a pressure seal: discussion. *Bulletin of the American Association of Petroleum Geologists,* **79,** 1075–1078.

HILL, R. 1950. *The mathematical theory of plasticity.* Clarendon Press, Oxford.

HOTTMAN, C. E. & JOHNSON, R. K. 1965. Estimation of formation pressures from log-derived shale properties. *Journal of Petroleum Technology,* **17,** 717–722.

HUBBERT, M. K. & RUBEY, W. W. 1959. Role of fluid-filled porous solids and its application to overthrust faulting. *Bulletin of the Geological Society of America,* **70,** 115–166.

JAMISON, W. R. 1992. Stress spaces and stress paths. *Journal of Structural Geology,* **14,** 1111–1120.

JENIKE, A. W, & SHIELD, R. T. 1959. On the plastic flow of Coulomb solids beyond original failure. *Journal of Applied Mathematics and Physics,* **23,** 599–602.

JONES, M. E. & ADDIS, M. A. 1985. On changes in porosity and volume during burial of argillaceous sediments. *Marine and Petroleum Geology,* **2,** 247–253.

—— & —— 1986. The application of stress path and critical state analysis to sediment deformation. *Journal of Structural Geology,* **8,** 575–580.

KÜMPLE, H.-J. 1991. Poro-elasticity: parameters reviewed. *Geophysical Journal International,* **105,** 783–799.

LUO, X. & VASSEUR, G. 1992. Contributions of compaction and aquathermal pressuring to geopressure and the influence of environmental conditions. *Bulletin of the American Association of Petroleum Geologists,* **76,** 1550–1559.

——, BAKER, M. R. & LEMONE, D. V. 1994. Distribution and generation of the overpressure system, eastern Delaware Basin, western Texas and southern New Mexico. *Bulletin of the American Association of Petroleum Geologists,* **78,** 1386–1405.

MAGARA, K. 1978. *Compaction and Fluid Migration.* Elsevier, Amsterdam.

MILLER, T. W. 1995. New insights on natural hydraulic fractures induced by abnormally high pore pressures. *Bulletin of the American Association of Petroleum Geologists,* **79,** 1005–1018.

—— & LUK, C. H. 1993. Contributions of compaction and aquathermal pressuring to geopressure and the influence of environmental conditions: discussion. *Bulletin of the American Association of Petroleum Geologists,* **77,** 2006–2010.

NEUZIL, C. J. 1995. Abnormal pressures as hydrodynamic phenomena. *American Journal of Science,* **295,** 742–786.

SCLATER, J. G. & CHRISTIE, P. A. F. 1980. Continental stretching: an explanation of the post-mid-Cretaceous subsidence of the central North Sea Basin. *Journal of Geophysical Research,* **85,** 3711–3739.

SHIELD, R. T. 1954. Stress and velocity fields in soil mechanics. *Journal of Applied Mathematics and Physics,* **18**, 144–156.

STEARNS, D. W., COUPLES, G. D., JAMISON, W. R. & MORSE, J. D. 1981. Understanding faulting in the shallow crust: contributions from selected experimental and theoretical studies. *In*: CARTER, N.

L., FRIEDMAN, M., LOGAN, J. M. & STEARNS, D. W. (eds) *Mechanical Behavior of Crustal Rocks (The Handin Volume)*, American Geophysical Union Monograph, **24**, 215–229.

SWARBRICK, R. E. 1995. Distribution and generation of the overpressure system, eastern Delaware Basin, western Texas and southern New Mexico. discussion. *Bulletin of the American Association of Petroleum Geologists,* **79,** 1817–1821.

—— & OSBORNE, M. J. 1996. The nature and diversity of pressure transition zones. *Petroleum Geoscience,* **2,** 111–116.

TERZAGHI, K. & PECK, R. B. 1948. *Soil Mechanics in Engineering Practice*. John Wiley and Sons, New York.

WAPLES, D. W. & OKUI, A. 1992. A sensitivity analysis of oil expulsion. *In*: DAWSON, M. (ed.) *Advances in Basin Modelling Techniques*. International Conference, Aberdeen, UK.

WILKINSON, M. W., DARBY, D., HASZELDINE, R. S. & COUPLES, G. D. 1997. Secondary porosity generation during deep burial associated with overpressure leak-off: Fulmar Formation, UKCS. *Bulletin of the American Association of Petroleum Geologists*, **81**, 803–813.

YASSIR, N. & BELL, J. S. 1994. Relationships between pore pressure, stresses, and present-day geodynamics of the Scotian Shelf, offshore eastern Canada. *Bulletin of the American Association of Petroleum Geologists,* **78,** 1863–1880.

An estimation of the intrinsic permeability of argillaceous rocks and the effects on long-term fluid migration

T. TOKUNAGA, S. HOSOYA[1], H. TOSAKA & K. KOJIMA

Department of Geosystem Engineering, Faculty of Engineering, University of Tokyo, 7-3-1 Hongo, Bunkyo-ku, Tokyo 113, Japan

[1] *Present address: Dia Consultants Co., Ltd., Kobunsha-Bldg., 3-1-2 Ikebukuro, Toshima-ku, Tokyo 171, Japan*

Abstract: The effect of intrinsic permeability of argillaceous rocks on long-term petroleum migration is discussed from results of both laboratory experiments and numerical sensitivity studies. Results of our experimental studies on one-dimensional mechanical compaction of both muddy slurries and mudstones show that porosity(ϕ)–permeability(K) relationships are linear on double logarithmic scales, i.e. ϕ and K satisfy the relationship: $K=K_0(\phi/\phi_0)^a$, where K_0 is initial permeability, ϕ_0 is initial porosity, and 'a' is a parameter depending on the samples used. Comparison with published ϕ–K relationships of mudstones indicates that the above mentioned relationship could be extrapolated to natural condition when ϕ>0.3. On the other hand, the relationship becomes deviated from measured data in the low porosity range.

Considering the importance of ϕ–K relationships in ϕ<0.3 on the timing of expulsion of petroleum from source rocks, this paper discusses the way to find out an appropriate ϕ–K relationship through trial and error matching of results of basin simulation using different ϕ–K relationships with observed pore pressure and porosity data. We also present one case study using an imaginary siliciclastic sedimentary basin and discuss the potential of a numerical study to establish one of the appropriate relationships, showing how calculated results become varied by choosing different relationships.

Integrated basin simulators have been developed by many organizations since the publications of physical and chemical models to express petroleum generation, migration, and accumulation in the basins (e.g. Welte & Yukler 1981; Ungerer *et al.* 1990). Applications of these models to real basins are also reported (e.g. Doré *et al.* 1993). We have also recently developed a three-dimensional two-phase basin simulator BASIN3D2P (Tokunaga *et al.* 1994*b*, 1996). Most of the models treat the compaction process using an effective stress concept, a two-phase Darcian fluid migration model for both primary and secondary petroleum migration, a conductive/convective heat flow model for heat distribution in the basin, and first order chemical kinetics for petroleum generation (e.g. Hermanrud 1993).

Numerical results of basin simulators depend strongly on the petrophysical parameters used, and hence, we should evaluate the behaviour and values of important and sensitive parameters through experimental, numerical, and theoretical investigations.

In this paper, we discuss the trend for reduction of intrinsic permeability of argillaceous rocks during compaction, which affects development of overpressured regions and thus affects timing of petroleum expulsion (Tokunaga *et al.* 1993), from results of both laboratory experimental studies on mechanical compaction (hereafter called consolidation) and numerical sensitivity studies using our simulator, BASIN3D2P.

Details of numerical treatments of geological processes, physical properties of fluids, solids, and solid–fluid systems, and methods to solve coupled and highly non-linear equations for BASIN3D2P are presented in Tokunaga *et al.*(1994*b*, 1996).

Permeability measurements during consolidation of muddy slurries

The one-dimensional consolidation apparatus, shown in Fig. 1, was used to carry out both consolidation and permeability measurements using muddy slurries as starting materials. Maximum compressive stress which can be applied by the apparatus is about 500 kg f cm^{-2}.

Artificially deposited or remoulded samples were inserted into the cylinder and sequentially

TOKUNAGA, T., HOSOYA, S., TOSAKA, H. & KOJIMA, K. 1998. An estimation of the intrinsic permeability of argillaceous rocks and the effects on long-term fluid migration. *In*: DÜPPENBECKER, S. J. & ILIFFE, J. E. (eds) *Basin Modelling: Practice and Progress*. Geological Society, London, Special Publications, **141**, 83–94.

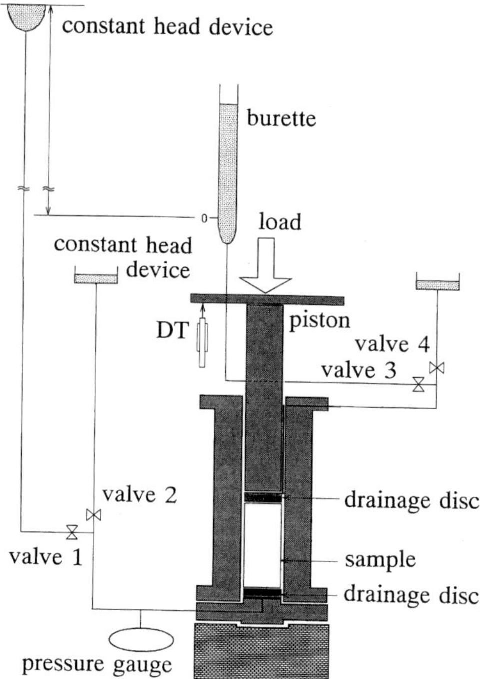

Fig. 1. Schematic diagram of the one-dimensional consolidation apparatus. DT is the displacement transducer.

loaded. Pore water was drained through both the upper and lower drainage discs. Hydraulic heads of both the upper and lower outlets were set at the same value during consolidation to avoid generation of vertical gradient of porosity in the sample. Falling head permeability measurements were conducted at the end of respective loading stages by injecting water from the lower side (Fig. 1). Porosities for respective loading stages were determined by back calculation from the sample weight and axial displacements recorded by displacement transducers attached in parallel with the loading piston (Fig. 1). The vertical stresses measured during the tests included the effects of sliding friction of both the sample and the seal along the cell wall. This side wall friction was minimized by thinly coating the cell walls with MoS_2 grease. In this paper, we do not discuss the stress–strain relationships during the experiments because it was difficult to estimate the effect of friction. Time effects on displacements after diffusion of excess pore pressure in the sample were not detected by displacement transducers (resolution of 0.01 mm) used in this study.

Samples used in the experiments were crushed siltstone from the Plio-Pleistocene Kiwada Formation of the Kazusa Group in Boso Peninsula, Japan; crushed Rochester shale in New York; and commercially available kaolinite powder. Fractions of clay size and silt size particles in the crushed samples and kaolinite powder are shown in Table 1 with the initial conditions of each experiment. Samples were set by either the following procedures:

(1) Large amounts of pure water and powder samples were mixed and poured into the cylinder. They were kept until the particles of minimum size settled down. Settling time of the smallest particles was estimated by the Stokes' law. This initial condition was used to obtain similar fabric to the naturally deposited sediments. This condition is referred to as 'depositional setting' here (Table 1).
(2) Samples were prepared as slurries, with relatively smaller amounts of pure water than method 1. These slurries had porosities of approximately 0.8 and were viscous enough to prevent segregation during filling of the cell. This condition is here named as 'slurry setting' (Table 1).

Measured vertical permeability data of one run (exp. 6) are plotted against porosity on double logarithmic axes in Fig. 2 as an example. The result was all obtained from the experiment on the normally consolidated sample for which the vertical effective stress was being steadily increased and porosity was steadily falling. Error bars shown in Fig. 2 were estimated from both errors derived from the measurements during experiments and errors derived from least square fitting to obtain permeability. Results of each run

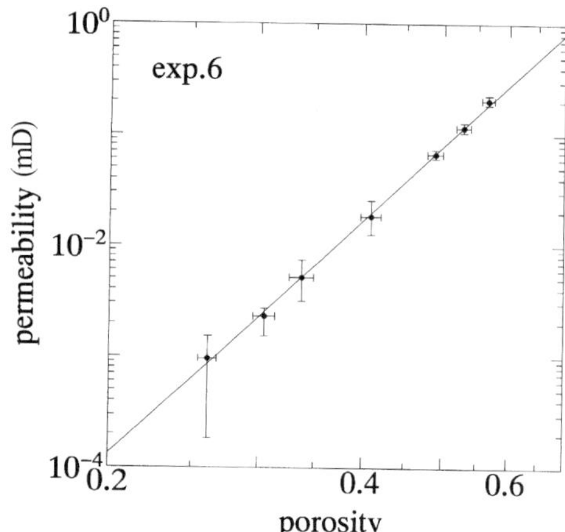

Fig. 2. An example of the ϕ–K relationships obtained from one-dimensional consolidation experiments.

Table 1. *Description of samples used for one-dimensional consolidation experiments*

Run no.	Sample name	Initial condition	Specific gravity	clay content (%)	silt content (%)	$D_{10}(\mu m)$	parameter '*a*'	Mineralogy
exp. 1	crushed Kiwada Fm	ds	2.56±0.01	10.9	82.4	3.71	11.89	quartz, chlorite, illite, calcite
exp. 2	crushed Kiwada Fm	ds	2.56±0.01	16.4	79.0	2.81	11.81	quartz, chlorite, illite, calcite
exp. 3	crushed Rochester shale	ds	2.77±0.01	30.9	69.1	2.40	8.26	illite, quartz
exp. 4	crushed Rochester shale	ds	2.77±0.01	29.0	71.0	2.38	9.94	illite, quartz
exp. 5	crushed Kiwada Fm	ss	2.56±0.01	22.5	76.0	2.52	10.27	quartz, chlorite, illite, calcite
exp. 6	crushed Rochester shale	ss	2.77±0.01	29.8	70.2	2.39	7.57	illite, quartz
exp. 7	kaolinite powder	ss	2.61±0.01	37.8	62.2	—	7.71	kaolinite, quartz
exp. 8	kaolinite powder	ss	2.61±0.01	37.8	62.2	—	8.84	kaolinite, quartz

ds: depositional setting, ss: slurry setting. See text for the description of each setting.

Table 2. *Physical properties and grain size distribution of the sample (Pleistocene Umegase Formation) used for consolidation experiments with transient pulse permeability measurements*

bulk density ($g\ cm^{-3}$)	grain density ($g\ cm^{-3}$)	uniaxial strength ($kg\ f\ cm^{-2}$)	initial porosity	clay content (%)	silt content (%)	sand content (%)	Mineralogy
1.94	2.62	166.0	0.425	15.6	77.8	6.6	quartz, chlorite, illite, calcite

show that the porosity(ϕ)–permeability(K) relationships (hereafter called ϕ–K relationships) are linear on double logarithmic scales. Thus, an empirical relationship

$$K = K_0\left(\frac{\phi}{\phi_0}\right)^a \quad (1)$$

is obtained from our experiments, where K_0 is initial permeability, ϕ_0 is initial porosity, and 'a' is a parameter depending on the samples used. The relationship was also confirmed by thorough review of published data (Tokunaga *et al.* 1994*a*) over the range $0.3<\phi<0.7$.

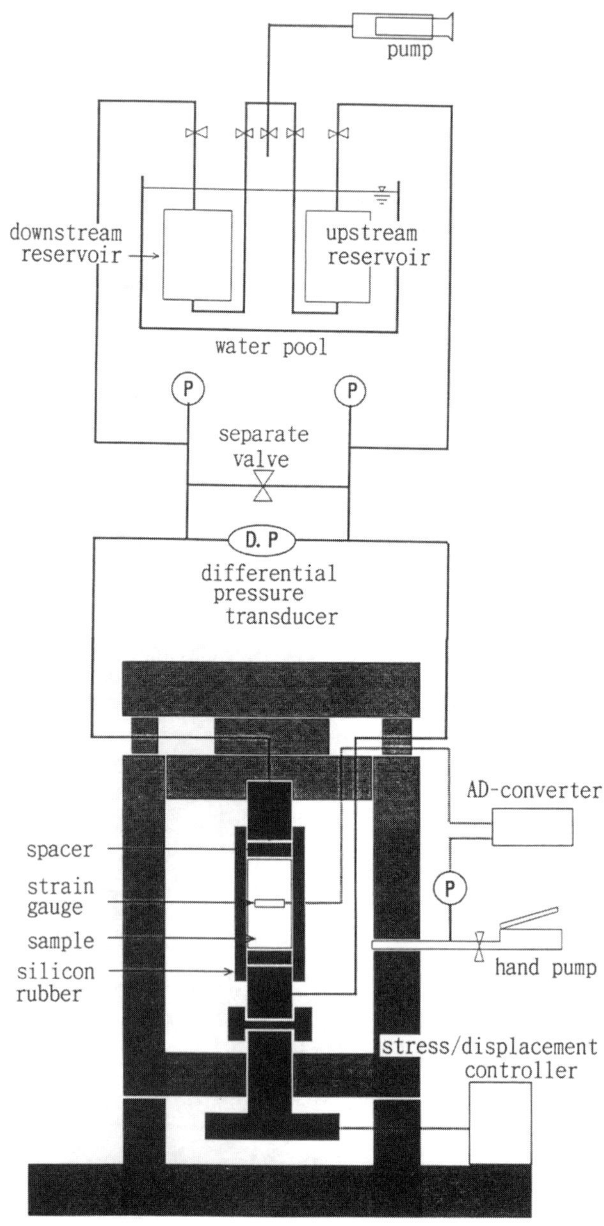

Fig. 3. Schematic diagram of triaxial compression apparatus with transient pulse permeability measurement system used in this study.

Permeability measurements during consolidation of mudstones

The microstructure or fabric of muddy slurries is possibly different from that compacted in nature, so consolidation experiments were carried out using naturally compacted mudstones to confirm whether the relationship obtained for muddy slurries was applicable to them. Triaxial compression apparatus, shown in Fig. 3, was used to carry out K_0-consolidation test, in which confining pressure was controlled to keep sample diameter constant during the consolidation/swelling processes. Maximum axial stress and maximum confining pressure of the apparatus are about 1000 kg f cm^{-2} and about 500 kg f cm^{-2}, respectively. A detailed description of the experimental procedure is presented in Hosoya *et al.* (1995). Permeability was measured using the transient pulse method (Brace *et al.* 1968) at the end of each loading/unloading stage. The system of the transient pulse method used in this study is similar to that presented in Takahashi *et al.* (1991). Permeability was calculated from pressure decay data according to the procedure described in Brace *et al.* (1968). Although the procedure ignores the effect of pore storage, the obtained permeability is good enough because the specific storage of the sample is prepared to be sufficiently small compared with the compressive storages of both the up-stream and down-stream reservoirs as suggested by Ishijima *et al.* (1991). Porosity of each loading stage was determined by the method described in the previous section. Time effects on consolidation were not observed during the experiments although we used the displacement transducer of which resolution was 0.001 mm.

Samples used in the experiments were siltstones from the Pleistocene Umegase Formation of the Kazusa Group in Boso Peninsula, Japan. These samples were cored perpendicular to the bedding surface from a massive block obtained from a river cliff. They were 30 mm in diameter and 60 mm in height. Physical properties and grain size distribution of the sample are shown in Table 2.

Three experiments were conducted using cored samples obtained from the same block by changing

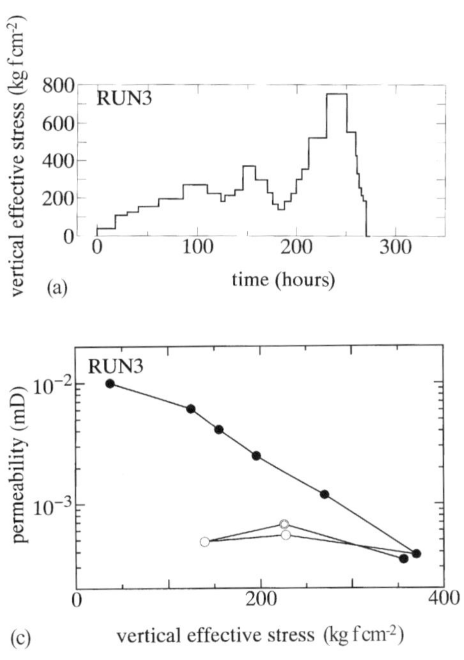

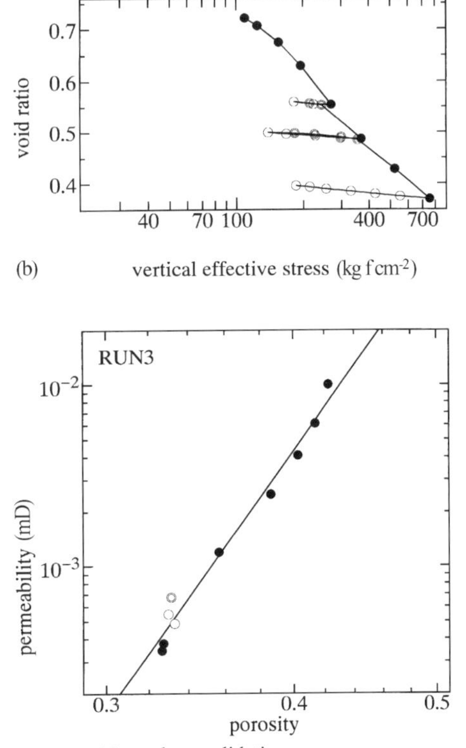

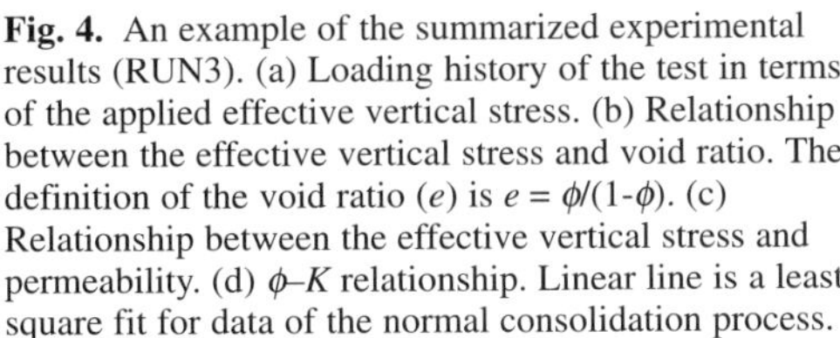

Fig. 4. An example of the summarized experimental results (RUN3). (a) Loading history of the test in terms of the applied effective vertical stress. (b) Relationship between the effective vertical stress and void ratio. The definition of the void ratio (e) is $e = \phi/(1-\phi)$. (c) Relationship between the effective vertical stress and permeability. (d) ϕ–K relationship. Linear line is a least square fit for data of the normal consolidation process.

Table 3. *Summary of the consolidation experiment with permeability measurements of mudstone (RUN3)*

Test no.	effective overburden pressure (kg f cm^{-2})	pore pressure (kg f cm^{-2})	step change of pressure in reservoir (kg f cm^{-2})	sample length (cm)	porosity	permeability (mD)
1	37.50	41.00	0.9045	6.349	0.4227	1.035×10^{-2}
2	124.74	40.79	0.9320	6.255	0.4141	6.324×10^{-3}
3	155.33	39.76	0.9420	6.139	0.4030	4.233×10^{-3}
4	195.45	41.03	0.9324	5.974	0.3865	2.588×10^{-3}
5	270.17	40.21	0.9203	5.696	0.3566	1.232×10^{-3}
6	369.97	40.91	0.9536	5.449	0.3274	3.912×10^{-4}
7	227.78	41.21	0.9479	5.468	0.3298	5.661×10^{-4}
8	139.11	41.20	0.9304	5.495	0.3330	5.030×10^{-4}
9	225.98	40.06	0.9287	5.480	0.3313	6.976×10^{-4}
10	356.22	39.88	0.9278	5.442	0.3265	3.571×10^{-4}

loading/unloading patterns. An example of the experimental results (RUN3) is summarized in Fig. 4 and Table 3. Fig. 4(d) shows a linear relationship between log K and log ϕ during normal consolidation, indicating that equation (1) is satisfied. Lapierre *et al.* (1990) reported that the remoulding of the sample did not modify the ϕ–K relationship for the Louiseville clay, Quebec, in their experiments on both intact and remoulded samples. They confirmed that the microstructures of both intact and remoulded samples were similar according to both the observation by scanning electron microscope and the measurements of pore size distribution by mercury-intrusion porosimetry. Both our results and results by Lapierre *et al.* (1990) indicate that the relationship obtained by us might be applicable for assessing the ϕ–K relationship of natural samples.

The ϕ–K data during unloading and reloading processes are also plotted adjacent to the line of least square fit of ϕ–K plots for the normal consolidation process (Figs 4(d) & 5). Although the deformation process is basically elastic for the unloading/reloading processes, which is different from that for the normal consolidation process, it is

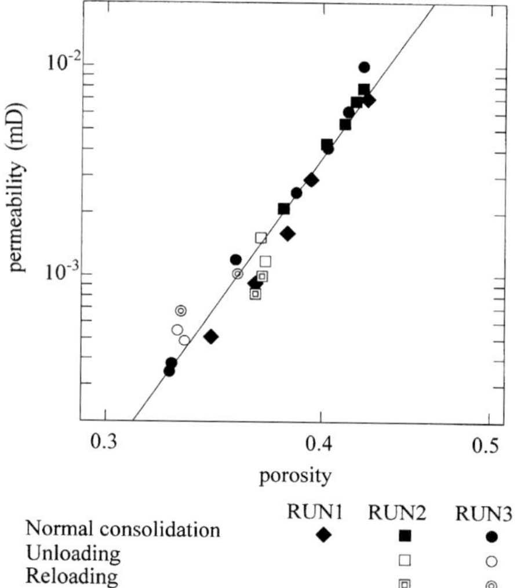

Fig. 5. ϕ–K relationships obtained from all the consolidation experiments. Linear line is a least square fit for data from a normal consolidation process.

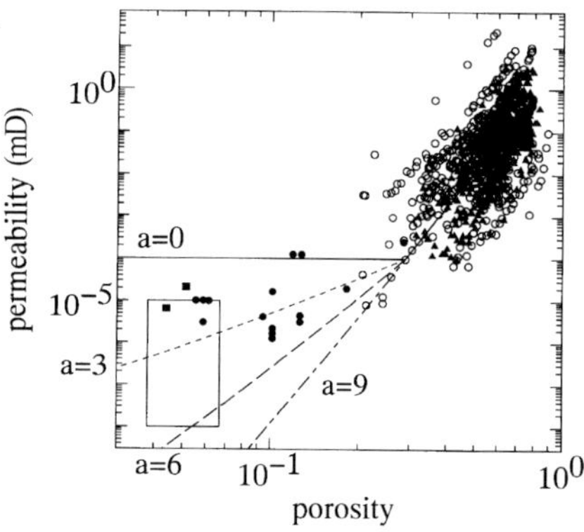

Fig. 6. ϕ–K plots of published data and results of our experiments. Data of Bryant *et al.* (1975), Dutta (1988), Katsube *et al.* (1991), and Lin (1978) after Neuzil (1994) are direct measurements of well core samples. Data sources of experimentally consolidated samples (except our data) are given in Tokunaga *et al.* (1994*a*). ϕ–K relationships used for this sensitivity study are also shown.

small enough that the small deviation from the line of normal consolidation is considered to be not so significant for basin simulation studies. Applicability of eqn (1) for both normal consolidation and during the unloading/reloading processes was also confirmed by using the data of Al- Tabbaa & Wood (1987) (Hosoya *et al.* 1994).

Published data for permeabilities of well core samples in a high porosity range (Bryant *et al.* 1975) are plotted with those obtained by laboratory consolidation experiments (Fig. 6). Their overall permeability reduction pattern is concordant with the previous discussions. Neuzil (1994) compared the ϕ–K relationships obtained from laboratory measurements of core samples and those obtained from inverse analyses of large-scale flow systems, and found that permeability of argillaceous rock is scale-independent. These observations indicate that eqn (1) is possibly extrapolated to the naturally compacted samples where the porosities of the samples are grater than about 0.3. Pore water salinity would affect the ϕ–K relationship; however, experimental results by Mesri & Olson (1971) and Lapierre *et al.* (1990) showed that it did not modify the relationship.

Relationship between parameter '*a*' and clay content of the sample

The parameter '*a*', which corresponds to the slope of the linear relationship between log ϕ and log K, varies depending on the samples used. Here, we try to correlate the value '*a*' in the ϕ–K relationship with clay content of the samples, from the results of experiments using muddy slurries. Figure 7 shows the results from experiments using muddy slurries of different clay content (see Table 1), in which a qualitative trend can be observed showing that decreasing the clay content causes '*a*' to increase, and vice versa.

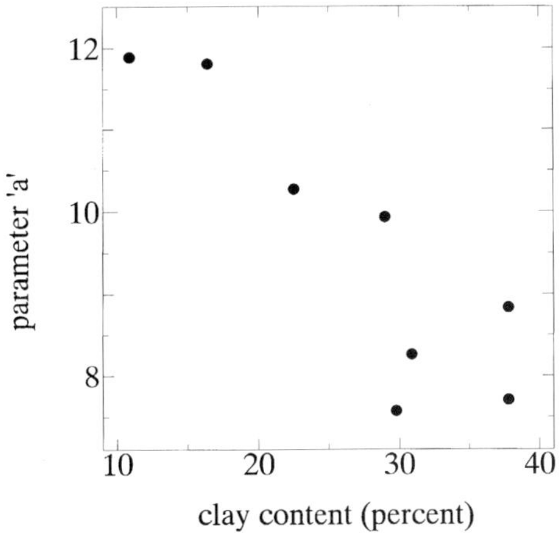

Fig. 7. Plots of the relationship between parameter '*a*' of equation (1) and clay content of the samples used in the one-dimensional consolidation experiments.

Assessing the ϕ–K relationships in a low porosity region

Published data for permeability of well core samples in the low porosity range (Dutta 1988; Katsube *et al.* 1991; Neuzil 1994) show considerable deviation from the equation (Fig. 6). This may be due to chemical effects including cementation, pressure-solution precipitation and clay transformation.

The transient pulse method (Brace *et al.* 1968) is the most appropriate for measuring very low permeability values. Trimmer *et al.* (1980) succeeded in measuring the permeability of granites and gabbros in the order of 10^{-12} Darcy. However, it took at least one week to obtain such data (Trimmer *et al.* 1980), and it is generally very difficult to maintain the same conditions during experiments of such a long duration. Furthermore, core samples obtained from deep boreholes suffer conditional changes such as the generation of microcracks caused by very fast pressure reduction during the coring processes. The experimental approach to quantify the ϕ–K relationships seems to be difficult in the highly compacted range. On the other hand, we know of no theoretical examination of permeability changes of low porosity mudstones. Thus, numerical studies would be useful in examining an appropriate ϕ–K relationship in a low porosity region as discussed in the next section.

Numerical sensitivity studies on the ϕ–K relationship

We conducted numerical sensitivity studies assuming that the change of permeability in the low porosity range followed the same form as eqn (1) but differed in the value of parameter '*a*' as shown in Fig. 6. This is because ϕ–K plots become deviated from eqn (1) when $\phi<0.3$ (Fig. 6), and because the trend of reduction of compressibility of argillaceous rocks as a function of porosity changes at around ϕ = 0.3 (Fig. 8), indicating that the mechanical behaviour of rocks changes from that of viscous fluids to that of plastic solids (Inami & Hoshino 1974). Aoyagi & Kazama (1980) summarized their results on consolidation experiments, porosity measurements, and mineralogical studies of cores and cuttings from deep wells in Japan, and concluded that compaction behaviour could be divided into two stages bounded at around $\phi = 0.3$.

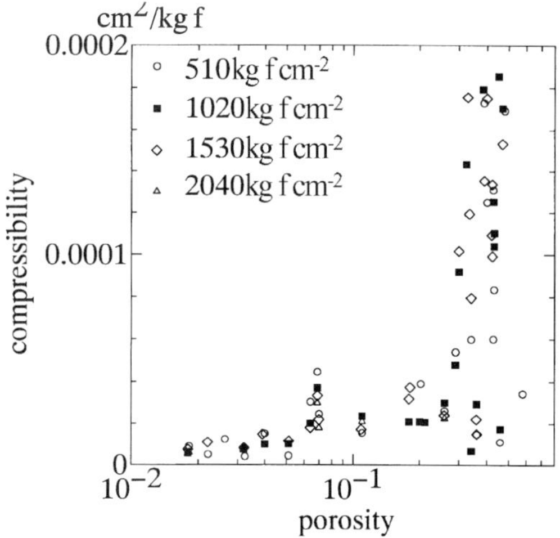

Fig. 8. Compressibility of argillaceous rocks at hydrostatic pressures of up to 2040 kg f cm^{-2}. The abscissa is porosity in logarithmic scale (after Inami & Hoshino 1974).

In this study, we initially chose the value $a = 0$, 3, 6, 9 and compared calculated results with one-another. Numerical studies were carried out using a cross-sectional model of an imaginary siliciclastic sedimentary basin. The input data are shown in Table 4. Initial and boundary conditions were set as follows:

(1) hydrostatic pressure, sea water temperature, and full water saturation were set as initial conditions for the deposited sediments at each time step;
(2) a no fluid flow boundary condition was set at the bottom of the basin;
(3) constant temperature was set at the sea bottom, and constant heat flow was given at the bottom of the basin;
(4) no water and no heat flow conditions were set at lateral boundaries.

Figure 9 shows input lithologies at 14.3 Ma after the initial deposition, and Fig. 10 presents the calculated results of porosity, pore pressure and temperature with respect to depth at the structural top. Figure 10 shows that porosity and pore pressure are highly sensitive to the given ϕ–K relationship. Thus, if we calculate using several 'a' values in the low porosity range and compare the

Table 4. *Input data for the imaginary sedimentary basin*

Number of time steps		22
Length of time steps		0.65 (Ma)*
Number of grids for X direction		8
Number of grids for Y direction		1
Width of grids for X direction		2000 (m)†
Width of grids for X direction		1000 (m)†
Initial porosity for mudstone		0.6
Initial porosity for sandstone		0.4
Constant of Athy's equation		
	for mudstone	6.0E–4 (m^{-1})
	for sandstone	3.0E–4 (m^{-1})
Heat flow between basement and sediments		1.9 (hfu)‡
Grain size of sandstone		medium§
Total organic carbon in mudstone		
	for Z = 1 & Z = 3	5.0 (wt%)
	for other grids	1.0 (wt%)
Composition of organic matters in mudstone		
	type I kerogen	30.0 (%)**
	type II kerogen	50.0 (%)**
	type III kerogen	20.0 (%)**
Amount of heat generation in sediments		0††

* Length of time step is to be constant for all time steps.
† Width of grids for X & Y directions is set to be constant for all grids.
‡ Heat flow is assumed to be constant for all time steps.
§ Grain size of sandstone is set to be constant for all sandstone grids
** Composition of organic matters in mudstone is assumed to be constant for all grids containing mudstone.
†† Heat generation in sediments is not considered for all time steps.

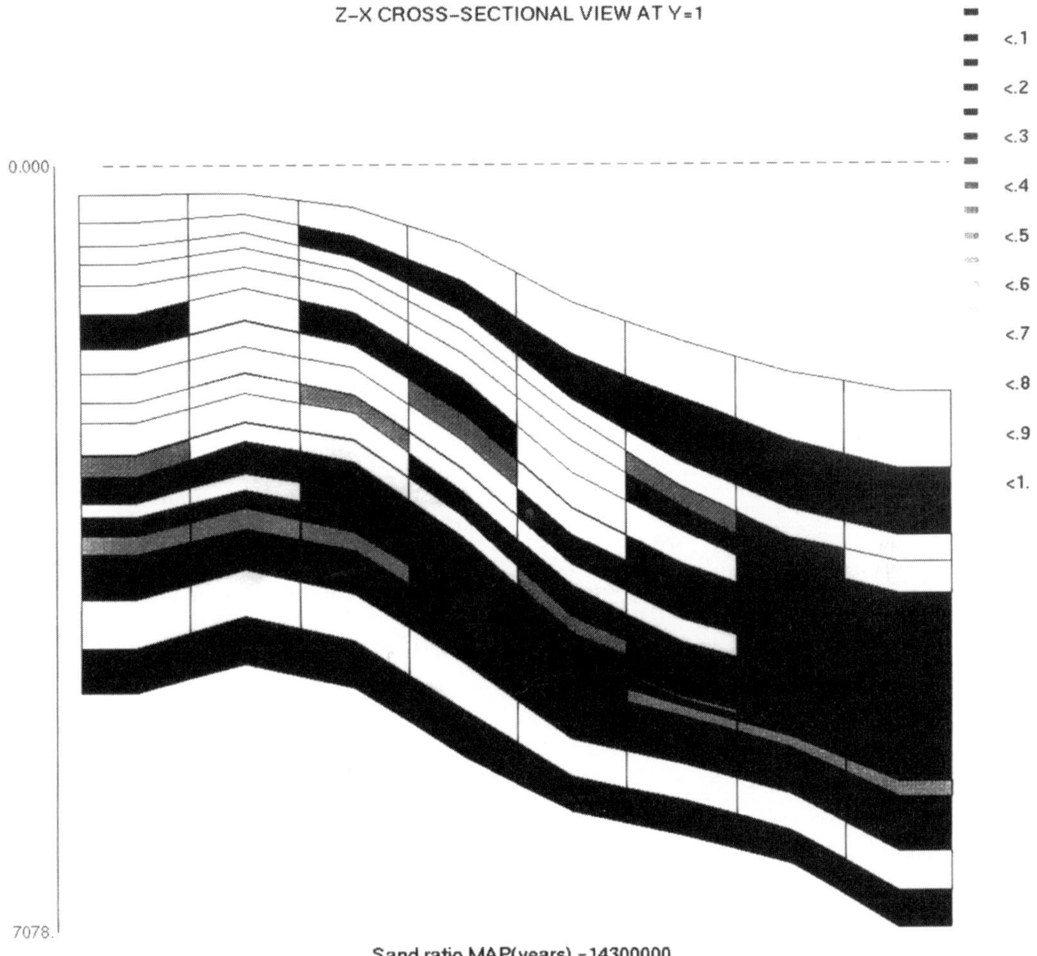

Fig. 9. Input lithology distribution (sand ratio map) of an imaginary sedimentary basin 14.3 Ma after the initial deposition.

results with measured porosity and pore pressure, we can get an appropriate value of 'a' for the basin model, which includes initial and boundary conditions, input values, and also constitutive equations. Note that the relationship obtained by this method does not always trace the real process that occurred in nature, however, for basin simulators like BASIN3D2P, it is one practical way to obtain the ϕ–K relationship in the low porosity region.

Calculated results of water saturation (Fig. 11) clearly indicate that the ϕ–K relationship for ϕ<0.3 strongly affects petroleum migration. If permeability does not change during the late compaction stage; an extreme case where $a = 0$, a certain amount of oil accumulation is found at the top of the anticline at 14.3 Ma after the initial deposition. However, if we choose $a = 3$, 6, 9, there is no accumulation at the equivalent time. This is caused by the delay of expulsion of oil from source rocks (Fig. 12). This is because development of overpressure in the source rock prevents reduction of porosity and permeability, and because the threshold oil saturation for petroleum expulsion from source rock is modelled as a decreasing function with increasing permeability in our simulator as suggested by Okui & Waples (1993) and Tokunaga *et al.* (1994*b*). However, if permeability reduction of mudstone with respect to porosity is fast enough ($a = 9$ in this case), overpressuring may cause hydraulic fracturing and expulsion of petroleum could be enhanced as we can see in Fig. 12 (compare a = 6 with $a = 9$).

The results obtained from the numerical case studies indicate that, at least for the simulators

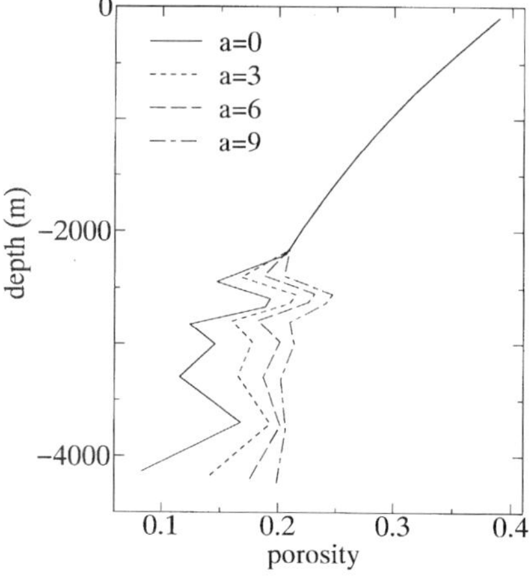

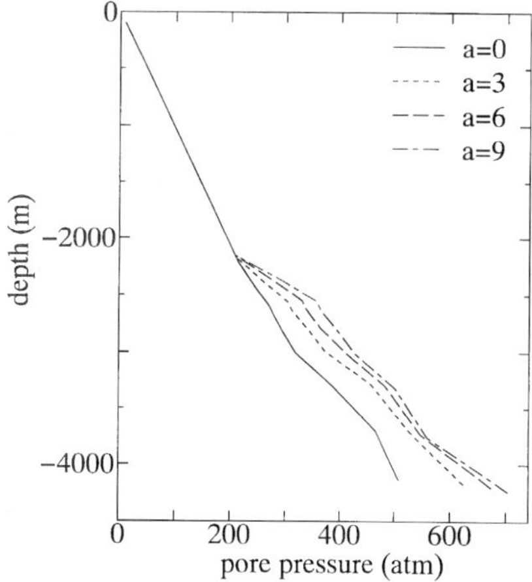

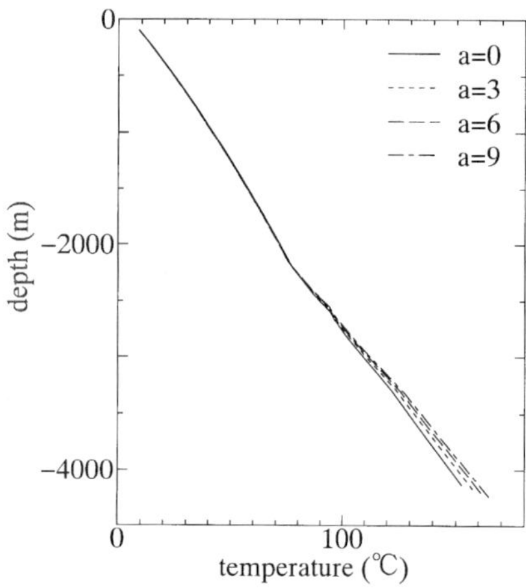

Fig. 10. Columnar displays of calculated results of porosity, pore pressure and temperature at the structural top 14.3 Ma after the initial deposition.

which adopt the concept of relative permeability for expulsion, an appropriate ϕ–K relationship of mudstone should be carefully chosen, especially for the low porosity region. A trial and error approach to match calculated performance with observed data is indispensable to make the basin simulation reliable and enhance its reproductivity. If pore pressure and porosity data are available, we could get proper trend for the ϕ–K relationship, and then, a reproduced migration pattern of petroleum might be more acceptable.

Conclusions

The following conclusions can be summarized from both laboratory studies on permeability measurements of muddy slurries and mudstones during consolidation, and from the results of numerical case studies:

(1) The ϕ–K relationship of muddy slurries and mudstones during loading, unloading, and reloading processes can be expressed by the equation:

$$K = K_0\left(\frac{\phi}{\phi_0}\right)^a$$

over the range $0.3<\phi<0.7$. The parameter 'a' of the equation could be correlated with the clay content of the sample. However, ϕ–K relationships deviate from the equation when porosities of the sample are less than about 0.3.

(2) Results of basin simulation, such as porosity, pore pressure, timing of expulsion/accumulation of oil, are sensitive to the ϕ–K relationship of mudstones. An appropriate form of the relationship can be obtained by trial and error matching of the numerical simulator's performance with observed data.

We thank S. Düppenbecker of BP Exploration and J. Iliffe of ERC Tigress Limited for providing us with an opportunity to submit this paper. We are also greatful to reviewers M. R. Giles and P. S. Ringrose for their thoughtful comments and helpful suggestions. Experiments on mudstones were carried out at the Geological Survey of Japan. We wish to thank M. Takahashi of the Geological Survey of Japan for his permission to use the apparatus and for his help during experiments. Discussions with C. J. Spiers and C. J. Peach of Utrecht University on a transient pulse method to obtain permeability of low permeable porous media was very helpful. We also thank S. Matsuo, Y. Saito, and K. Mogi of the University of Tokyo for their technical support. T. T. is grateful for a Grant-in-Aid for Encouragement of Young Scientists from the Ministry of Education, Science and Culture, Japan (Grant No. 05750833).

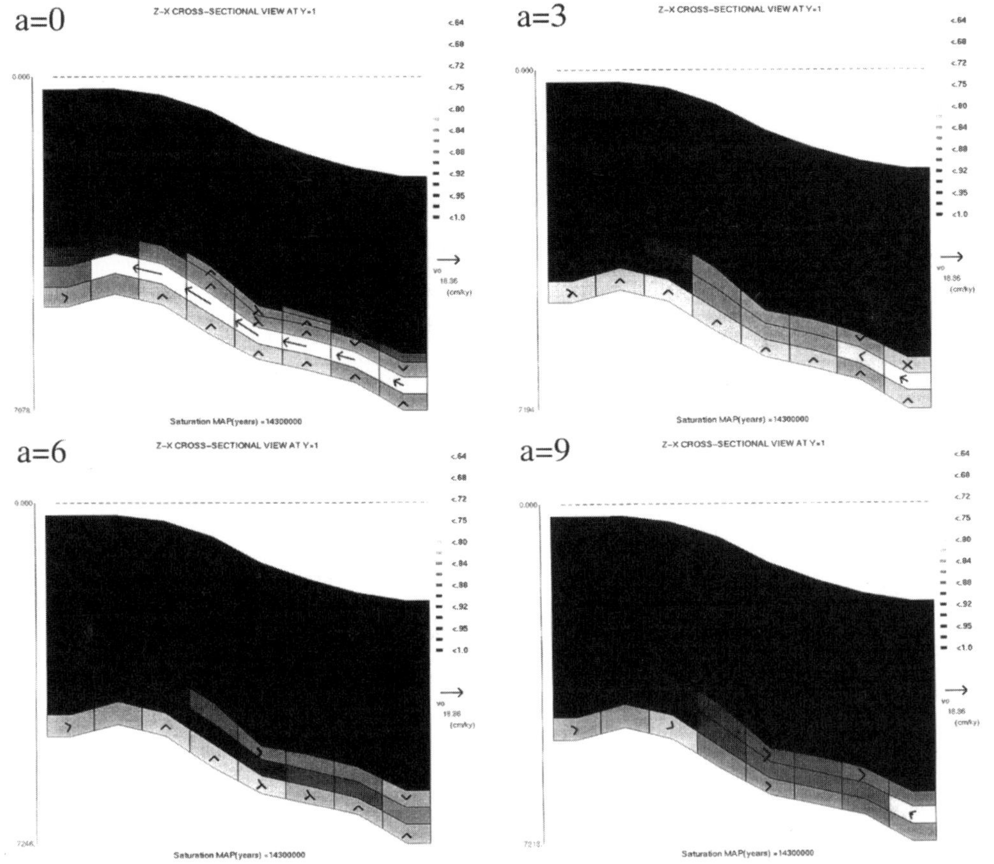

Fig. 11. Calculated results of water saturation distribution of the basin 14.3 Ma after the initial deposition.

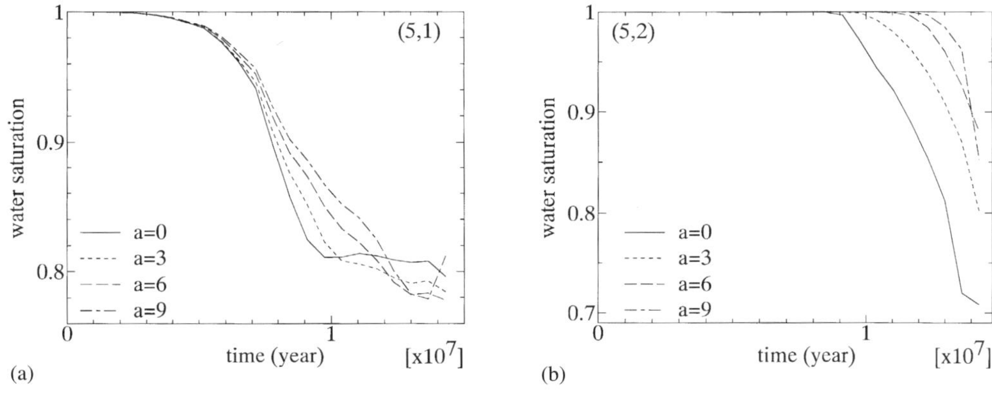

Fig. 12. Examples of the history of water saturation in the grids. (a) Water saturation history of grid (5,1) (source horizon). (b) Water saturation history of grid (5,2) (carrier horizon).

References

AL-TABBAA A. & WOOD, D. M. 1987. Some measurements of the permeability of kaolin. *Geotechnique*, **37**, 499–503.

AOYAGI, K. & KAZAMA, T. 1980. Transformational changes of clay minerals, zeolites and silica minerals during diagenesis. *Sedimentology*, **27**, 179–188.

BRACE, W. F., WALSH, J. B. & FRANGOS, W. T. 1968. Permeability of granite under high pressure. *Journal of Geophysical Research*, **73**, 2225–2236.

BRYANT, W. R., HOTTMAN, W. & TRAVANT, P. 1975. Permeability of unconsolidated and consolidated marine sediments, Gulf of Mexico. *Marine Geotechnology*, **1**, 1–14.

DORÉ, A. G., AUGUSTSON, J. H., HERMANRUD, C., STEWART, D. J. & SYLTA, Ø. 1993. *Basin Modelling: Advances and Applications.* Elsevier, Amsterdam.

DUTTA, N. C. 1988. Fluid flow in low permeable, porous media. *Revue de L'Institut Français du Pétrole*, **43**, 165–180.

HERMANRUD, C. 1993. Basin modelling techniques —an overview. *In*: DORÉ, A. G. *et al.* (eds) *Basin Modelling: Advances and Applications*. Elsevier, Amsterdam, 1–34.

HOSOYA S., TOKUNAGA, T., KOJIMA, K., TOSAKA, H. & TAKAHASHI, M. 1994. Evaluation of permeability of muddy deposits during consolidation–expansion processes: an experimental approach. *Proceedings of 1994 Meeting of the Japan Society of Engineering Geology*, 135–138 (in Japanese).

——, ——, TAKAHASHI, M., KOJIMA, K. & TOSAKA, H. 1995. Trends of changes in K_0 and permeability of soft rocks obtained from high-stress K_0 consolidation tests. *Proceedings of 26th Symposium on Rock Mechanics in Japan*, 271–275 (in Japanese with English abstract).

INAMI, K. & HOSHINO, K. 1974. Compressibility and compaction of clastic sedimentary rocks. *Journal of the Japanese Association for Petroleum Technology*, **39**, 366–374. (in Japanese with English abstract).

ISHIJIMA, Y., XUE, Z. & TAKAHASHI, M. 1991. Some basic problems on measurements of hydraulic properties of rock by the transient pulse method. *Shigen-to-Sozai*, **109**, 511–516 (in Japanese with English abstract).

KATSUBE, T. J., MUDFORD, B. S. & BEST, M. E. 1991. Petrophysical characteristics of shales from the Scotian shelf. *Geophysics*, **56**, 1681–1689.

LAPIERRE, C., LEROUEIL, S. & LOCAT, J. 1990. Mercury intrusion and permeability of Louiseville clay. *Canadian Geotechnical Journal*, **27**, 761–773.

LIN, W. 1978. Measuring the permeability of Eleana Argillite from area 17, Nevada Test Site, using the transient method. *Report UCRL-52604.* Lawrence Livermore Laboratory, Livermore.

MESRI, G. & OLSON, R. E. 1971. Mechanisms controlling the permeability of clays. *Clays and Clay Minerals*, **19**, 151–158.

NEUZIL, C. E. 1994. How permeable are clays and shales? *Water Resources Research*, **30**, 145–150.

OKUI, A. & WAPLES, D. W. 1993. Relative permeabilities and hydrocarbon expulsion from source rocks. *In*: DORÉ, A. G. *et al.* (eds) *Basin Modelling: Advances and Applications*. Elsevier, Amsterdam, 293–301.

TAKAHASHI, M., XUE, Z. & KOIDE, H. 1991. Permeability characteristics of Inada granite, Shirahama sandstone, Kimachi sandstone, and Neogene argillaceous rock. *Bulletin of the Geological Survey of Japan*, **42**, 305–331 (in Japanese with English abstract).

TOKUNAGA T., HOSOYA, S., TOSAKA, H. & KOJIMA, K. 1993. Change of physical properties of sediments during compaction and its importance on basin simulation. *Chikyu Monthly*, **15**, 601–607 (in Japanese).

——, ——, —— & —— 1994*a*. Change of hydraulic properties of muddy deposits during compaction: assessment of mechanical and chemical effect. *Proceedings of 7th Congress, International Association of Engineering Geology*, 635–643.

——, TOSAKA, H. & KOJIMA, K. 1994*b*. Modelling petroleum field formation processes and development of a three dimensional basin simulator. *Journal of the Japanese Association for Petroleum Technology*, **59**, 519–530 (in Japanese with English abstract).

—— 1996. Development of a three-dimensional basin simulator and its application to an actual sedimentary basin. *Unpublished doctors thesis, University of Tokyo.* (in Japanese with English abstract).

TRIMMER, D., BONNER, B., HEARD, H. C. & DUBA, A. 1980. Effect of pressure and stress on water transport in intact and fractured gabbro and granite. *Journal of Geophysical Research*, **85**, 7059–7071.

UNGERER, P., BURRUS, J., DOLIGEZ, B., CHÉNET, P. Y. & BESSIS, F. 1990. Basin evaluation by integrated two-dimensional modeling of heat transfer, fluid flow, hydrocarbon generation, and migration. *American Association of Petroleum Geologists Bulletin*, **74**, 309–335.

WELTE, D. H. & YUKLER, M. A. 1981. Petroleum origin and accumulation in basin evolution -a quantitative model. *American Association of Petroleum Geologists Bulletin*, **65**, 1387–1396.

Central North Sea overpressures: insights into fluid flow from one- and two-dimensional basin modelling

D. DARBY[1], R. S. HASZELDINE[2] & G. D. COUPLES[3]

Centre for Research in Applied Geoscience, Department of Geology and Applied Geology, University of Glasgow, Glasgow G12 8QQ, UK

[1] Present address: Department of Marine Chemistry and Geochemistry, Woods Hole Oceanographic Institution, Woods Hole, MA 02540, USA

[2] Present address: British Gas Research and Technology Centre, Loughborough LE11 3GR, UK

[3] Present address: Department of Petroleum Engineering, Heriot-Watt University, Edinburgh EH14 4AS, UK

Abstract: Jurassic and Triassic reservoirs in the Central North Sea are highly overpressured (>40 MPa above hydrostatic pressure). Simulation of the interplay between rapid Tertiary subsidence, seal permeability and fluid flow allows insight into the geological controls on the distribution and magnitude of the overpressure. One-dimensional models demonstrate that, unlike other basins, the overpressure developed in the Graben is not determined by the thickness and permeability of the shale pressure seal. A two-dimensional model simulating lateral flow beneath the pressure seal provides an accurate simulation of the overpressure distribution. Disequilibrium compaction of shale-dominated off-structure regions forms the principal overpressuring mechanism. Lateral flow in the permeable Fulmar sandstones leads to high overpressure and focused vertical escape on an axial high, where the seal is thinner above a subcropping Fulmar Fm. A layered hydrogeological regime is suggested, with shallow Tertiary pressure cells separated from deep Cretaceous–Triassic pressure cells by normally pressured, permeable Palaeocene sandstones. The pre-Cretaceous rift-associated configuration of the Graben, in combination with 3 km of Tertiary subsidence, controls the distribution of overpressure.

Jurassic and Triassic reservoirs in the Central Graben of the North Sea are characterized by great depths (>4000 m) and extreme overpressures (in excess of 40 MPa above hydrostatic pressure). Overpressure in the Graben provides an intellectual challenge to geologists: as a problem for safe drilling of deep prospects; a clue to fluid flow; and a control of porosity distribution. An increased understanding of the processes controlling the magnitude and distribution of overpressure may lead to wider insights into the processes of fluid flow in sedimentary basins. This paper presents a model of the origin and distribution of overpressure in the Central North Sea based on the results of one- and two-dimensional simulations of fluid flow.

The Central Graben is the area of greatest fault-related subsidence in the North Sea sedimentary basin. The Central Graben is divided into two depocentres (Fig. 1): a western graben, forming the 'true' Central Graben; and the East Forties Basin, a half-graben whose crest forms the axial Forties–Montrose High (Roberts *et al.* 1990). Thick, massive Upper Jurassic sandstones in the Central Graben are referred to as the Fulmar Formation (Stewart 1986) and they are the target for the present high-pressure/high-temperature exploration play. These sandstones are overlain and sealed by a sequence of low-permeability, Jurassic–Lower Cretaceous shales and Upper Cretaceous chalks, including the Kimmeridge Clay Formation, the source rock for most of the Graben's hydrocarbons. Overlying the chalk is a basin-wide sheet of permeable Palaeocene sandstones, which form the reservoir to giant oil accumulations such as the Forties field. Accelerating subsidence of the Graben has led to the deposition of over 3000 m of Tertiary shales, of which 1000 m may have been deposited in the past 5 Ma. A representative subsidence curve is presented in Fig. 2.

The rapid rate of Tertiary deposition, coupled with the presence of low-permeability chalks and shales, has led to restricted fluid flow and consequent overpressuring in the sediments of the

Darby, D., Haszeldine, R. S. & Couples, G. D. 1998. Central North Sea overpressures: insights into fluid flow from one- and two-dimensional basin modelling. *In*: Düppenbecker, S. J. & Iliffe, J. E. (eds) *Basin Modelling: Practice and Progress*. Geological Society, London, Special Publications, **141**, 95–107.

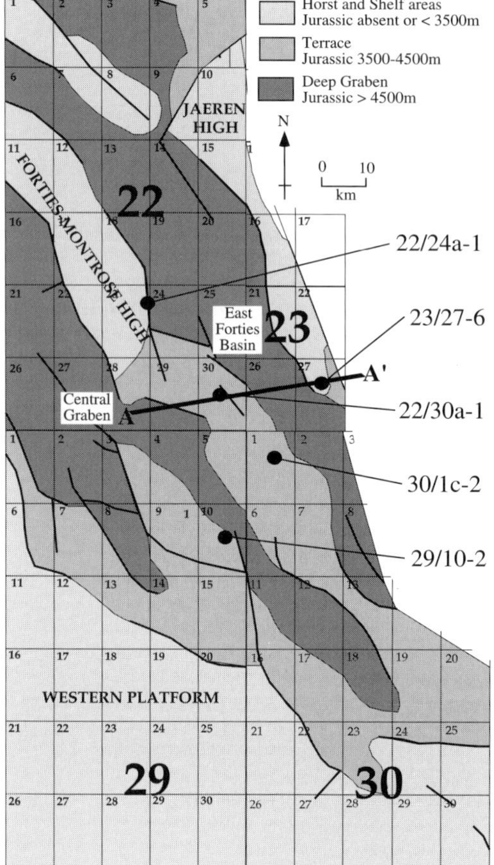

Fig. 1. Location Map of the modelled wells and section.

Graben. Repeat Formation Tests (RFTs) in Fulmar sandstones have revealed that the Fulmar Formation is highly compartmentalized, and divided into a series of pressure cells. The magnitude of pressure within these pressure cells can approach the minimum horizontal stress (Engelder & Fischer 1994) and shows a general increase with the depth of the cell, increasing towards the centre of the Graben (Gaarenstroom *et al.* 1993). However, the highest degree of overpressure is encountered in relatively shallow wells (e.g. wells 22/30a-1 and 30/1c-2) located on the axial Forties–Montrose High.

The upper boundaries of pressure cells (pressure seals) are marked by rapid rises in pore pressure, termed transition zones. These transition zones are encountered within the regional chalk–shale interval. Lateral boundaries to pressure cells are implied to be faults (Darby *et al.* 1996). Numerous theories have recently been advanced to explain the position of vertical pressure seals and the magnitude of pressure. These aspects have been suggested to be controlled by temperature (Hunt 1990); organic–inorganic interactions (Hunt *et al.* 1994; Whelan *et al.* 1994); or the porosity and permeability of the sediment forming the seal (Mello *et al.* 1994).

The complex interplay between the dynamical physical processes of subsidence, compaction of sediments and fluid flow lends itself to investigation by computer simulation (Dewers & Ortoleva 1994). Basin models have been employed in deciphering the hydrogeology of many overpressured basins throughout the world, including the Northern North Sea (England *et al.* 1987) and the Central North Sea (Mudford *et al.* 1991).

This paper presents the results of one- and two-dimensional modelling of overpressures in the Central North Sea. The aim of this work is to identify the controls on the origin, distribution, magnitude and retention of abnormal fluid pressures developed in the Central North Sea by using quantitative basin models, which provide a method of building and testing hypotheses regarding fluid flow.

From one-dimensional modelling of the system we conclude that the overpressure developed in the Central Graben today is the product of a dynamic system. The presence of low-permeability shales has restricted vertical compactional fluid flow, and the rapid Tertiary subsidence of the Graben has produced disequilibrium compaction of the shale-rich Graben sediments. Importantly, however, we find that the Central Graben, unlike other basins world-wide, does not show a relationship between the magnitude of overpressure developed and the thickness and permeability of the pressure seal. We present a two-dimensional model to suggest that lateral flow, into structurally-elevated permeable sandstones subcropping below the pressure seal, controls the distribution and magnitude of overpressure in the region. The model provides an accurate description of the present-day hydrogeological system of the Central Graben, and may provide insight into the palaeo-hydrogeology that has controlled hydrocarbon migration and diagenesis in the region.

Modelling methodology

In this paper we report the modelling of five government-released wells from Quadrants 22,23, 29 and 30 of the UK sector of the Central North Sea. The locations of the study wells are shown on Fig. 1.

The simulations in this paper have used the BasinMod™ finite-difference one- and two-dimensional basin modelling packages. Details of these models are proprietary, but they are based on well-

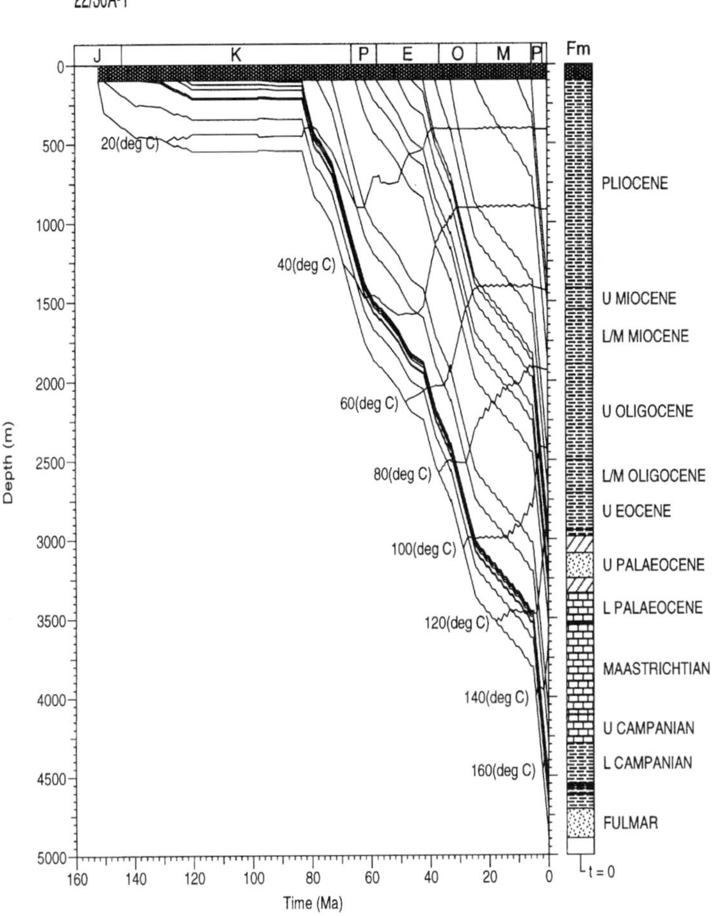

Fig. 2. Subsidence curve of well 22/30a-1. The Graben has undergone accelerating subsidence during the Tertiary, leading to the deposition of 3000 m of shales. 1000 m of these shales may have been deposited in the Pliocene.

known primary relationships: the law of conservation of solid mass; Darcy's Law; and equations of state (Bethke 1985; Lerche 1990). The approach taken by BasinMod™ is to ensure that a porosity–depth relationship is obtained, and then to solve for other variables. We have used the Sclater and Christie (1980) porosity-decline curve, and a power-law relationship between porosity and permeability (Lerche 1990). Table 1 lists the values selected for the necessary lithological constants that arise in the equations which define these functions.

The modelling methodology employed falls into two parts (Mudford *et al.* 1991). The first phase is the construction of the model using lithological and stratigraphic information derived from well composite logs. However, we do not have complete knowledge of the system under investigation: many parameters that influence fluid flow are rarely measured during drilling and coring. The most critical of these parameters is the permeability of the shales that form the barrier to vertical fluid flow (Mudford *et al.* 1991).

Accordingly, we are faced by an inverse problem common to hydrogeological modelling: the independent variable in the equation of fluid flow, the rock permeability, is poorly known, whilst the distribution of the dependent variable in the equation (the fluid pressure) is the most well known aspect of the system under investigation. Thus the second phase of the model investigation is a confirmatory check by calibration of the assumed porosity and permeability against observed pressures. Sources of pressure data include direct measurements such as reservoir RFTs and drill stem tests, and indirect measurements such as mudweight.

Table 1. *Lithological parameters used in the models*

	Sandstone	Chalk	Shale	Siltstone
One-dimensional model: high-permeability case				
Initial porosity	0.45	0.10	0.60	0.55
Initial permeability (mD)	27 000.00	27 000.00	0.01	0.10
Compaction factor (m^{-1})	0.45	–0.60	–0.50	–0.80
Permeability power	5.50	5.50	5.50	5.50
One-dimensional model: low-permeability case				
Initial porosity	0.45	0.10	0.40	0.55
Initial permeability (mD)	27 000.00	0.0001	0.01	0.10
Compaction factor (m^{-1})	0.45	–0.50	–0.60	–0.80
Permeability power	5.50	5.50	5.50	5.50
Two-dimensional model				
Initial porosity	0.45	0.60	0.60	0.55
Initial permeability (mD)	27 000.00	50 000.00	100.00	5000.00
Compaction factor (m^{-1})	0.27	–0.41	–0.51	–0.22
Permeability power	7.00	6.00	6.00	6.00
Permeability anisotropy	0.50	0.50	0.20	0.20

Calibration of a dynamic model of geohistory faces problems of comparing the computed results against data localized in time (at the present day) and in space (most data is focused on structural highs, within the reservoir interval). This study maximizes the use of spatially-continuous data such as sonic logs and mudweight. Mudweight provides a continuous indirect record of pressure. Direct calibration of model pressure against mudweight is not valid for the Central North Sea, as mudweight is always greater than formation fluid pressure. However, the mudweight recorded from a specific drilling depth provides a maximum pressure that cannot be exceeded by a valid model. We note that calibration of a physical model to reservoir porosity and permeability is not valid in this region of the Central Graben due to geochemical influences: important volumes of quartz and albite cement; and localized creation of up to 15% secondary porosity (Wilkinson & Haszeldine 1996; Wilkinson *et al.* 1997), have modified the primary porosity in sandstone reservoirs.

Quantitative models of geological processes are being employed in many fields of geology, from the exploration for oil and gas to the disposal of nuclear waste. As the results of such models may be used to make decisions regarding the safety and economics of a hydrocarbon prospect, we feel that it is necessary to consider the underpinning (Oreskes *et al.* 1994) of the models employed in this paper. The models are based on simple mathematical descriptions of complex geological processes. Where possible, the models are checked against observed data to allow decisions to be taken as to whether the model represents an accurate description of the geological reality. However, as the data for calibration is necessarily localized in time and space, reproduction of observed data is insufficient to demonstrate that the model is correct throughout geohistory. Quantitative models are a form of complex scientific hypothesis. They are subject to improvement or refutation as new data are acquired by additional wells in the Graben, or as a fuller understanding of the geological processes being simulated is achieved.

We have used one-dimensional models to allow rapid evaluation of initial hypotheses, to investigate the importance of variation in unobserved model parameters, to focus on parameters and data important for the model calibration, and to guide the construction of secondary hypotheses. We have then tested these secondary hypotheses using two-dimensional models.

One-dimensional modelling

One-dimensional models of gravitational compaction can adequately reproduce the observed pressure–depth profiles from many overpressured basins world-wide, including the northern North Sea (England *et al.* 1987), the Sable Basin of offshore Nova Scotia (Williamson & Smyth 1992), the South Caspian Sea (Audet & McConnell 1992), the Louisiana Gulf Coast (Mello *et al.* 1994), and southern regions of the Central North Sea (Mudford *et al.* 1991) adjacent to the area examined in this paper. In all these regions, the vertical flow of compactional fluid is impeded by the presence of low-permeability rocks, whose thickness, porosity and permeability govern the magnitude of overpressure developed in the underlying sediments. Thus the initial hypothesis to be examined by this

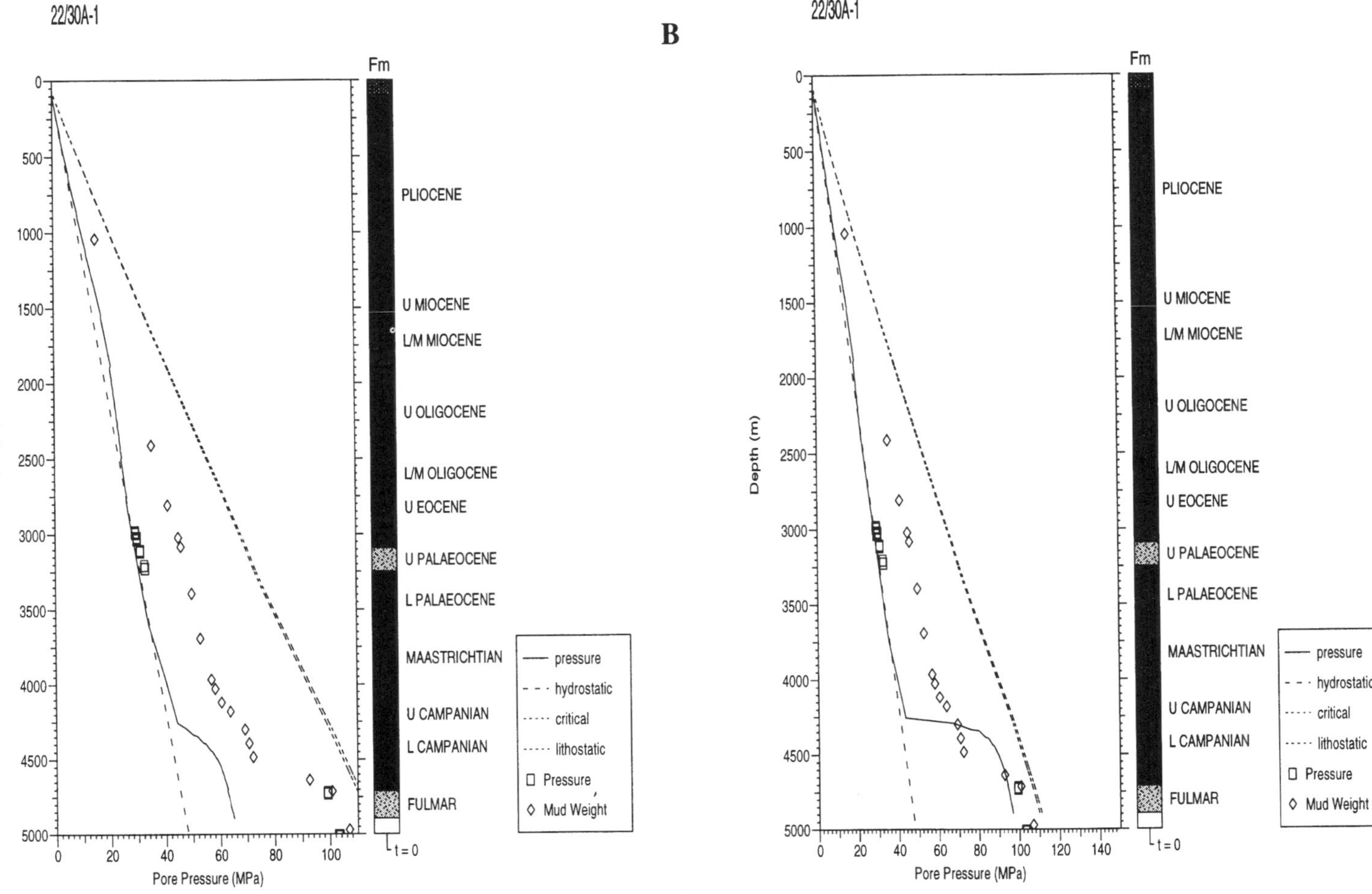

Fig. 3. Calibration of modelled fluid pressures in the Fulmar Fm in well 22/30a-1 to observed mudweight and RFT pressures. The low-permeability case (b) provides an adequate reproduction of reservoir pressures, but pressures higher in the section are in excess of the measured mudweight. Increasing seal permeability (case a) will honour the mudweight profile but forces a shortfall between modelled and observed pressures. (**A**) Chalk porosity 0.03, shale porosity 0.05, chalk permeability 10^{-6} mD, shale permeability 10^{-7} mD. (**B**) Chalk porosity 0.02, shale porosity 0.1, chalk permeability 10^{-6} mD, shale permeability 10^{-8} mD.

investigation is that the magnitude of overpressure developed in the Fulmar sandstones in the Central Graben is proportionate to the thickness and permeability of the overlying pressure seal, and the rate of subsidence of the basin. That is, the physical properties of the pressure seal will restrict vertical fluid flow during rapid sedimentation and lead to disequilibrium compaction, causing overpressuring in the Fulmar sandstone. This hypothesis can be tested by examining the interplay of sedimentation rate and seal permeability evolution and the resultant pore pressures using a one-dimensional model.

The model results are constrained by fluid pressures measured by RFT in the reservoir (Fig. 3a). However, we find that to attain such pressures requires extremely low shale permeabilities. This low permeability produces pressures that exceed the mudweight measured in the Lower Cretaceous shales at the top of the pressure seal, and so these simulations cannot be considered valid. The mudweight provides an important constraint to the minimum seal permeability in the Central Graben. If the mudweight-derived pressure profile is honoured, calculated pressures in the Upper Jurassic sandstone are too low (Fig. 3b). The model is extremely sensitive to variations in the permeability of shale. Although no direct measurements of shale permeability are available in the Central Graben, the values required (10^{-6} mD – 10^{-8} mD) are consistent with permeabilities derived from shales globally (Katsube *et al.* 1991). We cannot reconcile all pressure measurements in the region at this stage of the modelling. The divergence between the maximum model pressure and the observed pressure occurs across the region (Fig. 4), and is most marked where the pressure seal is thin and sharp gradients in pore pressure occur, for example well 22/30a-1.

From the initial, one-dimensional investigation of the region we conclude that overpressure does not show a relationship to seal thickness and permeability. The permeability of these seals cannot be modelled to be sufficiently low to reproduce the reservoir pressures without simultaneously violating other constraints. The initial hypotheses are disproved.

We infer from this result that the studied region of the Central North Sea is an unusual overpressured environment: the hydrogeological system expressed by the distribution of overpressure must differ from the Northern North Sea and even from

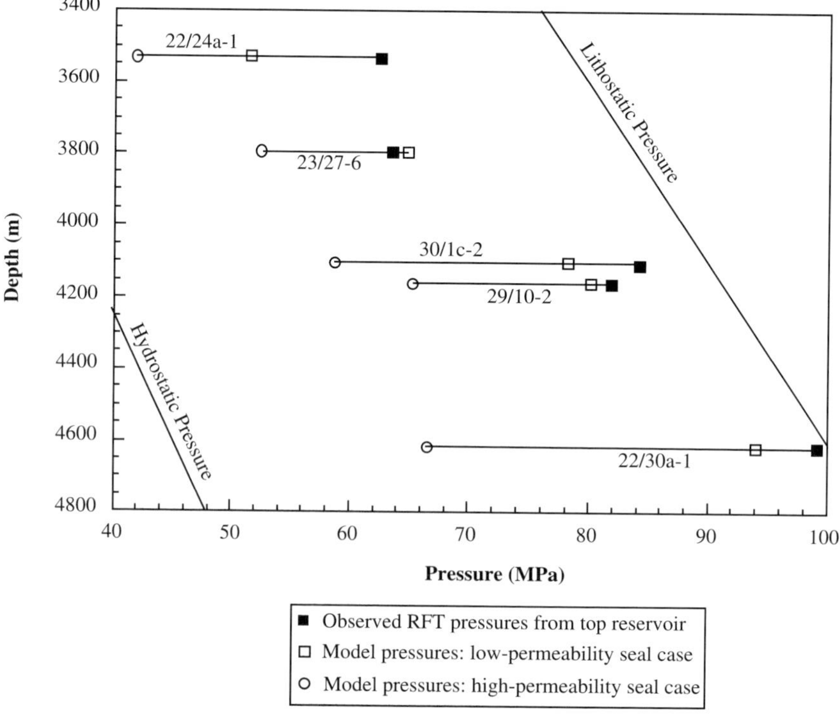

Fig. 4. One-dimensional modelling provides a mismatch between modelled and observed reservoir pressures across the region if pressures in the seal are honoured.

adjacent regions of the Central North Sea. The one-dimensional model cannot simulate the magnitude of pressure in the Central Graben, or correctly reproduce its distribution with depth. It is now necessary to develop this result to construct new hypotheses. The one-dimensional models, while not an adequate approximation to observed pressures, can nevertheless serve as a useful framework for the building of new hypotheses.

A similar shortfall in modelled pressures has been observed for the Sable Basin of Nova Scotia (Williamson & Smyth 1992). This shortfall has been attributed to additional overpressuring mechanisms such as hydrocarbon generation that are not addressed by simple physical models based on sediment compaction. The Kimmeridge Clay Formation is thin or absent in the wells we have studied, but may thicken greatly in off-structure kitchen regions. Thus one hypothesis that can be constructed from the one-dimensional model is that overpressures resulting from gravitational compaction disequilibrium are supplemented by hydrocarbon generation in the Central Graben.

The high pressures in the permeable Upper Jurassic sandstones suggest that the aquifer system beneath the regional pressure seal may be continuous and allows lateral connectivity to deep, highly overpressured off-structure regions. If this is the case, the hydrogeological behaviour of the sandstones in the Central Graben cannot be accurately simulated using a one-dimensional model. The hydrogeological role of the sandstones may be important in controlling the distribution and magnitude of overpressure in the region. Thus to support or disprove this hypothesis we require a model capable of simulating this behaviour. It is necessary to employ two-dimensional models. Although more expensive in terms of the computer time and hardware required to simulate the processes, the use of sophisticated two-dimensional models allows insight into the controls on the distribution and magnitude of pressure in the Central Graben,

Two-dimensional modelling

To test the hypotheses using a two-dimensional model, we have modelled a basin-wide section across the East Forties Basin (Fig. 5). A simplified geometry derived from published interpreted regional seismic data (Roberts *et al.* 1990) has been simulated. The section crosses the Forties–Montrose High, the central tilted fault block penetrated by well 22/30a-1. The crest of this structure is flanked by deeper regions of the Graben, with Jurassic and Lower Cretaceous sediments deepening and thickening eastwards into the East Forties Basin. An eastern shelf zone, the Norwegian–Danish Platform, is penetrated by well 23/27-6. The seismic-derived geometry of the model has been calibrated against the well stratigraphy. The section allowed us to study the

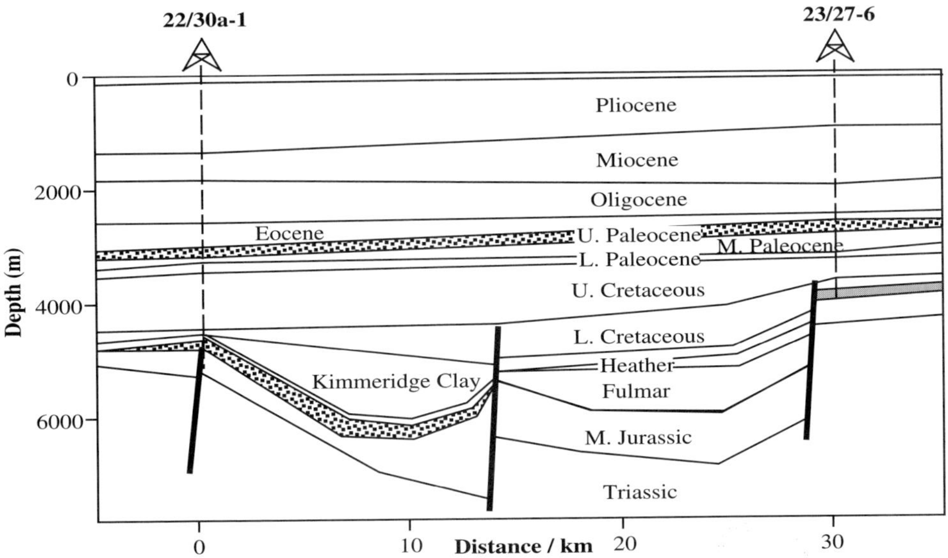

Fig. 5. Model geometry of the two-dimensional model (after Roberts *et al.* 1990). The axial Forties–Montrose High penetrated by well 22/30a-1 is formed by the crest of the NE-dipping fault block that deepens into the East Forties Basin. Well 23/27-6 is sited on the Norwegian–Danish High. Stippled pattern denotes sandstones.

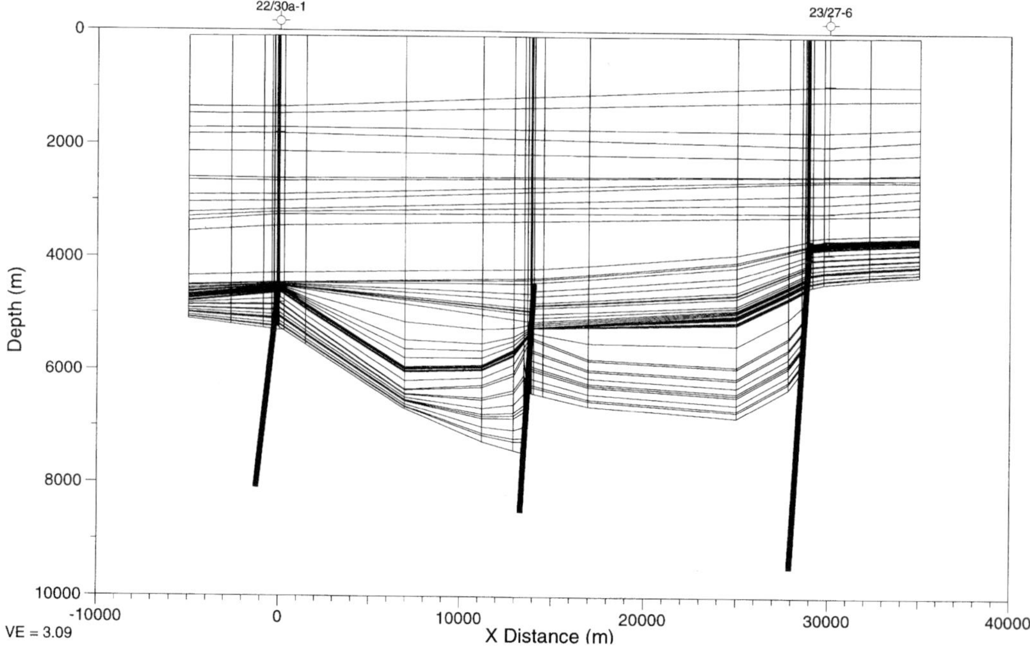

Fig. 6. Grid used in calculation. 2448 cells have been used, with calculation density greatest in zones of hydrogeological interest such as fault block crests.

potential of up-dip transfer of pressure from the deep Graben towards the crestal wells located on the tops of the generally NE-dipping fault blocks in the area (Roberts *et al.* 1990).

The geometry of the section was discretized onto a grid of 2448 cells (Fig. 6). As the model calculates flow values for the centre of each cell through time, grid density is non-uniform to allow greater density of calculation in crestal zones. Accurate reproduction of the thin transition zones in the Central Graben requires a dense array of cells on the crest of the modelled fault block. Calculation density decreases in the Triassic hydrogeological basement to the model. We have assumed that the Permo-Triassic salt that underlies the Central Graben (Roberts *et al.* 1990) will impede downward flow, and so have assigned lateral and basal no-flow boundary conditions.

The model reproduces observed RFT pressures in the Fulmar sandstones and Palaeocene sandstones in both wells (Fig. 7); it is also consistent with the mudweight profile in the Cretaceous–Jurassic pressure cell. As the observed hydrogeological data are replicated by the model, we suggest that this model is accurately describing the processes controlling overpressure in the region.

The model produces high overpressure (40 MPa above hydrostatic pressure) and a sharp transition zone in the crestal well 22/30a-1 due to lateral transmission of pressure from deeper, off-structure regions where the sandstones are encased in thick, highly overpressured, undercompacted shales. Modelled pressures on the crests of the fault block are close to the minimum horizontal stress, and vertical fluid flow through the pressure seal is focused at the crest of the fault block. These results support the hypothesis that lateral flow in permeable sandstones beneath the pressure seal is controlling the pressure distribution. This lateral flow is controlled by the geometry of the sandstones, which in turn is controlled by the pre-Cretaceous structure of the Graben.

Discussion

The model rests on the undercompaction of deep shales in the off-structure regions of the Graben. Is there evidence that Graben shales are undercompacted? Sonic logs may provide a clue. Increasing undercompaction as the pressure cell is penetrated has been reported from overpressured regions world-wide (Schmidt 1973) and sonic log porosity in the Central Graben shows a similar pattern across a range of depths (Fig. 8). The model reproduces the observed porosity–depth profile derived from sonic logs (Fig. 9). Although the relationship of sonic transit time to porosity has

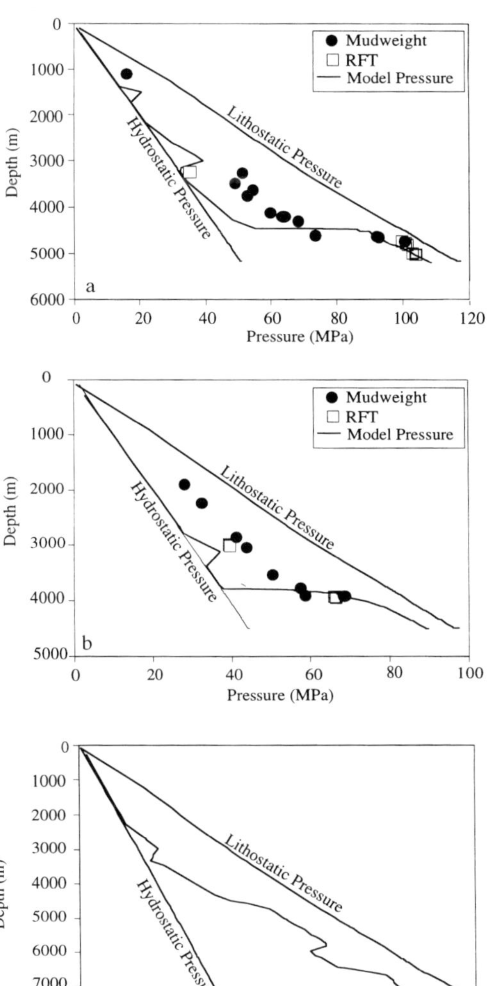

Fig. 7. Pressure–depth plots for wells (**a**) 22/30a-1 and (**b**) 23/27-6 demonstrate that the two-dimensional model provides an adequate match to observed data. A modelled off-structure region (**c**) at $x = 10\ 000$ m is shown for comparison. A Tertiary pressure cell is decoupled from the Cretaceous–Jurassic pressure cell by the hydropressured/low-pressured Palaeocene sandstones.

been questioned (Wensaas *et al.* 1994), it forms the only available porosity data for model calibration in the Central Graben. High sonic transit times implying undercompaction have also been noted from seismic investigation of the shales in this region (Japsen 1993). Modelled shale and chalk permeability is low (Fig. 9), and is in the same range as that of the one-dimensional model.

Our two-dimensional model also addresses the hydrogeological behaviour of the Palaeocene sandstone overlying the pressure seal, which forms a basin-wide, high-permeability aquifer. This hydropressured unit acts as a pressure 'drain' and serves to decouple the Tertiary hydrogeological regime from the underlying Cretaceous–Triassic pressure cell regime. An upper and lower pressure cell regime is implied, separated by the hydropressured Palaeocene aquifer. The decoupling effect of the hydropressured Palaeocene sandstones plays an important role in controlling the details of the pressure–depth profiles. We suggest that regions to the south of our study region, where the Palaeocene sandstones become thin and discontinuous, will demonstrate different pressure regimes from the region we have modelled (Mudford *et al.* 1991).

Moving into two dimensions poses intricate problems for model calibration, as the geometry of the system distal to wells is speculative and the hydrogeological role of faults is unclear. This problem is common to all similar basin modelling exercises (Burrus *et al.* 1991). In the Central Graben our models suggest that fluid flow must be generally restricted by poorly permeable faults. With permeable Graben-bounding faults, overpressures in the Fulmar sandstones of the Norwegian–Danish Platform can leak laterally and model pressures provide a poor match to observed pressures (Fig. 10). Additionally, the present generation of models cannot simulate the halokinetic movements that determined the geometry of the Jurassic sediments in the region. The effect of Tertiary halokinesis and salt piercement structures on the hydrogeology of the Central Graben cannot yet be approached using a quantitative modelling methodology.

Hydrocarbon generation pressures are calculated assuming source rock properties from Cayley (1987). The Kimmeridge Clay Formation is simulated as having oil-prone Type I kerogen at 8–15 % TOC, whilst the Upper Jurassic Heather Fm has gas-prone type III kerogen at 3% TOC. Overpressure resulting from hydrocarbon generation is minor compared to overpressure resulting from disequilibrium compaction of the shale-rich Graben sediments (Fig. 11). Although playing an important role at the pore-scale in the fine-grained source rocks, in particular for the expulsion of petroleum (Duppenbecker *et al.* 1991), we infer that hydrocarbon generation plays a minor role in regulating the observed pressures in the Central Graben. Although our model of hydrocarbon generation is geochemically simplistic, our rapid analysis allows us to neglect hydrocarbon generation pressures for the aims of this paper. The model suggests that

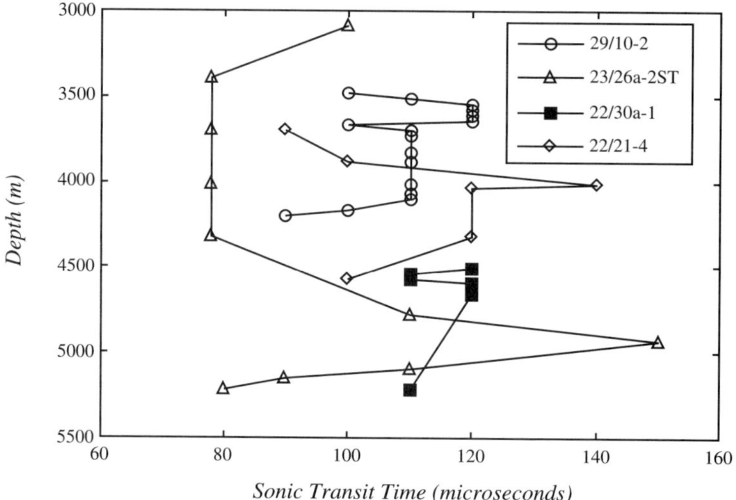

Fig. 8. Sonic log transit time from wells across the region suggests increasing undercompaction with depth in Central Graben shales.

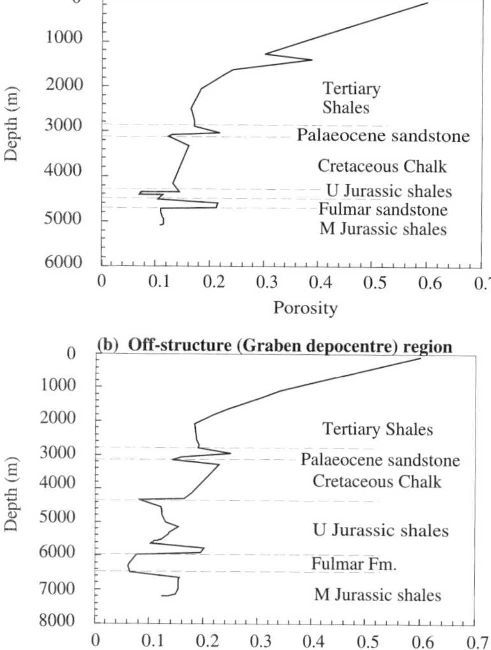

Fig. 9. Model porosity shows increasing undercompaction with depth and major undercompaction of shales in deep off-structure regions. The trend qualitatively replicates the sonic transit time pattern shown in Fig. 8.

disequilibrium compaction of thick shale sequences off-structure in the Eastern and Western Graben is the principal overpressuring mechanism affecting the region. This conclusion will ultimately be supported or disproved by the acquisition of direct porosity measurements in shale cores retrieved from deeper wells, and by advances in our understanding of the compaction processes of clay-rich sediments. We have demonstrated that the over-

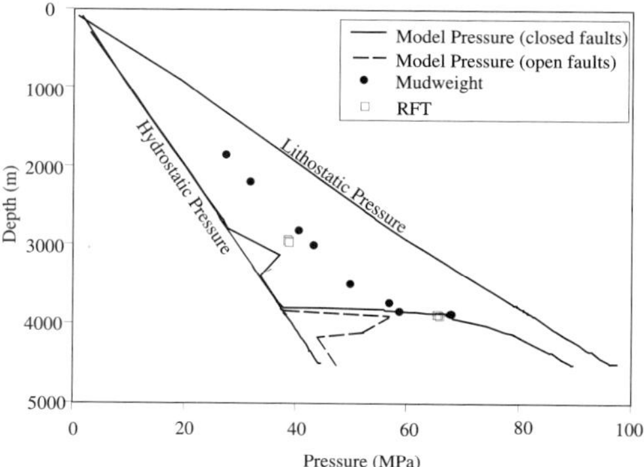

Fig. 10. Fault hydrogeological behaviour in well 23/27-6. Simulating faults as closed, poorly permeable zones provides a better match to observed pressures on the Norwegian–Danish High than if faults are considered as open, permeable zones.

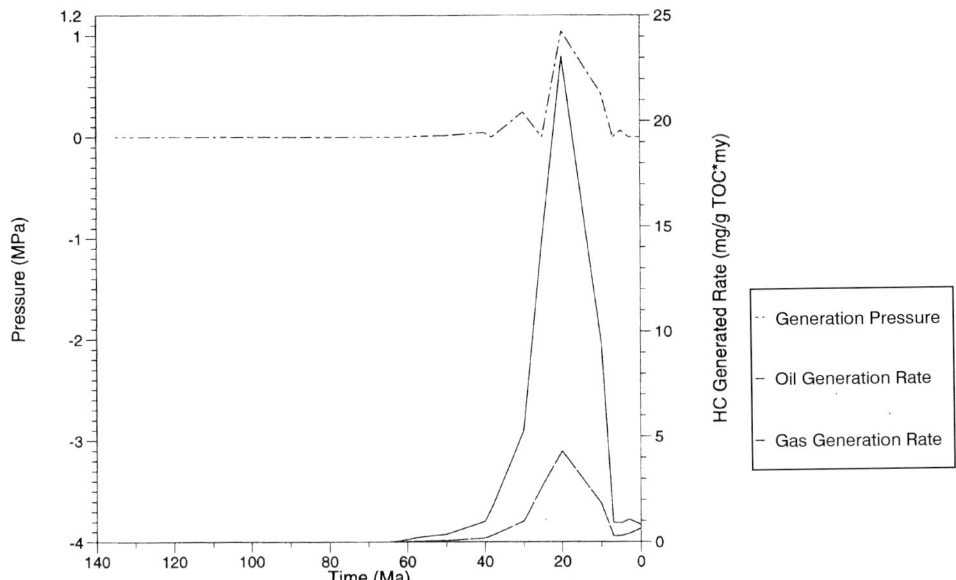

Fig. 11. Generation pressures in the deep kitchen regions of the Graben. The graph shows the generation pressure in the Kimmeridge Clay Formation at a present-day depth of 5500 m at cross-section position x = 7000 m. Pressures resulting from hydrocarbon generation are low compared to those arising from gravitational compaction disequilibrium.

pressures observed in wells penetrating structural highs may be controlled by the hydrogeological behaviour of adjacent, off-structure regions which are currently unobserved. We suggest that maximizing observed data from shales in deep regions of the Graben will allow testing of the hypotheses presented.

The two-dimensional model presented in this paper accurately replicated the pressure data observed in the Central Graben at the present day. We infer that it provides an accurate depiction of the present-day hydrogeological system, and is adequately simulating the processes controlling the system. Our inference is made even though we recognise that our model is only two-dimensional, and that three-dimensional effects may be important. In the case examined here, these effects would be related to the increase in flow-path length over what we have depicted in two dimensions. Such increases would have to be extremely large to substantially alter the behaviour of the system we have modelled.

Given the apparent validity of our model, it may provide a basis for formulating hypotheses regarding those historical basin processes controlled by the palaeo-hydrogeological environment of the Graben, such as hydrocarbon migration and entrapment, and the diagenesis of the Fulmar sandstones. These hypotheses can be tested against observed data such as fluid inclusion palaeobarometry (Swarbrick 1994) and the distribution of hydrocarbons in the basin.

Conclusions

1. The magnitude and distribution of overpressure in the Central Graben is controlled by lateral flow beneath the regional pressure seal. One-dimensional modelling of the region shows that the distribution of pressure in the region does not show a relationship to the thickness and permeability of the vertical seal. This is unlike other basins world-wide. Two-dimensional modelling allows recognition that the pre-Cretaceous structure of the Graben controls the flow of pore fluids, leading to elevated pressures and sharp transition zones on axial fault blocks where the regional pressure seal is thin (Fig. 12).
2. The models suggest that disequilibrium compaction of thick shale sequences in off-structure regions is the principal overpressuring mechanism in the Graben. This disequilibrium

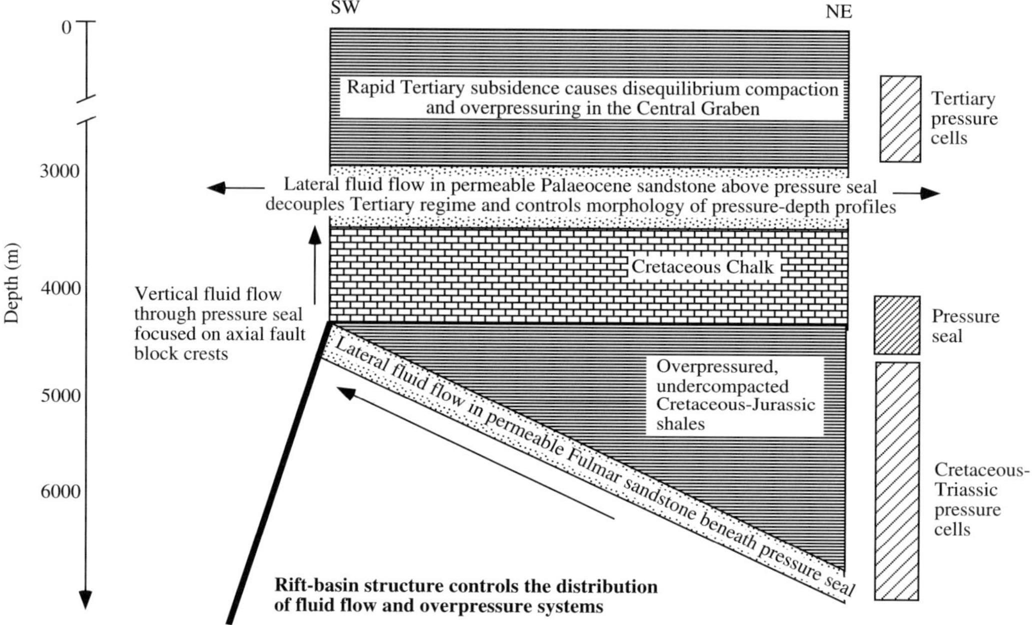

Fig. 12. Summary cartoon of the overpressure system in the Central Graben.

compaction has arisen due to restricted fluid flow in low-permeability lithologies, coupled with the high Tertiary sedimentation rate.

3. Free lateral flow of porewater in the hydrostatically pressured Palaeocene sandstones above the pressure seal leads to a layered hydrogeological regime in the Central Graben. An upper Tertiary pressure cell is decoupled from the deeper pre-Cretaceous pressure cell by this hydropressured zone.
4. Basin models provide an invaluable tool in building and testing hypotheses regarding the complex dynamic interplay of the basin processes that control fluid flow. The detailed models resulting from computer simulation need to be improved by additional porosity–permeability data acquisition from deeply buried shales and an improved understanding of the processes controlling shale petrophysical behaviour.

This work was supported by the Natural Environment Research Council Grant No. GT4/92/164/G. Platte River Associates, Inc. are thanked for providing the BasinMod™ software. Elf Enterprise Caledonia are thanked for pressure data. All data presented in this paper is released. We appreciate the helpful reviews provided by T. Dodd and B. Mudford, and G. Holm and P. Ringrose are thanked for many useful comments.

References

AUDET, J. M. & MCCONNELL, J. D. C. 1992. Forward modelling of porosity and pore pressure evolution in sedimentary basins. *Basin Research,* **4**, 147–162.

BETHKE, C. M. (1985) A numerical model of compaction-driven groundwater flow and heat transfer and its application to the palaeohydrology of intracratonic sedimentary basins. *Journal of Geophysical Research,* **90**, B8, 6817–6828.

BURRUS, J., KUHFUSS, A., DOLIGEZ, B. & UNGERER, P. 1991. Are numerical models useful in reconstructing the migration of hydrocarbons? A discussion based on the Northern Viking Graben. *In:* ENGLAND, W. A. & FLEET, A. J. (eds) *Petroleum Migration,* Geological Society, London, Special Publications, **59**, 89–109.

CAYLEY, G. T. 1987. Hydrocarbon migration in the central North Sea. *In:* BROOKS, J. & GLENNIE, K. (eds) *Petroleum Geology of North West Europe,* Graham & Trottman, London, 549–555.

DARBY, D., HASZELDINE, R. S. & COUPLES, G. D. 1996. Pressure cells and pressure seals in the UK Central Graben. *Marine and Petroleum Geology,* **13**, 865–878.

DEWERS, T. & ORTOLEVA, P. 1994. Nonlinear dynamical aspects of deep basin hydrology: fluid compartment formation and episodic fluid release. *American Journal of Science,* **294**, 713–755.

DUPPENBECKER, S. J., DOHMEN, L. & WELTE, D. H. 1991. Numerical modelling of petroleum expulsion in two areas of the Lower Saxony Basin, Northern Germany. *In:* ENGLAND, W. A. & FLEET, A. J. (eds) *Petroleum Migration* , Geological Society, London, Special Publications, **59**, 47-64.

ENGELDER, T. & FISCHER, M. P. 1994. Influence of poroelastic behaviour on the magnitude of minimum horizontal stress in overpressured parts of sedimentary basins. *Geology,* **22**, 949–952.

ENGLAND, W. A., MACKENZIE, A. S., MANN, D. M. & QUIGLEY, T. M. 1987. The movement and entrapment of petroleum fluids in the subsurface. *Journal of the Geological Society,* **144**, 327–347.

GAARENSTROOM, L., TROMP, R. A., DE JONG, M. C. & BRANDENBURG, A. M. 1993. Overpressures in the Central North Sea: implications for trap integrity and drilling safety. *In:* PARKER, J. (ed.) *Petroleum Geology of Northwest Europe: Proceedings of the 4th Conference,* Geological Society, London, 1305–1313.

HUNT, J. M. 1990. Generation and migration of petroleum from abnormally pressured fluid compartments. *Bulletin of the American Association of Petroleum Geologists,* **74**, 1–12.

——, WHELAN, J. K., EGLINTON, L. B. & CATHLES, L. M. 1994. Gas generation— a major cause of deep Gulf Coast overpressures. *Oil and Gas Journal,* **92**, 18–21.

JAPSEN, P. 1993. Influence of lithology and Neogene uplift on seismic velocities in Denmark: implications for depth conversions of maps. *Bulletin of the American Association of Petroleum Geologists,* **77**, 194–211.

KATSUBE, T. J., MUDFORD, B. S. & BEST, M. E. 1991. Petrophysical characteristics of shales from the Scotian shelf. *Geophysics,* **56**, 1681–1689.

LERCHE, I. 1990. *Basin Analysis— Quantitative Methods*, Volume 1, Academic Press Inc., San Diego.

MELLO, U. T., KARNER, G. D. & ANDERSON, R. N. 1994. A physical explanation for the positioning of the depth to the top of overpressure in shale-dominated sequences in the Gulf Coast basin, United States. *Journal of Geophysical Research,* **99**, B2, 2775–2789.

MUDFORD, B. S., GRADSTEIN, F. M., KATSUBE, T. J. & BEST, M. E. 1991. Modelling 1-D compaction-driven flow in sedimentary basins: a comparison of the Scotian Shelf, North Sea and Gulf Coast. *In:* ENGLAND, W. A. & FLEET, A. J. (eds) *Petroleum Migration.* Geological Society, Special Publications, **59**, 65–85.

ORESKES, N., SHRADER-FRECHETTE, K. & BELITZ, K. 1994. Verification, validation and confirmation of numerical models in the earth sciences. *Science,* **263**, 641–646.

ROBERTS, A. M., PRICE, J. D. & OLSEN, T. S. 1990. Late Jurassic half-graben control on the siting and structure of hydrocarbon accumulations: UK/Norwegian Central Graben. *In:* HARDMAN, R. & BROOKS, J. (eds) *Tectonic Events Responsible For Britain's Oil and Gas Reserves.* Geological Society, London, Special Publications, **55**, 229–257.

SCHMIDT, G. W. 1973. Interstitial water composition and geochemistry of deep Gulf Coast shales and sandstones. *Bulletin of the American Association of Petroleum Geologists,* **57**, 321–337.

SCLATER, J. G. & CHRISTIE, P. A. F. 1980. Continental stretching: an explanation for the post-mid-Cretaceous subsidence of the Central North Sea Basin. *Journal of Geophysical Research,* **85**, 3711–3739.

STEWART, D. J. 1986. Diagenesis of the shallow marine Fulmar formation in the Central North Sea. *Clay Minerals,* **21**, 537–564.

SWARBRICK, R. E. 1994. Reservoir diagenesis and hydrocarbon migration under hydrostatic palaeopressure conditions. *Clay Minerals,* **29**, 463–474.

WENSAAS, L., SHAW, H. F., GIBBONS, K., AAGAARD, P. & DYPVIC, H. 1994. Nature and causes of over-pressuring in mudrocks of the Gullfaks area, North Sea. *Clay Minerals,* **29**, 139–450.

WHELAN, J. K., KENNICUTT, M. C., BROOKS, J. M., SCHUMACHER, D. & EGLINTON, L. B. 1993. Organic geochemical indicators of dynamic fluid flow processes in petroleum basins. *Organic Geochemistry,* **22**, 587–615.

WILKINSON, M., DARBY, D., HASZELDINE, R. S. & COUPLES, G. D. 1997. Secondary porosity generation during deep burial associated with overpressure leak-off: Fulmar Formation, UKCS. *Bulletin of the American Association of Petroleum Geologists*, **81**, 803–813.

—— & HASZELDINE, R. S. 1996. Aluminium loss during sandstone diagenesis. *Journal of the Geological Society, London*, **153**, 657–660.

WILLIAMSON, M. A. & SMYTH, C. 1992. Timing of gas and overpressure generation in the Sable Sub-basin, offshore Nova Scotia. *Bulletin of Canadian Petroleum Geology,* **40**, 151–169.

A comparison between 1-D, 2-D and 3-D basin simulations of compaction, water flow and temperature evolution

T. THRONDSEN & M. WANGEN

Institute for Energy Technology, PO Box 40, N-2007, Kjeller, Norway

Abstract: Is 3-D basin modelling really necessary? The answer depends on the type of problem under consideration. The Institute for Energy Technology (IFE) 3-D Basin Simulator offers the possibility to investigate certain aspects of this issue. It is a prototype 1-D, 2-D or 3-D basin simulator for reconstruction of the time-dependent pressure and temperature evolution of compacting sedimentary basins. Comparison of results, from 1-D, 2-D and 3-D simulations confirms that there may be significant differences due to water flow and pressure behaviour. The advantage of 2-D simulations as compared to 1-D is seen in overpressured configurations where 2-D allows more realistic calculations, while 1-D is less predictable. When going from 2-D to 3-D advantages are evident in configurations with pronounced 3-D basin geometries in combination with rapid processes and overpressure. The differences in water flow velocities and consequently the pressure calculations, may be quite significant. The rates in 3-D may be several orders of magnitude higher than in 2-D due to lateral focusing effects, while the overpressure still remains high. One consequence is, for example, reduced compaction in 3-D as compared to 2-D, with notable effects on basin geometry and temperature evolution of individual layers. Temperature anomalies due to increased convection in 3-D seem to be of less importance, unless processes of lateral focusing of water flow reach extreme levels.

The importance of three-dimensional basin modelling as compared with one- and two-dimensional models has been a theme of many informal discussions among geologists during the last ten years. The key issues include whether 3-D is really necessary, difficulties in obtaining necessary input data, resolution with respect to basin details and limitations in computer time. Despite these discussions, surprisingly few documented evaluations have been put forward, and very little have so far been published about 3-D basin modelling (Welte & Yükler 1981; Novelli *et al.* 1988). Moreover, documented benefits and drawbacks of 3-D as compared to 1-D and 2-D are lacking.

The Institute for Energy (IFE) 3-D Basin Simulator (hereinafter called IFE-3-D) offers the possibility to evaluate certain aspects of this issue, of course limited by its current capabilities. It is a combined 1-D, 2-D and 3-D basin simulator for reconstruction of the time-dependent water flow, pressure and temperature evolution of compacting sedimentary basins.

The purpose of the present study is to provide a comparison between 1-D, 2-D and 3-D modelling obtained from simulations in the same basin. The basin is synthetic and specifically designed to enhance differences between 1-D, 2-D and 3-D. Although being synthetic, it offers a realistic situation with respect to natural conditions.

The basin simulator

IFE-3-D is a prototype 3-D basin simulator that can also be run as a 1-D and 2-D simulator. The various elements and capabilities are described by Wangen (1993; 1994). It bears many similarities to existing 2-D fluid flow simulators with respect to geological capabilities, input requirements, functionality and output.

The current version also has certain limitations with respect to geological capabilities, grid configuration and resolution.

IFE-3-D makes use of a 3-D hexahedral FEM-grid with vertical walls. The horizontal geometry of the grid is kept fixed during the simulations, while it is allowed to change vertically due to compaction and erosion. The fixed horizontal geometry implies that horizontal movements within the basin are not allowed, for example lateral stretching. Moreover, there is a lower resolution limit of the grid which introduces certain restrictions on the discretization of the basin. This limit has not yet been investigated, but successful tests have been performed on a DEC Alpha workstation with approximately 10 000 grid cells without spending to much

THRONDSEN, T. & WANGEN, M. 1998. A comparison between 1-D, 2-D and 3-D basin simulations of compaction, water flow and temperature evolution. *In*: DÜPPENBECKER, S. J. & ILIFFE, J. E. (eds) *Basin Modelling: Practice and Progress*. Geological Society, London, Special Publications, **141**, 109–116.

computer time (13.5 hours). Between approximately 3 000 and 10 000 cells the computer time has been found to increase approximately linearly with cell numbers.

The main objective of IFE-3-D is the calculation of the dynamic evolution of sediment compaction, heat flow, temperature, pressure and water flow. This is based on physical laws and empirical assumptions, defined boundary conditions, rock and water properties, and knowledge of the process history. The simulator allows laterally varying lithologies, together with temporarily and laterally varying boundary conditions, such as sea-bottom temperature and heat flow into the bottom of the basin. The lateral boundaries of the system (e.g. sediment basin) are assumed isolated with respect to heat, and may be either open or closed for water flow. An open lateral boundary is at hydrostatic pressure, whilst a closed one is impermeable. The sea-level and the sediment–water interface are used as top boundaries for the pressure and temperature calculations, respectively. In the present study the base of the sedimentary basin is used as bottom boundary for both pressure and temperature calculations. However, an underlying crust and mantle may be included in the temperature solution.

Heat capacity, thermal conductivity and permeability are porosity-dependent parameters, and a variety of functions are available in the simulator. The actual functions being used in the present study are:

Porosity (Korvin 1984):

$$\Phi = \Phi_0 \times \exp(-\alpha \times ps_{max});$$

Permeability (Bryant *et al.* 1984):

$$k = a \times \exp(b \times \Phi);$$

Thermal conductivity (Lewis & Rose 1970):

$$\lambda = \lambda_w^{\Phi} \times \lambda_s^{(1-\Phi)};$$

where Φ is porosity, Φ_0 is initial (depositional) porosity, α is rock pore volume compressibility, ps_{max} is maximum effective stress (i.e. sediment grain-to-grain pressure), k is permeability, a and b are constants, λ is bulk rock thermal conductivity, and λ_w and λ_s are heat conductivities of the water and rock matrix, respectively.

Anisotropic rock properties are permitted in the simulator, but the option is not used in this study.

The water density is assumed to be pressure- and temperature-dependent according to the following formulation:

Water density (Wangen 1994):

$$\rho_w = \rho_0 \times [1 + \alpha_0 \times (p_w - p_{w0}) - \beta_0 \times (T - T_0)]$$

Where ρ_w is water density, ρ_0 is water density (998 kg m^{-3}) at a reference water pressure $p_w = 0$ Pa and a reference temperature of $T_0 = 0$ °C. α_0 is water compressibility (4.5×10^{-10} Pa^{-1}) and β_0 is the coefficient of thermal expansion of water (2×10^{-4} K^{-1}).

The viscosity of water is approximated as a function of temperature alone, and is represented in the simulator by a table (Wangen 1994).

The simulator works forward in time and the following parameters are calculated for each time step:

(a) Compacted formation thickness;
(b) Porosity;
(c) Permeability (in $x = y$ and z directions);
(d) Water flow (Darcy velocity in x, y and z direction);
(e) Water (pore) pressure;
(f) Hydrostatic pressure;
(g) Excess pressure;
(h) Total pressure;
(i) Effective stress (sediment grain-to-grain pressure);
(j) Temperature;
(k) Heat capacity;
(l) Thermal conductivity (in $x = y$ and z direction);
(m) Heat flow (in x, y and z direction).

Synthetic test basin

A three-dimensional synthetic test basin has been constructed specifically to enhance differences between 1-D, 2-D and 3-D basin modelling, whilst still being realistic with regard to natural conditions. The basin configuration is illustrated in Figs 1, 2 and 3.

The basin consists of five formations (Fm 1–Fm 5): a thin highly permeable sandstone formation (Fm 2) in the lower part, surrounded above and below by two less permeable siltstone formations (Fm 1 and Fm 3), and two low-permeable shale formations (Fm 4 and Fm 5) at the top capable of building up considerable overpressure (Fig. 1). The rock parameters used are listed in Table 1. The formations are gently dipping from relatively shallow depths in the eastern parts of the basin to more than 6500 m in the west. An impermeable vertical barrier crosses the basin in a north to south direction. It has a small opening in the middle, only where it intersects the sandstone formation (Fm 2). The horizontal size of the basin is 18×15 km^2.

The burial history of the basin is simple, ending with a fairly rapid burial during the last five million years. The burial history for pseudowell 1 located in the deepest part of the basin is illustrated in Fig. 2 (for location see Fig. 1).

The boundary conditions for the temperature calculations are: constant heat flow of 50 mW m^{-2} into the base of the basin and constant temperature

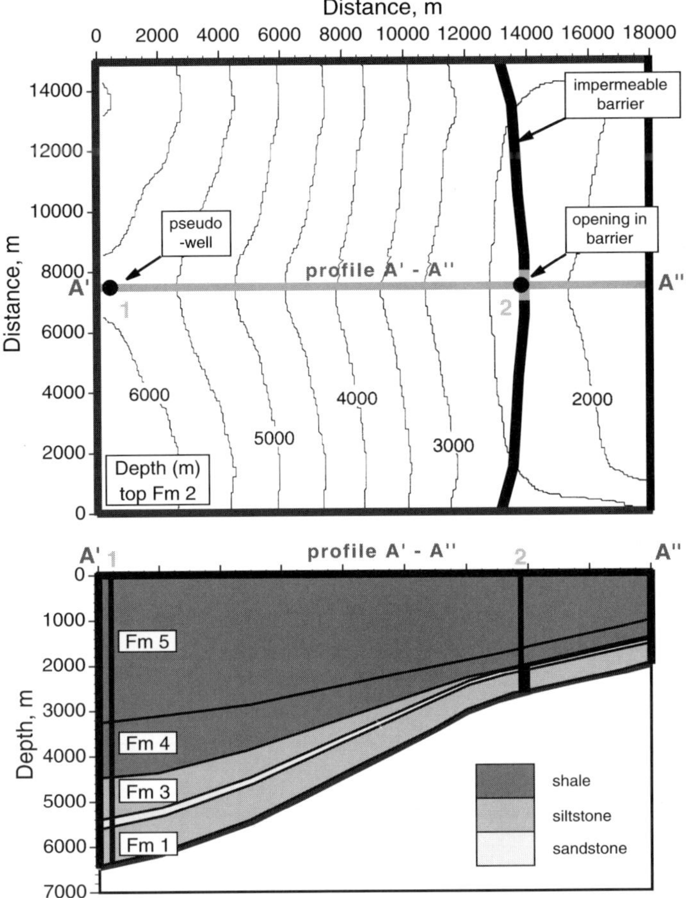

Fig. 1. Configuration of the synthetic test basin showing position of the barrier, opening in the barrier, pseudowells 1 and 2 and profile A′–A″.

of 5 °C at the sediment surface. For the water flow and pressure calculations, the base as well as the northern, southern and western boundaries of the sedimentary basin are kept closed (impermeable) to water flow. The eastern boundary, however, is kept open, i.e. at hydrostatic pressure. The sea-level is kept constant and equal to zero. The grid configuration for the calculations is shown in Fig. 3.

Simulation results

Three-dimensional basin simulation has been performed for the entire basin, 2-D has been run along profile A′–A″, whilst 1-D has only been performed in pseudowells 1 and 2 (Fig. 1). Note that profile A′–A″ passes through the opening in the barrier across the basin.

Water flow distribution

Figure 4 shows the horizontal velocity of water flow at the present day in the permeable sandstone formation (Fm 2) as calculated by 3-D simulations. There is a very strong lateral focusing effect on the water flow, from less than 100 m Ma^{-1} in the westernmost part of the basin to more than 30 000 m Ma^{-1} in the opening in the barrier. Note that this flow is driven by compaction only.

Figure 5 shows a comparison of the lateral water flow in the permeable sandstone formation (Fm 2) at present as calculated by 3-D simulation for the entire basin, and by 2-D simulation along profile A′–A″. In the western and central parts of the profile 2-D and 3-D simulations give approximately similar water flow within 200 m Ma^{-1}. In the position of the barrier, however, there is a

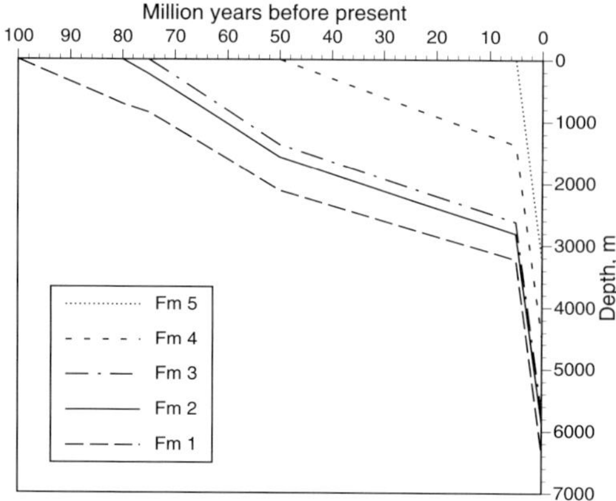

Fig. 2. Burial history for pseudowell 1.

dramatic difference, with 3-D reaching flow rates more than 30 times higher than in 2-D.

Figure 6 shows water flow velocities at present in pseudowells 1 and 2, and provides a comparison between 1-D, 2-D and 3-D fluid flow modelling. These diagrams also include the component of water flow in the z-direction (vertical). In pseudowell 1 in the deepest part of the basin, both the horizontal (x-direction) and vertical water flow is very similar in 2-D and 3-D, both with respect to magnitude and direction. Above approximately 3500 m and below 5700 m the vertical component of the water flow is positive upwards, whilst in the intermediate section the water is drained downwards to the permeable sandstone formation (Fm 2). The 1-D simulations are different since compaction water is only allowed to escape upwards, and no lateral water flow is permitted. The restrictions causes somewhat higher vertical flow rates in 1-D as compared to 2-D and 3-D. In

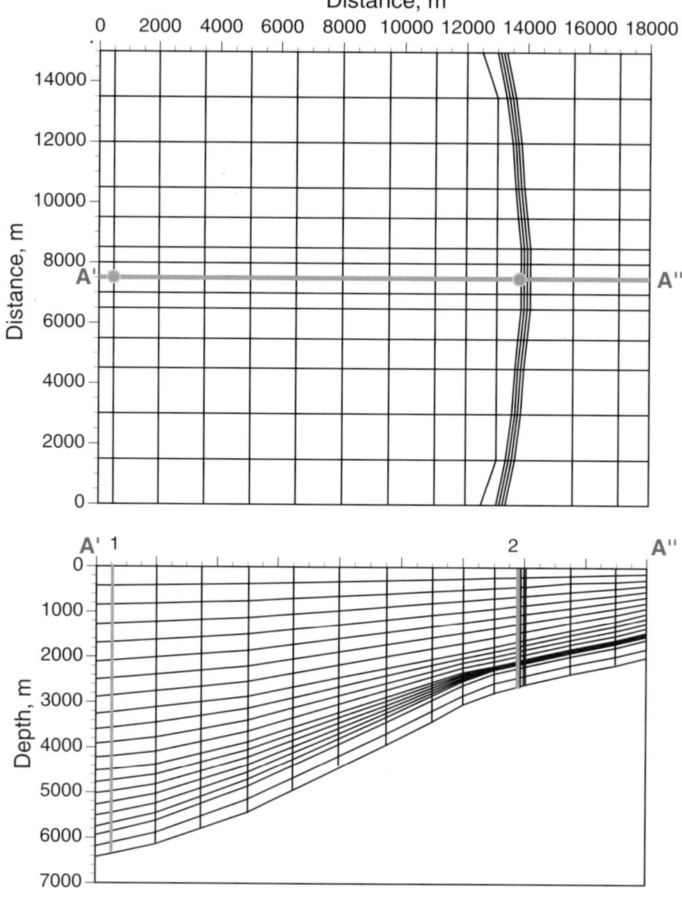

Fig. 3. Grid configuration used in the simulations.

Table 1. *Rock properties used in the simulations*

Formation	Lithology	Initial porosity (%)	Compressibility (1/Pa)	Permeability a (mDarcy)	Permeability b	Matrix density ($kg\ m^{-3}$)	Matrix heat capacity ($J\ kg^{-1}\ K$)	Matrix thermal conductivity ($W\ m^{-1}\ K$)
Fm 5	Shale	50	2E-08	1E-07	20	2600	1000	2.00
Fm 4	Shale	50	2E-08	1E-07	20	2600	1000	2.00
Fm 3	Siltstone	50	2E-08	1E-04	17	2600	1000	3.00
Fm 2	Siltstone	50	2E-08	1E-02	17	2600	1000	4.00
Fm 1	Siltstone	50	2E-08	1E-04	17	2600	1000	3.00
Water							4180	0.63

pseudowell 2 the 3-D simulations show extremely high lateral flow rates due to the focusing effect, whilst 2-D gives more moderate values.

Pressure

Figure 7 shows a comparison between the overpressure distribution at present, as calculated by 3-D simulation for the entire basin, and by 2-D simulation along profile A′–A″. Both models give significant overpressures in the middle part of the shale formations (Fm 4–Fm 5). However, the absolute values for the overpressure are significantly higher in 3-D than in 2-D. A consequence of this is that 3-D predicts higher porosity than 2-D.

Temperature distribution

Figure 8 shows the results of temperature calculations at present in the sandstone formation (Fm 2) as calculated by 3-D simulation for the entire basin, 2-D simulation along profile A′–A″, and by 1-D simulations in the pseudowells. Note that two models are used for the 1-D calculations, one including pressure calculations, and the other assuming hydrostatic pressure.

There are noticeable differences between the various models. The 3-D simulation gives the highest temperatures, about 10 °C higher than analogous 2-D simulations in the deepest part of the basin and progressively less in the shallower parts. The difference is mainly due to higher porosity in 3-D and accordingly lower thermal conductivity of

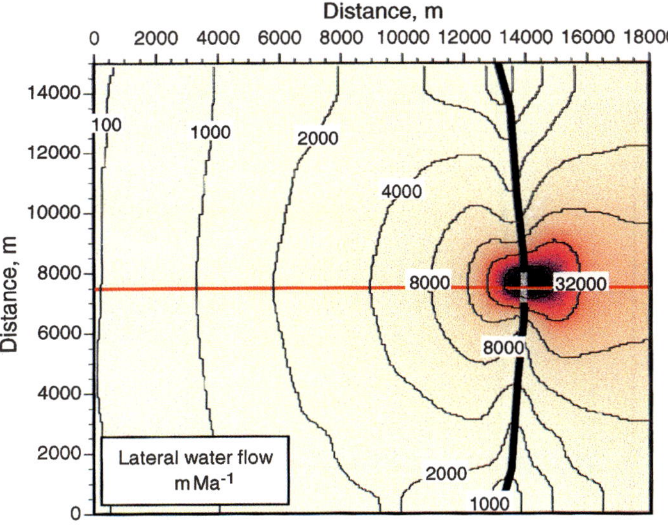

Fig. 4. Lateral water flow velocities in the permeable sandstone formation (Fm 2) at present, as calculated by 3-D basin modelling.

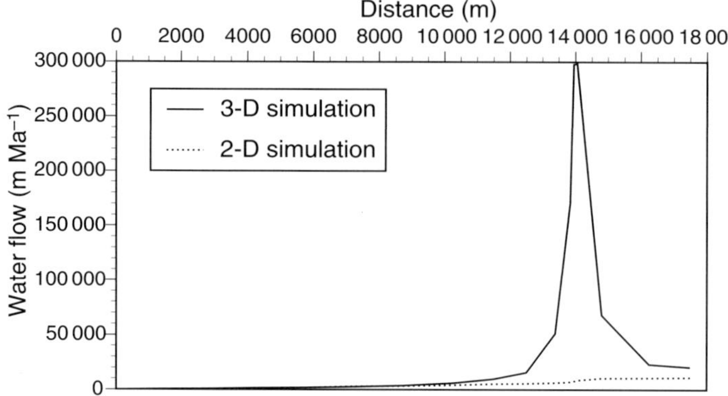

Fig. 5. Lateral water flow velocities in the permeable sandstone formation (Fm 2) at present along profile A′–A″ as calculated by 2-D and 3-D basin modelling.

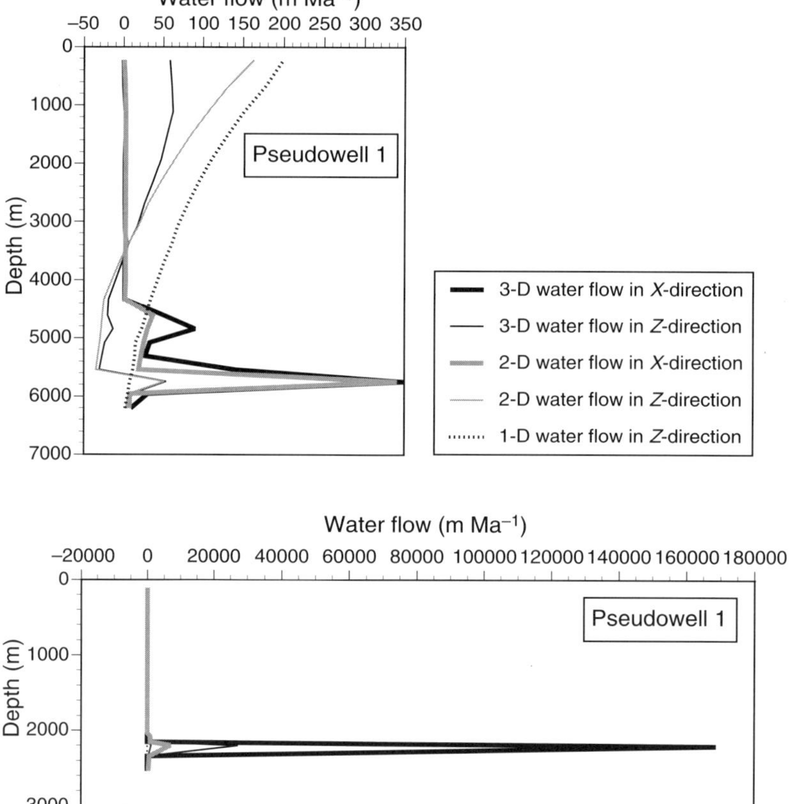

Fig. 6. Lateral and vertical water flow velocity at present in pseudowells 1 and 2 as calculated by 1-D, 2-D and 3-D basin modelling.

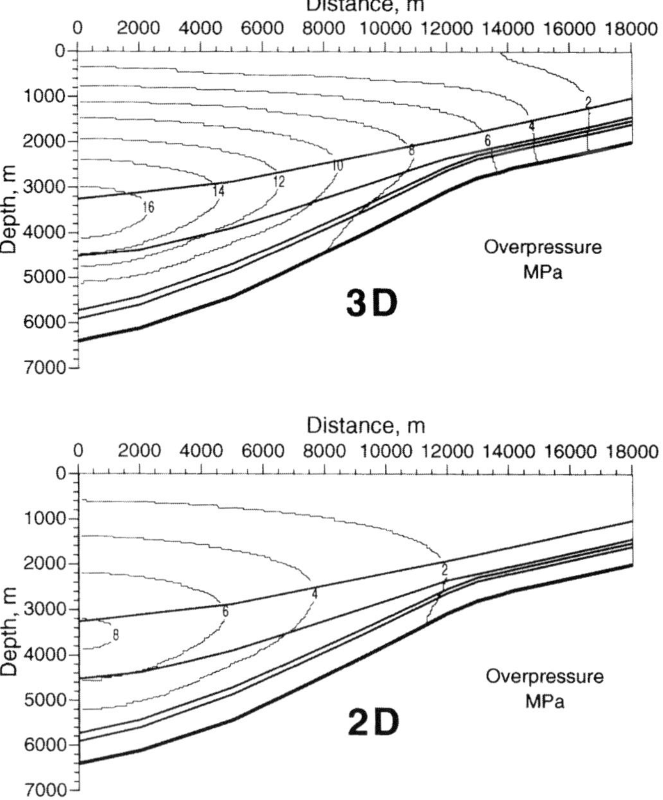

Fig. 7. Overpressure distribution at present along profile A′–A″ as calculated by 2-D and 3-D basin modelling.

the sediments than in 2-D. The strong focusing of the water flow in 3-D in the position of the barrier gives only a slight effect (increase) on the temperature field due to convective heat transport (Fig. 8).

The temperature calculations in 1-D depend strongly on the model assumptions and the geological setting. Simulations assuming hydrostatic pressure give temperatures that are similar to or lower than 1-D, 2-D and 3-D with pressure calculations. The 1-D model with hydrostatic pressure gives more compaction, higher thermal conductivities and lower temperatures than the models that account for overpressure. In pseudo-well 1 the difference between the two 1-D models is approximately 10 °C.

At the eastern edge of the basin where the pressure, by definition, is hydrostatic, the temperature calculations are nearly identical irrespective of 1-D hydrostatic, 1-D fluid flow, 2-D or 3-D. Very minor differences are observed and can be ascribed to minor and slightly different lateral convective contributions in 2-D and 3-D.

Conclusion

This study offers a comparison between 1-D, 2-D and 3-D basin modelling in a rather simple setting, and can only highlight certain aspects of this issue. However, even this simple case demonstrates noticeable differences between 1-D, 2-D and 3-D simulations.

Calculated water flow velocities may locally reach much higher levels in 3-D as compared to 2-D — up to several orders of magnitude higher. This is due to lateral focusing effects in 3-D taking into account both *x*- and *y*-direction flows. 3-D basin modelling is definitely very important if water flow is an essential element of the problem under consideration. For example, simulation of diagenesis where transport of chemical species dissolved in water is an important mechanism.

Convective heat transport seems to be of minor importance for the temperature field, even at high levels of water flow. Note that effects of meteoric water flow and topography above sea-level are not discussed here.

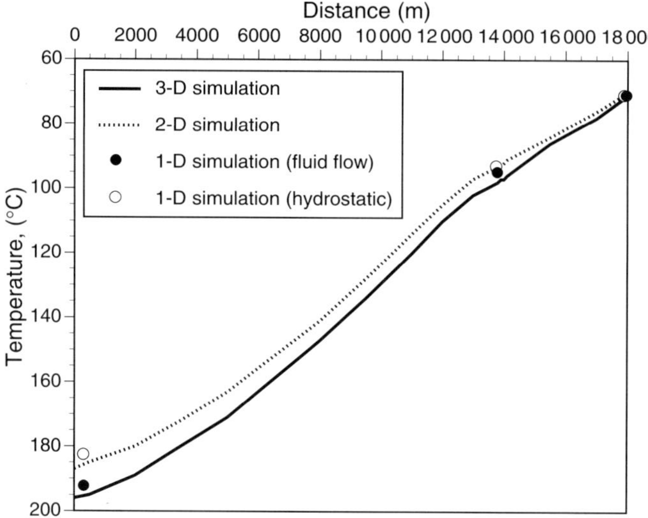

Fig. 8. Temperature distribution at present in the permeable sandstone formation (Fm 2) as calculated by 1-D, 2-D and 3-D basin modelling.

Two-dimensional and 3-D simulation may give significantly different pressures, with 3-D yielding higher overpressures, and presumably being the more correct. A consequence is that 2-D tends to over-estimate compaction relative to 3-D. This increases the thermal conductivity of the sediments and causes under-estimation of temperature. This under-estimation may reach significant levels, up to 10 °C or more, but the effect is restricted to basins with overpressure.

One-dimensional simulations with pressure calculations are somewhat spurious with respect to natural process because severe restrictions on the pressure and fluid flow calculations must be assumed, unless water is allowed to escape laterally in a second dimension. Such an escape mechanism has to be arbitrary, and is not included in the present study. The tendency in overpressured situations is to under-estimate the compaction and thereby over-estimate the temperature. One-dimensional hydrostatic simulations will in overpressured situations always over-estimate compaction and consequently under-estimate temperature relative to 1-D fluid flow, 2-D and 3-D.

The authors wish to thank the Norwegian Reasearch Council who has financed this study and the development of the IFE 3-D Basin Simulator as part of a strategic research programme (STP) at IFE (Modelling of geochemical processes and petroleum basins). We are also grateful to P. Ringrose and J.-L. Rudkiewics for their valuable reviews which improved the manuscript.

References

BRYANT, W. R., HOTTMAN, W. & TRABANT, P. 1975. Permeability of unconsolidated marine sediments, Gulf of Mexico. *Marine Geotechnology*, **1**, 1–14.

KORVIN, G. 1984. Shale compaction and statistical physics. *Geophysical Journal of the Royal Astranomical Society*, **78**, 35–50.

LEWIS, C. R. & ROSE, S. C. 1970. A theory relating high temperatures and over-pressures. *Journal of Petroleum Technology*, **22**, 11–16.

NOVELLI, L., WELTE, D. H., MATTAVELLI, L., YALÇIN, M. N., CINELLI, D. & SCHMITT, K. J. 1988. Hydrocarbon generation in southern Sicily. A three dimensional computer aided basin modelling study. *Organic Geochemistry*, **13**, 153–164.

WANGEN, M. 1993. *User's guide to Bas7. IFE/KR/F-93/206*. IFE report, available upon request.

—— 1994. Numerical simulation of thermal convection in compacting sedimentary basins. *Geophysical Journal International*, **119**, 129–150.

WELTE, D. H. & YÜKLER, A. 1981. Petroleum origin and accumulation in basin evolution – A quantitative model. *American Association of Petroleum Geologists Bulletin*, **65**, 1387–1396.

A review of techniques used to determine geological and thermal history in the Southern North Sea

G. ARCHARD, J. STAFFORD[1], K. BARDWELL[1] & M. BAGGE

GeoQuest Reservoir Technologies, Highlands Farm, Greys Road, Henley on Thames, Oxfordshire RG9 4PS, UK

[1] *PGS Reservoir (UK) Limited, Chapel House, Liston Road, Marlow, Bucks SL7 1XJ, UK*

Abstract: The Southern North Sea has been a prolific source of gas to the UK economy over the last 30 years. As many of the producing fields are now becoming depleted, attention has switched to exploration in more marginal areas of the basin. In order to evaluate suitable structures in these areas, an understanding of the complex structural and thermal history of the Southern North Sea Basin is required. Particular emphasis must be placed upon the prediction of source rock maturity, the timing of generation and migration of hydrocarbons, and the thermal controls exerted on potential reservoir sand units.

Over large parts of the basin the Carboniferous source rocks are not at their maximum depth of burial at present day. A variety of techniques utilizing sonic velocity, vitrinite reflectance and apatite fission track analysis have been used to quantify the magnitude of uplift that the Carboniferous section has undergone through geological time. With the exception of apatite fission track analysis, these techniques provide only an estimate of 'net uplift', i.e. the amount of uplift relative to maximum depth of burial or maximum temperature. They record neither the timing and effects of individual uplift events nor the timing of maximum burial prior to net uplift. In order to accurately reconstruct the burial and thermal history of the basin these techniques must be integrated with a detailed knowledge of the geological history of the Southern North Sea.

The use of these techniques in isolation, together with a lack of appreciation of the limitations of the data has in the past resulted in spurious results, leading to criticism and mistrust of the techniques. This paper attempts to show that when correctly applied and in appropriate geological circumstances, each analytical technique can provide useful data for burial and thermal history reconstruction. Examples are shown to highlight the effects of incomplete data interpretation and the impact of inaccurate determination of present day temperatures. Particular attention is paid to determining the burial and thermal history of the economically important Carboniferous source and Permian reservoir units, and the interpretation of uplift data is restricted to the post-Zechstein section.

The Southern North Sea is a major gas province where gas generated from Carboniferous source rocks is now trapped in Carboniferous, Permian (Rotliegendes and Zechstein) and Triassic (Bacton) reservoirs. At the present day the source rocks over the majority of the basin are not at their maximum depth of burial or maximum temperature, having been affected by tectonic events at several times in the past which resulted in uplift and cooling of the Carboniferous source rock section. Establishing the nature and effects of these tectonic events has been the focus of much work over the past two decades and as the science of basin modelling has developed, quantifying the magnitude of the events has become critical. The ultimate aim in quantifying uplift is to determine where and when Carboniferous source rocks have generated gas in the past and where this gas is likely to be trapped. Furthermore, the study of reservoir quality in the context of diagenesis requires a knowledge of the timing and duration of the thermal regimes the reservoir has passed through in the geological past.

Exploration in the Southern North Sea is currently focused on the margins of the basin where the presence of a mature gas source rock cannot be taken for granted. Maturity modelling techniques are therefore of great value to the explorationist in assessing the likelihood that structures contain gas. As in the centre of the basin, maturities in these marginal areas cannot be estimated from present day depth of burial, and information is required on maximum depth of burial and maximum temperature.

Many analytical techniques have been applied in the Southern North Sea to quantify the degree of uplift from maximum burial utilising sonic derived velocities (Marie 1975; Glennie & Boegner 1981; Bulat & Stoker 1987), vitrinite reflectance (Barnard & Cooper 1983; Cope 1986), and apatite fission track analysis (Green 1989). The theme of this

ARCHARD, G., STAFFORD, J., BARDWELL, K. & BAGGE, M. 1998. A review of techniques used to determine geological and thermal history in the Southern North Sea. *In*: DÜPPENBECKER, S. J. & ILIFFE, J. E. (eds) *Basin Modelling: Practice and Progress*. Geological Society, London, Special Publications, **141**, 117–136.

paper is to outline the limitations of these uplift determination techniques and the possible pitfalls of using them in isolation. We display how the techniques can be refined when integrated with the geological framework to produce more realistic simulations of basin history. One of the main points we emphasize is the need to include as many different techniques as possible in an evaluation and to ensure data from individual wells is placed in the correct geological context.

Location and stratigraphy

The area of study is shown in Fig. 1 and a stratigraphic column for the area in Fig. 2. Illustrated on the latter diagram are the significant source and reservoir units. It is important to highlight that the key source and reservoir horizons in the Carboniferous and Early Permian are overlain by the Late Permian Zechstein Group, which comprises evaporites, including mobile salt.

Figure 2 also shows the position of major unconformities in the stratigraphic column. The amount of eroded section at each of these unconformities varies considerably, giving rise to radically different burial history profiles in different parts of the basin. Quantification of the eroded section is therefore a high priority in all burial and thermal modelling studies. The amount of erosion at unconformities above the Zechstein is strongly influenced by diapiric movement of the mobile salt.

Burial history modelling

An example burial history diagram for the Sole Pit Trough area of the Southern North Sea is shown in

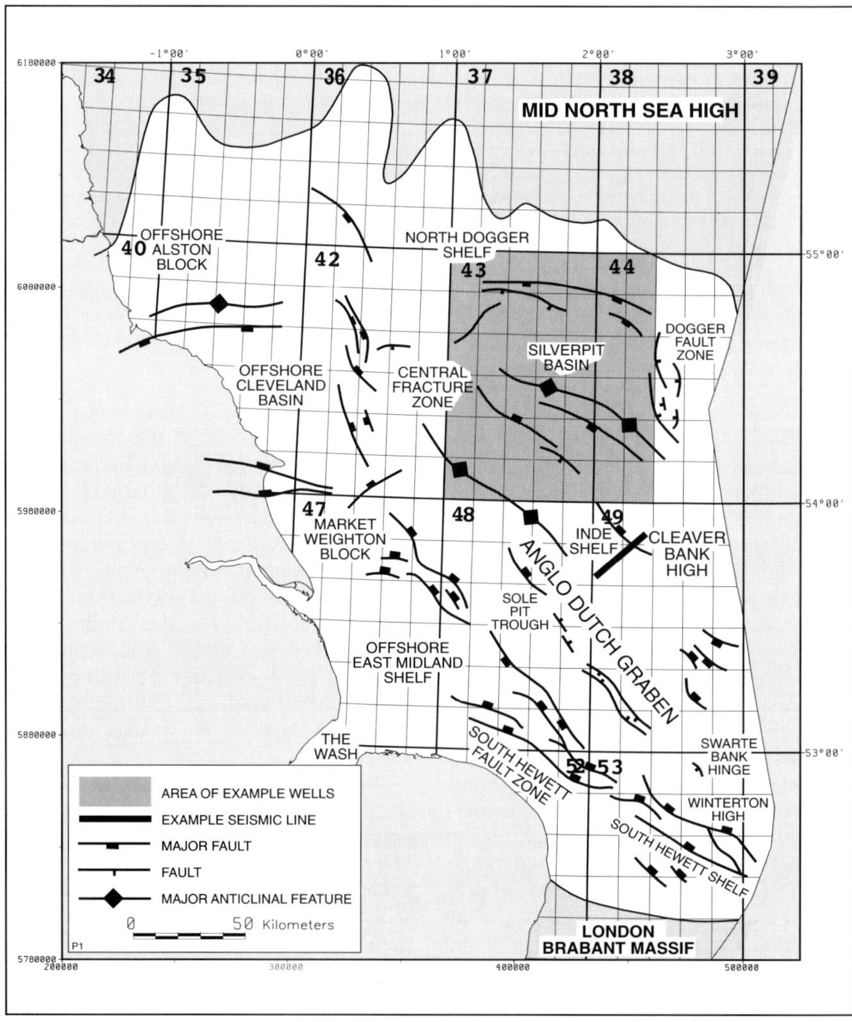

Fig. 1. Location map and present day structural elements.

Fig. 3. This diagram shows the particular importance of the Laramide uplift event (Late Cretaceous to Early Tertiary). Over large areas of the Gas Basin the principal source and reservoir rocks (Carboniferous and Rotliegendes) are considered to have attained maximum temperatures immediately prior to this event (Green 1989). Quantification of uplift associated with this event has been the focus of many exploration studies.

An early perception was that Laramide uplift was

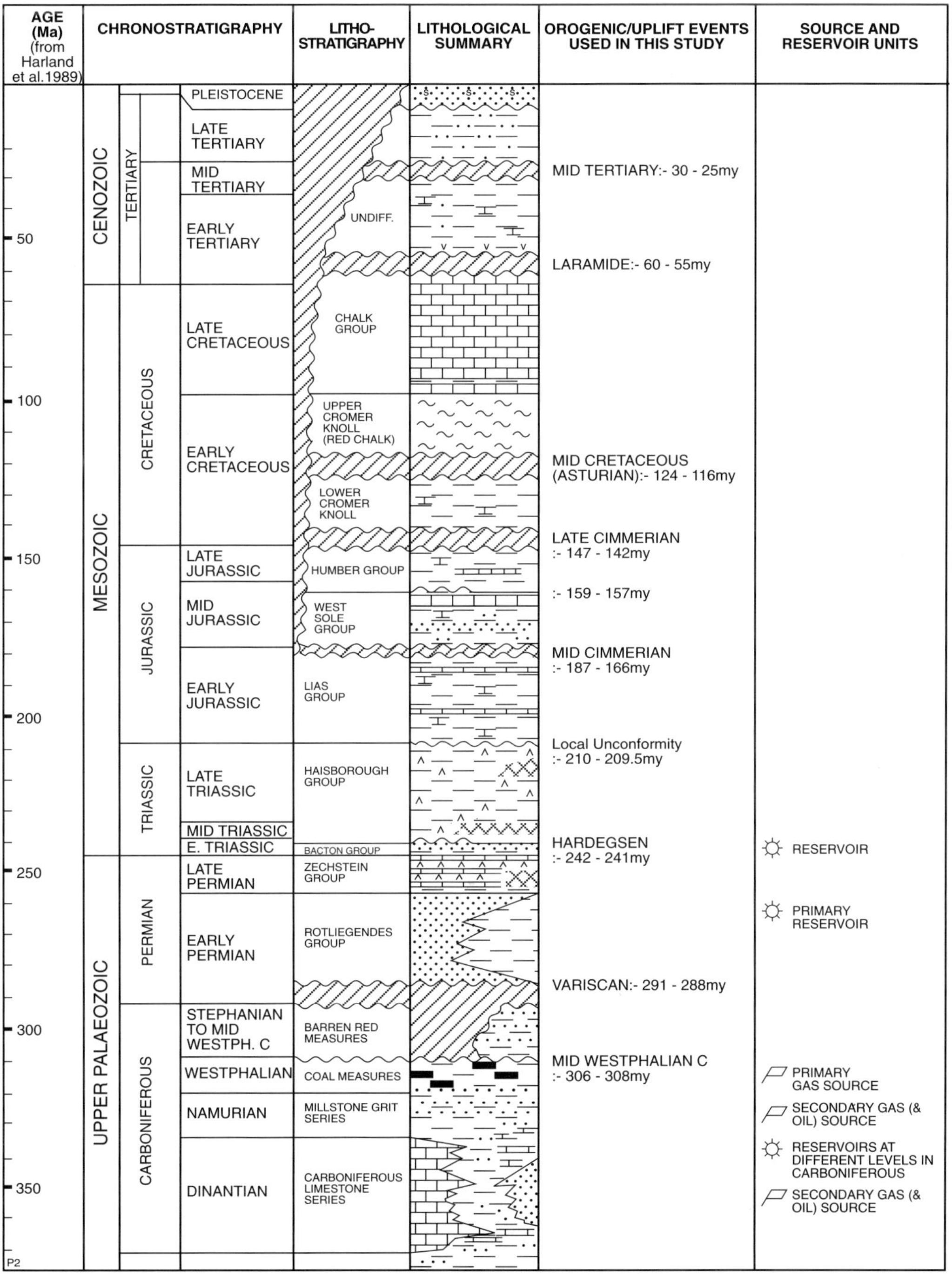

AGE (Ma) (from Harland et al.1989)	CHRONOSTRATIGRAPHY			LITHO-STRATIGRAPHY	OROGENIC/UPLIFT EVENTS USED IN THIS STUDY	SOURCE AND RESERVOIR UNITS
	CENOZOIC	TERTIARY	PLEISTOCENE			
			LATE TERTIARY			
			MID TERTIARY		MID TERTIARY:- 30 - 25my	
50			EARLY TERTIARY	UNDIFF.	LARAMIDE:- 60 - 55my	
	MESOZOIC	CRETACEOUS	LATE CRETACEOUS	CHALK GROUP		
100			EARLY CRETACEOUS	UPPER CROMER KNOLL (RED CHALK)	MID CRETACEOUS (ASTURIAN):- 124 - 116my	
				LOWER CROMER KNOLL	LATE CIMMERIAN :- 147 - 142my	
150		JURASSIC	LATE JURASSIC	HUMBER GROUP	:- 159 - 157my	
			MID JURASSIC	WEST SOLE GROUP	MID CIMMERIAN :- 187 - 166my	
200			EARLY JURASSIC	LIAS GROUP	Local Unconformity :- 210 - 209.5my	
		TRIASSIC	LATE TRIASSIC	HAISBOROUGH GROUP		
			MID TRIASSIC			
			E. TRIASSIC	BACTON GROUP	HARDEGSEN :- 242 - 241my	RESERVOIR
250	UPPER PALAEOZOIC	PERMIAN	LATE PERMIAN	ZECHSTEIN GROUP		
			EARLY PERMIAN	ROTLIEGENDES GROUP	VARISCAN:- 291 - 288my	PRIMARY RESERVOIR
300		CARBONIFEROUS	STEPHANIAN TO MID WESTPH. C	BARREN RED MEASURES	MID WESTPHALIAN C :- 306 - 308my	
			WESTPHALIAN	COAL MEASURES		PRIMARY GAS SOURCE
			NAMURIAN	MILLSTONE GRIT SERIES		SECONDARY GAS (& OIL) SOURCE
350			DINANTIAN	CARBONIFEROUS LIMESTONE SERIES		RESERVOIRS AT DIFFERENT LEVELS IN CARBONIFEROUS; SECONDARY GAS (& OIL) SOURCE
P2						

Fig. 2. Generalized stratigraphy and position of unconformities. Principal source and primary reservoir sections lie below the Zechstein Salt.

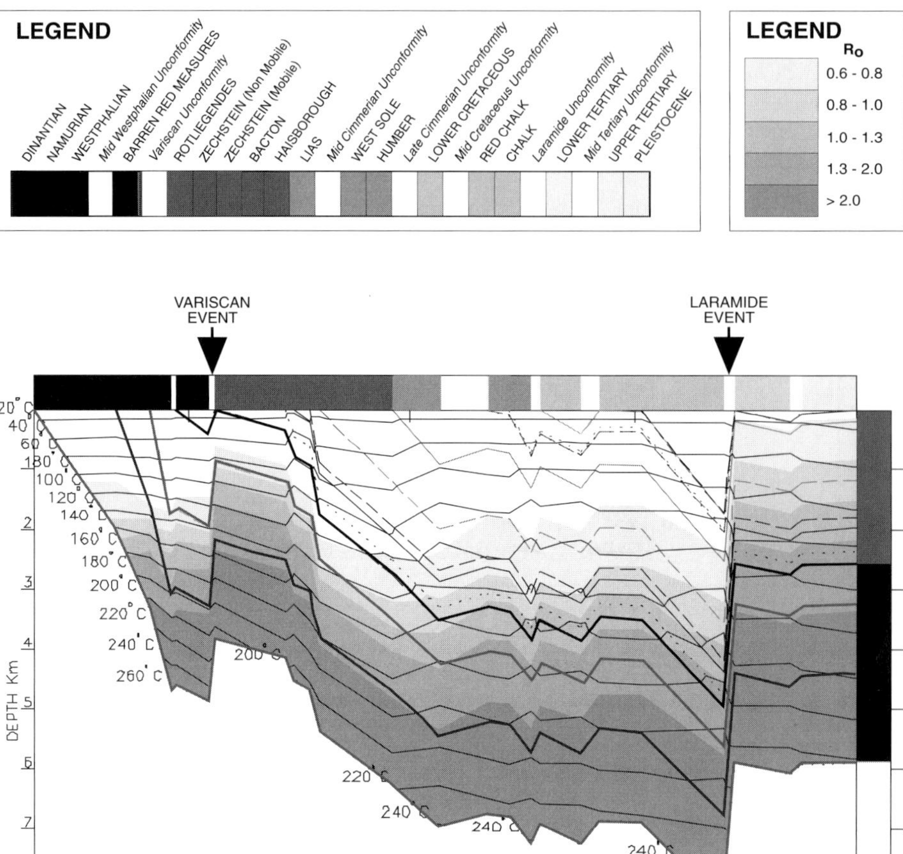

Fig. 3. Burial history diagram for Sole Pit trough well. Maximum temperatures in carboniferous reached prior to laramide uplift.

related to local inversion axes, such as the Sole Pit–Cleveland Basin axis. Away from these axes, it was believed that uplift decreased to zero (Kent, 1980; Zeigler 1982; Hemingway & Riddler 1982). Early attempts at quantifying uplift employed shale velocity techniques (Marie 1975; Glennie & Boegner 1981; Bulat & Stoker 1987) but uplift estimates were comparative and required a baseline area where uplift was assumed to be zero. Wells on the East Midlands Shelf were frequently used as 'baseline' sections. Later studies employing vitrinite reflectance data (Cope 1986) were also used to display uplift, but again relative to wells assumed to currently be at or near their maximum depth of burial i.e. at or near zero uplift (Cope 1986; fig. 2).

The development of the apatite fission track analysis technique revised the notion that uplift was restricted to local inversion axes. Fission track data implied that regional warping involving kilometre scale uplift had occurred with the recognized inversion axes representing local areas of maximum uplift and erosion (Green 1989 and Green *et al.* 1993). The apatite fission track data also provided information on the timing and rate of cooling from maximum palaeotemperatures.

By providing this additional information the apatite fission track analysis technique became popular among basin modelling workers in the Southern North Sea Gas Basin. Several oil industry studies were performed using this technique to calibrate burial and thermal models. However, in some areas of the basin (e.g. southern flank of Quadrant 43) uplift values derived from apatite fission track analysis conflicted with existing data derived from sonic velocities or vitrinite reflectance. Uplift estimates derived from the apatite technique were generally higher, resulting in the exclusion of one or other set of data during burial and thermal model calibration.

When data generated from the different techniques are considered in the correct geological context it becomes apparent that the datasets are often complementary, with variations largely due to sampling differences and the geological conditions related to the sample points. It is particularly important to understand the nature of uplift in the Southern Gas Basin and how the distribution of sample points effects the determination of uplift. Within the basin, uplift can be either the result of regional tectonism, or related to salt mobilization or a combination of both. The important pre-Zechstein source and reservoir sections (Carboniferous and Rotliegendes) are affected only by tectonic uplift. Therefore uplift determinations derived from samples from the post-Zechstein section may be in error in areas where salt mobilization has occurred and cannot be routinely used to derive burial history for the pre-Zechstein section (Fig. 4). A seismic line from Quadrant 49 is shown in Fig. 5, the location of the line is shown on Fig. 1. This line illustrates the geological structure in the basin and shows the effects of salt movement. It should be noted from this seismic display that the expansion of salt does not depress the pre-Zechstein section, so care must also be taken during modelling to ensure the process of increasing the salt volume does not result in increased burial of this section.

Techniques to determine uplift

The techniques commonly applied to determine uplift are as follows:

(i) sonic velocity analysis;
(ii) vitrinite reflectance determinations;
(iii) apatite fission track analysis.

The two latter techniques are largely based on the analysis of an individual well. In the following section a review of each technique will be given and its potential shortcomings outlined. Examples will be shown to illustrate how integration of the various techniques can provide a more realistic estimate of uplift for both the pre-Zechstein and post-Zechstein sections.

Sonic velocity determinations

Several published papers detail the application of sonic velocity analysis to derive uplift in the Southern North Sea (Marie 1975; Glennie & Boeger 1981; Bulat & Stoker 1987; Hillis 1993). Marie's early work involved the calculation of uplift of the Sole Pit–Cleveland Basin axis by analysing the sonic transit time of the Triassic Brockelschiefer (base of Bunter Shale), and comparing it to other offshore wells (assumed to be removed from the main inversion axis). A sonic velocity vs depth plot was made and a curve drawn through wells assumed to be at maximum depth of burial at present day. Uplift was calculated for wells not at maximum depth of burial at present day by measuring their deflection from the 'normal burial curve'. A net uplift value of 1200–1800 m was derived for the Sole Pit–Cleveland Basin Axis.

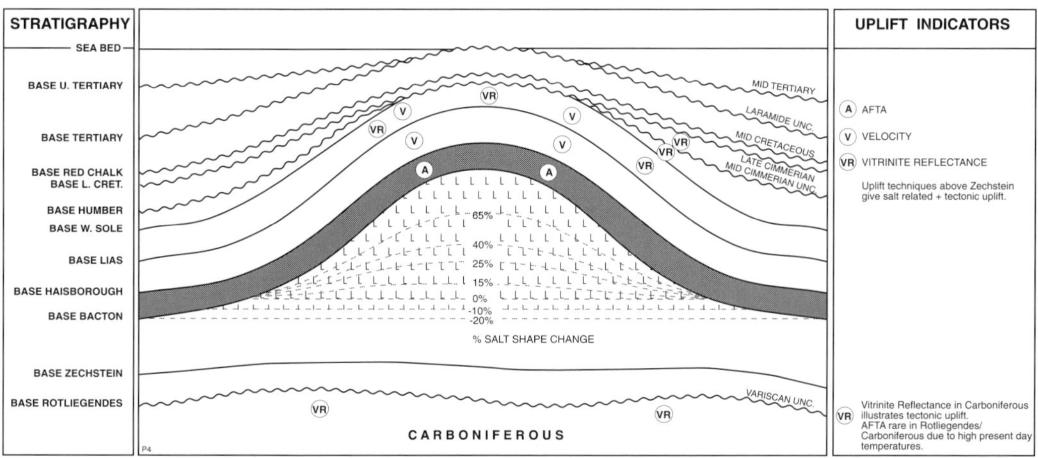

Fig. 4. Typical sample locations for AFTA, vitrinite reflectance and velocity analyses. Samples located above salt can give misleading assessment of tectonic uplift of Carboniferous.

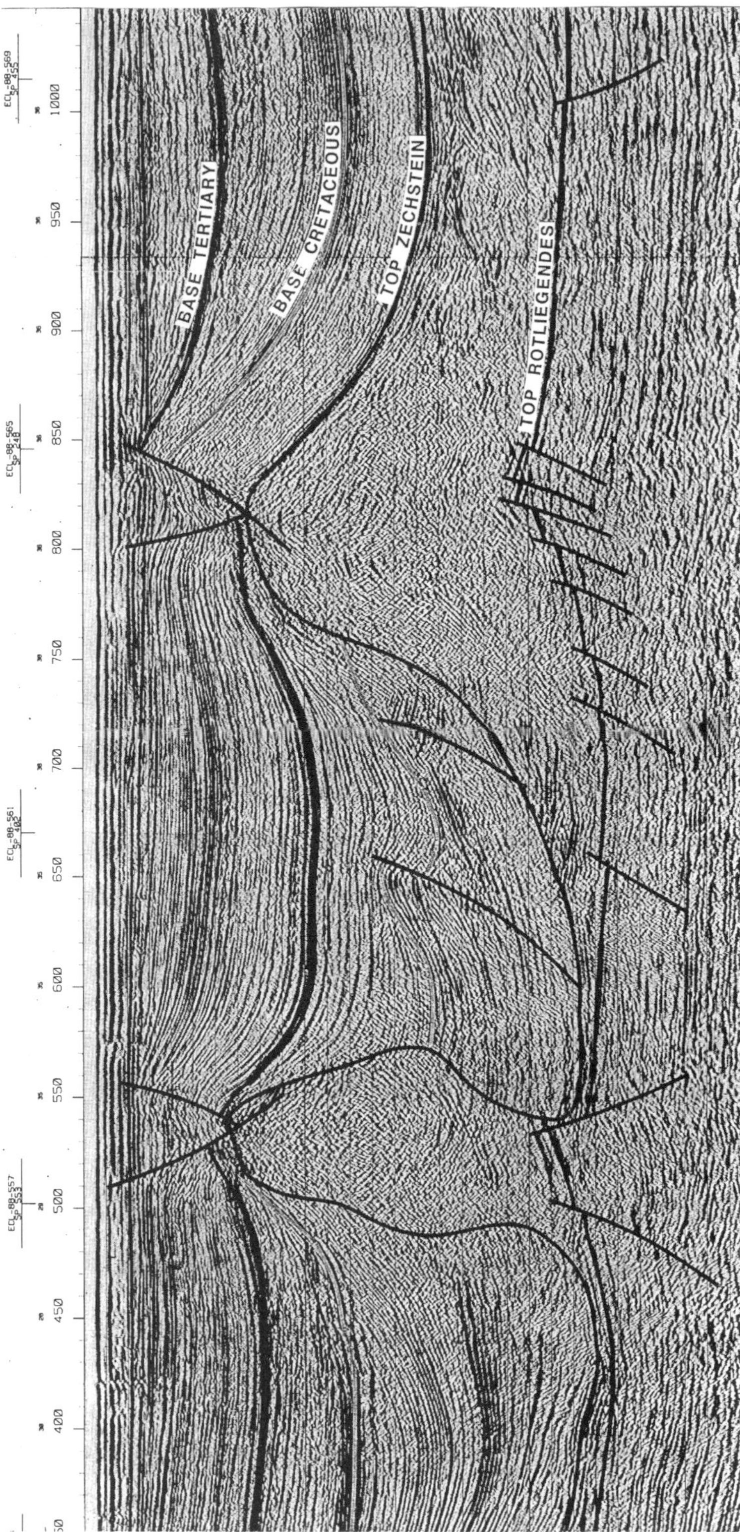

Fig. 5. Example seismic section illustrating salt effects in the basin.

Bulat & Stoker's (1987) approach was to use a range of stratigraphic intervals from the Carboniferous upwards. In all wells a plot was made of interval velocity vs mid point depth for each interval. A line was drawn through values from wells assumed to represent a 'normal' compaction trend (i.e. with zero uplift). Uplift values were then calculated for each individual unit. Lithological variations, undercompaction, fracturing or mineralization were not taken into account. The work produced some contradictory results, although values derived from shales yielded more consistent results. The finite values for uplift derived from the study relied on the assumption that wells on the East Midlands Shelf had suffered no uplift, which later work demonstrated was not the case (Green 1989).

Following construction of a large database of well velocities and stratigraphies, Bulat and Stoker's interpretations have been reviewed. When plotted, many data points fall below the Bulat and Stoker 'normal compaction curve' suggesting that the curve may not represent zero uplift. Uplift values derived using this curve will therefore be underestimated. To generate a more realistic 'zero uplift' curve, data from wells on the Mid North Sea High and North Dogger Shelf have been used, these are believed to be at or near maximum depth of burial at present day (an assumption supported by apatite fission track analysis and vitrinite reflectance data). Figures 6 and 7 illustrate interval velocity vs depth data for wells in different structural settings and show how selected Mid North Sea High/North Dogger Shelf wells may be used as 'zero uplift' cases.

To minimize the effects of lithological variation, individual mid-point velocities were extracted for the argillaceous facies from the Bacton, Haisborough, Jurassic and Lower Cretaceous and a 4-point weighted average uplift value based on these layers was calculated. Thicker sections were weighted more heavily to reduce the effects of interval velocities measured from thin sections. Layers thinner than 30 m were excluded from the analyses.

Uplift values calculated using this technique generally display similar ranges to the apatite fission track and vitrinite reflectance derived uplift values in areas where thick Late Cretaceous and or Tertiary sediments are present (Silverpit Basin). However, they produce markedly different results in areas with high uplift, such as the Sole Pit or Cleveland Basin, where velocities consistently underestimate uplift with respect to the other two techniques. The difference is generally in the order of 25 % with differences approaching 50 % in some wells. The reasons for the anomalous uplift values in these areas are unclear. However, it is noted that the intervals measured are often close to the sea bed, suggesting that the interval velocities could be effected by surface related processes. The anomalies could be due to either:

(i) relaxation of compaction following uplift or;
(ii) fracturing and groundwater circulation in the upper 1500 m of sediment (Oxburgh and Andrews-Speed 1981).

Depth vs velocity plots show considerable variation in the top 1500 m of sediment and

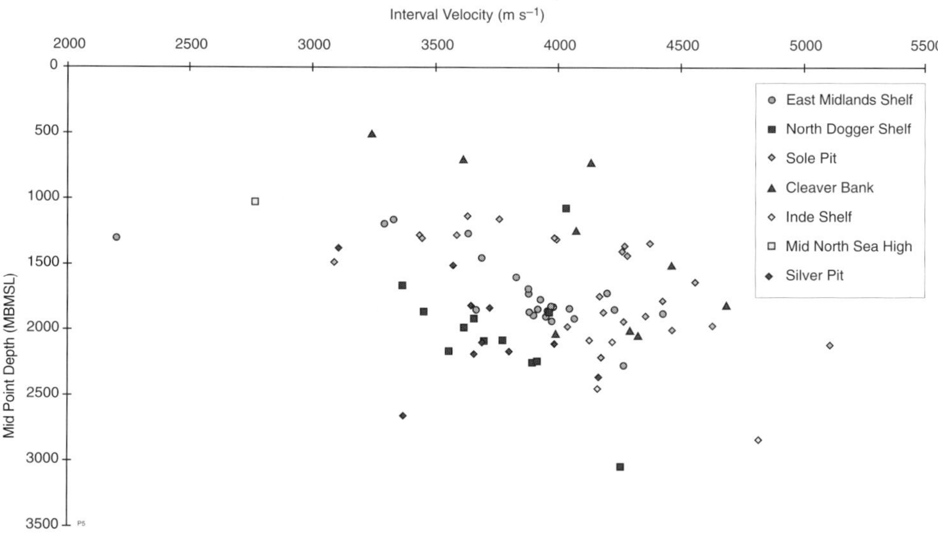

Fig. 6. Regionally divided mid-point depth interval velocity plot — Bacton Group.

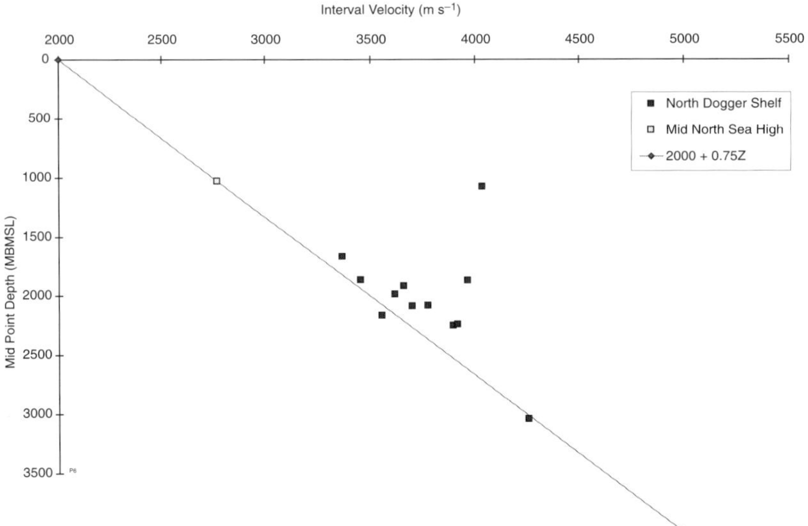

Fig. 7. Mid North Sea High and North Dogger Shelf interval velocity vs mid point depth plot — Bacton Group.

velocity data from these near surface intervals have been largely ignored.

Caution must also be exercised when interpreting velocity derived uplift data in areas of salt movement. Uplift values derived from the Bacton, Haisborough, Jurassic and Lower Cretaceous are all relevant for the post-Zechstein section and therefore give a value which is a sum of the tectonic and salt-related uplift. These uplift values cannot be translated directly to the Carboniferous source rocks.

Vitrinite reflectance

Vitrinite reflectance data are particularly important in the analysis of thermal histories in the Southern North Sea as they provide a direct assessment of the maturation of the Carboniferous gas source rock interval. Vitrinite reflectance data are commonly used to calibrate basin models by comparison with predicted values calculated using kinetic schemes (Burnham & Sweeney 1989). On their own, the data do not indicate when the maximum depth of burial and probable maximum temperatures were reached, so cannot be used as a direct indicator of timing of gas generation. Furthermore, the data cannot distinguish multiple burial and uplift events and lower magnitude events (compared to maximum depth of burial) are not resolved. When used to determine uplift, the data can however be a valuable indicator of net uplift i.e. the difference between maximum depth of burial and present day depth of burial.

The Westphalian section is rich in coals and dispersed terrestrial organic matter which provide a reliable source of vitrinite for measurement. This is particularly true of core samples containing coals where individual types of vitrinite can be distinguished with caved and low reflecting vitrinite populations being excluded. This rich and reliable supply of vitrinite does not occur in many stratigraphic units overlying the Westphalian, so vitrinite reflectance analysis outside the Carboniferous section has to rely on fewer and generally poorer quality sample points. In addition large sections such as the Rotliegendes, Triassic and the Chalk do not yield vitrinite suitable for analysis. Determining an accurate vitrinite reflectance profile outside the Carboniferous to calibrate burial and thermal modelling can therefore be difficult.

The work of Cope (1986) in outlining variation in uplift using vitrinite reflectance analysis has been a major source of reference for Southern Gas Basin studies. Cope's methodology was to derive an average vitrinite reflectance value for the top 500 feet (150 m) of the Carboniferous in 68 onshore and offshore wells. A depth vs %Ro plot was constructed (Fig. 8) and used to derive a curve drawn through well points considered to have zero or near zero uplift. PercentRo values from many wells were displaced from this trend indicating they had experienced uplift and therefore higher temperatures in the past. Uplift values were calculated

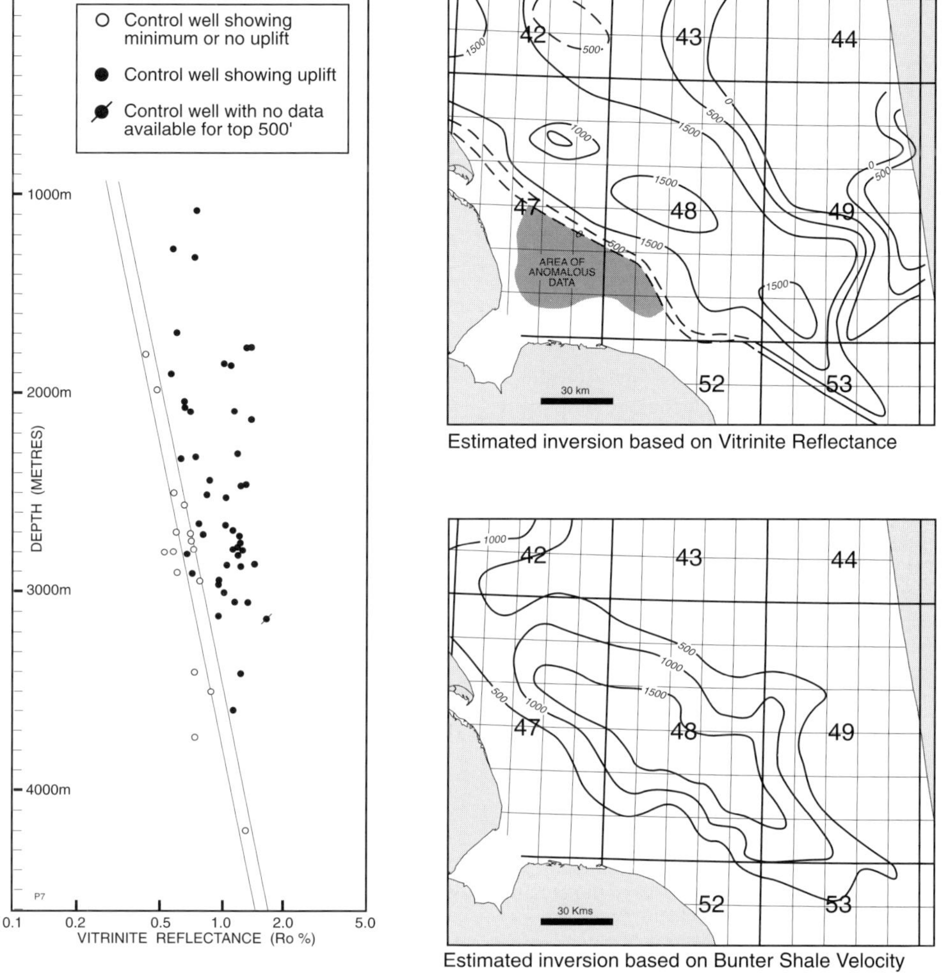

Fig. 8. Estimates of uplift using Vitrinite Reflectance and comparison to velocity derived uplift (Cope 1986; figs 2, 9, 10). Discrepancies between VR derived uplift (UVR) and velocity derived uplift (Uvel) are possibly due to salt effects (higher Uvel) or variable geothermal gradients (higher UVR). 'Area of Anomalous Data' on East Midlands Shelf later shown by AFTA to be uplifted by approx. 1 Km.

using this displacement assuming a single constant geothermal gradient throughout the basin.

Cope's estimates of uplift reached a maximum of 1500 m in the Sole Pit trough and 2000 m in the Cleveland Basin. The uplift values reduced to zero on the East Midlands Shelf and the Quad 44 and northern Quad 49 areas (Fig. 8). The uplift estimations were then compared with those using the sonic velocity technique proposed by Marie (1975). Although the work was in broad agreement with the sonic velocity derived uplift values, in some areas a discrepancy was observed between the two determinations. It was recognized that salt effects in some areas may explain higher velocity derived uplift values from the Bacton compared to those derived from vitrinite reflectance in the Carboniferous. The explanation that higher Carboniferous vitrinite derived uplift compared to velocity derived uplift in the Bacton was due to Hercynian uplift (i.e. maximum burial was obtained in the Carboniferous) is less convincing. One area, reported to have higher values of Carboniferous uplift than Bacton uplift is close to the Hewett gas field. Carboniferous subcrop maps (Leeder & Hardman 1990; Cameron *et al.* 1992) show the occurrence of Stephanian Barren Red Measures in this area, making it difficult to imagine the removal of a thickness in excess of the current post-Carboniferous section during the Hercynian Orogeny. A possible explanation is that these differences could be explained by higher geothermal gradients which increased the vitrinite reflectance values but are not recorded in the sonic velocities.

Cope's 'area of anomalous data' on the East Midlands Shelf can be partly explained by background uplift in the order of 1 km as later indicated by apatite fission track analysis. Increased heat flow in the Carboniferous due to the earlier emplacement of the Wash Granite could also contribute to this anomaly.

For the purposes of this study several refinements have been made to the technique employed by Cope. These include (i) the adoption of better 'zero uplift' reflectance curves, (ii) the recognition that geothermal gradients are variable over the basin, and (iii) the use of a consistent 'Top Carboniferous' reflectance value for each well.

By using a single 'zero uplift' line through wells considered to be currently at their maximum depth of burial, Cope's calculation of uplift for those wells that fall to the right of this line is based on their deviation from the line (Cope 1986; fig. 2). This line represents a single geothermal gradient. As the stratigraphic succession in the Southern Gas Basin displays both vertical and lateral variability in lithology the heat flow and geothermal gradient throughout the basin are likely to have been variable (Oxburgh & Andrew-Speed 1981).

In a refinement of Cope's technique a series of %Ro vs depth curves were calculated based on a Burnham & Sweeney (1989) algorithm for different geothermal gradients assuming continued burial (Fig. 10). The %Ro vs depth data were then plotted for each well and the range of uplift values within a reasonable range of geothermal gradients were determined by the displacement of each point away from the reference line. By integrating areal trends, the geology and the range of uplift values from the other techniques a 'best case' geothermal gradient reference line for each well was established from which the uplift was calculated. In addition, the top Carboniferous vitrinite reflectance value for each well was derived by extrapolating the %Ro vs. depth trend to the Top Carboniferous (Fig. 9). The values derived in this way exclude the influences of leaching on the Variscan unconformity that can

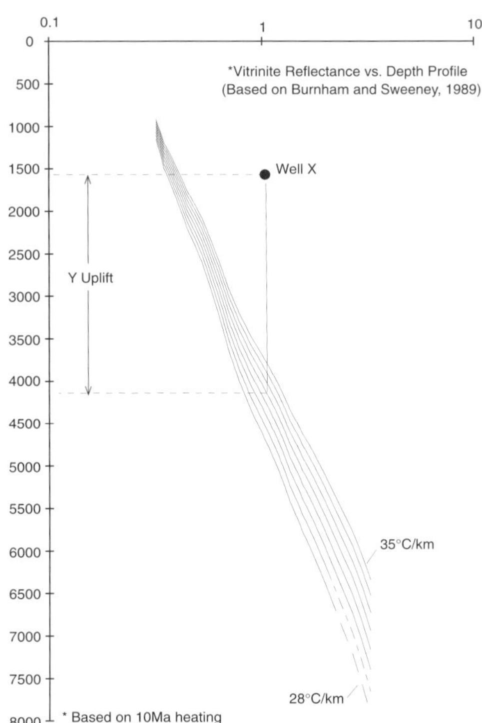

Fig. 10. Uplift calculation method from vitrinite reflectance in Carboniferous. Derivation of uplift. See text for discussion.

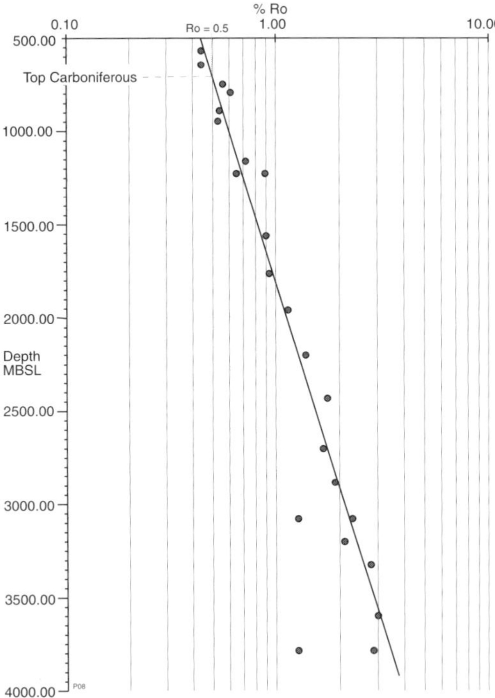

Fig. 9. Uplift calculation method from vitrinite reflectance in Carboniferous. Calculation of Top Carboniferous Ro value.

affect %Ro determinations at the top of the Carboniferous section.

The uplift values calculated using this methodology give an estimate of the tectonic uplift that has directly affected the Carboniferous section. As the uplift is measured directly from the Carboniferous it is unaffected by salt movement.

Apatite fission track analysis

The use of apatite fission track analysis has become commonplace in the Southern Gas Basin in recent years. Apatite fission track analysis provides estimates of both the timing of cooling from maximum palaeotemperature and in ideal circumstances, the 'palaeogeothermal gradient' at the time of maximum palaeotemperature. This information is used to interpret the amount of uplift and erosion. When complemented by vitrinite reflectance data it can also provide information on palaeogeothermal gradients and the style of cooling from maximum palaeotemperature.

In the Southern North Sea, apatite fission track analysis has been used to distinguish between different heating scenarios, particularly whether high palaeotemperatures prior to Laramide uplift were due to (i) increased heatflow or (ii) heating due to increased burial. This differentiation is made by a comparison of present day geothermal gradients and palaeogeothermal gradients. Green *et al.* 1993 (and references therein) used evidence from apatite fission track analysis to conclude that geothermal gradients prior to Laramide uplift were approximately the same as present day. The interpretation of apatite fission track analysis data has therefore played a large part in establishing the fact that elevated temperatures during the Late Cretaceous in the Southern North Sea were due to regional burial and subsequent uplift under 'normal' geothermal gradients and not due to a local source of heat, such as igneous activity (Green *et al.* 1993). The methodology used in the palaeogeothermal gradient estimation is shown in Fig. 11.

One drawback of apatite fission track analysis is that it provides data over a limited temperature range. In most of the Southern Gas Basin temperatures are currently too high in the Rotliegendes and Carboniferous section to permit analysis as all fission tracks are annealed above 110 °C. Furthermore, temperatures relating to the main stage of gas generation are higher than the annealing temperature for apatite. Therefore, in most areas of the Southern Gas Basin, maximum temperatures in the Carboniferous source rock cannot be derived using apatite fission track analysis alone.

High temperatures in the Carboniferous mean that sampling for apatite fission track analyses in many areas of the Southern Gas Basin are restricted to the post-Zechstein section. The palaeotemperatures derived from the analyses are valid in terms of calibrating thermal models but uplift calculations are often unreliable for three main reasons:

(i) measurement of present day temperature is often unreliable ;
(ii) temperature gradients are assumed to be linear;
(iii) interpretation of palaeogeothermal gradients and uplift often takes no account of Zechstein salt movement.

Temperature data from exploration wells rarely allow us to constrain temperatures accurately. This

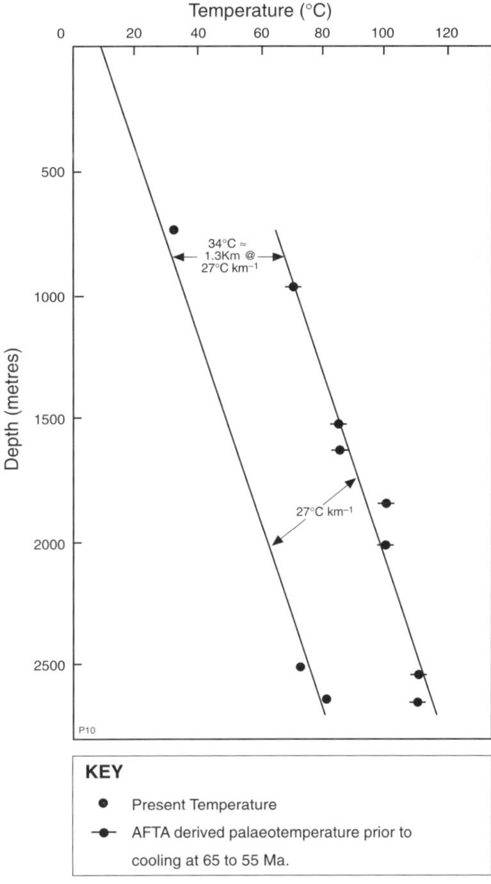

Fig. 11. Derivation of uplift and palaeogeothermal gradient from AFTA. Gradient prior to cooling is roughly equivalent to that at present day; hence cooling due to uplift is implied. Uplift calculated by displacement of palaeotemperature trend to present day temperature profile.

is primarily due to the fact that measured temperatures recorded from logging runs under-estimate true formation temperatures as a result of the cooling effect of the circulating drilling mud. No definitive correction method for this effect exists. Poor temperature recording practices and sparse measurements add to the poor quality of the data. A variety of correction methods can be applied to attempt to ascertain true formation temperature but even the most sophisticated and established such as the Horner Plot method can give false comfort as the data required to perform the correction are usually incomplete. Furthermore, techniques that are proven to give realistic corrections in some circumstances can be unrealistic in others. For example, a good suite of drill stem test derived temperatures can provide a very detailed and accurate profile of formation temperature, but when

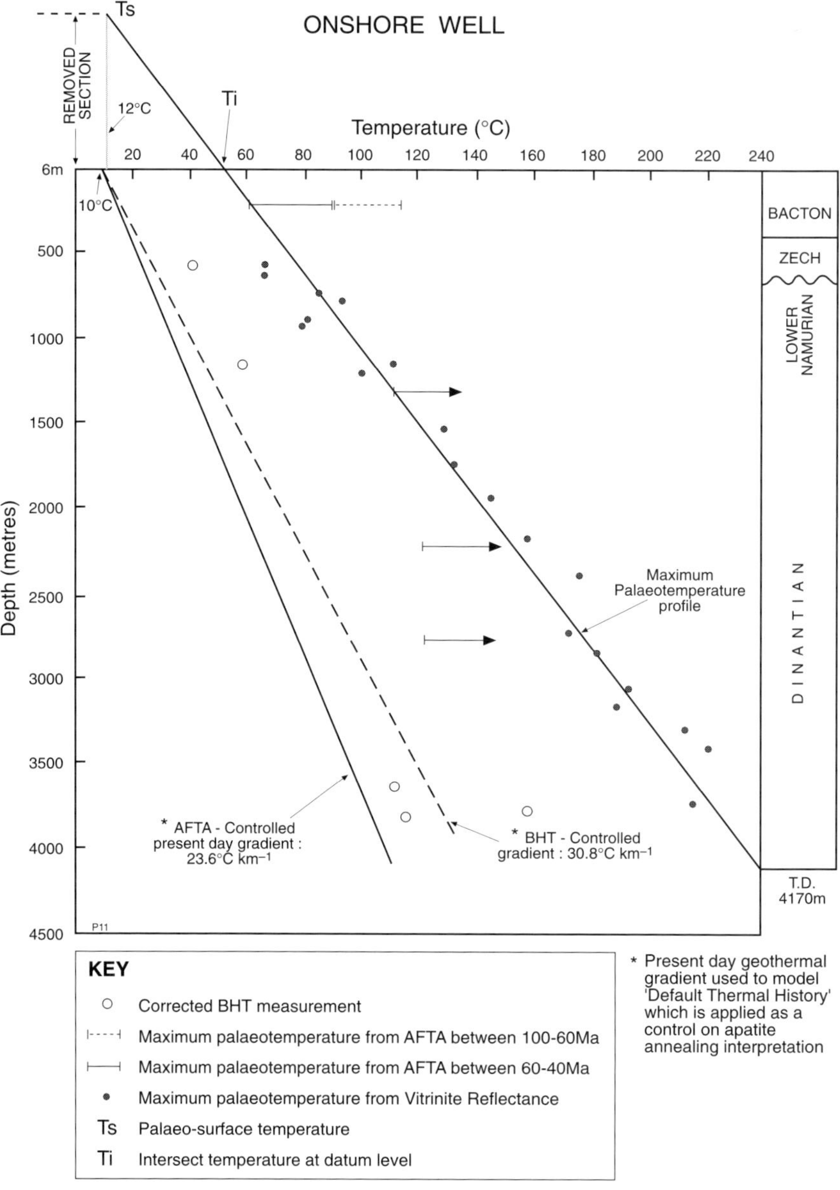

Fig. 12. Derivation of uplift from AFTA and vitrinite refectance for an onshore UK well.

taken, these are largely restricted to the reservoir sections in the pre-Zechstein. Data from the Zechstein salt and post-Zechstein sections where heat flow is likely to be variable are restricted to the logging run measurements. The end result is a poor estimate of present day temperatures. A technique such as apatite fission track analysis which uses present day temperature data as a fundamental part of its interpretation is therefore affected.

The interpretation of uplift from apatite fission track analysis uses a comparison of geothermal gradients that are linear. However, an examination of temperature profiles from the Southern North Sea clearly illustrates that present day temperature profiles cannot be described in terms of linear gradients. Temperature gradients vary with depth and are often described as 'dog leg'. This phenomenon is thought to be real and not an artefact resulting from post-drilling influences, as drill stem test temperatures follow the same profile. Non-linear temperature gradients are an understandable feature considering the contrast in thermal conductivity of sediments (for example halite and shale). Most modelling software can accommodate these variations and there is usually no need to describe the temperature in terms of a single geothermal gradient in the model calibration process. Caution must be exercised when using temperature gradients in the derivation of uplift as the non-linearity can lead to inaccurate extrapolations.

Perhaps the greatest possible error in apatite fission track analysis derived uplift is due to failure to separate differential uplift of the pre-Zechstein and post-Zechstein sections in areas of salt movement. For example, a Bacton sample will give an incorrect measurement of pre-Zechstein (basement) uplift in an area of salt movement, whilst a Rotliegendes or Carboniferous sample will be unaffected. If this effect is not recognized, uplift values calculated from apatite fission track analysis can be spuriously high, resulting in inaccurate interpretations of the burial history of the Carboniferous and Rotliegendes. This situation is illustrated in the example wells in the following section.

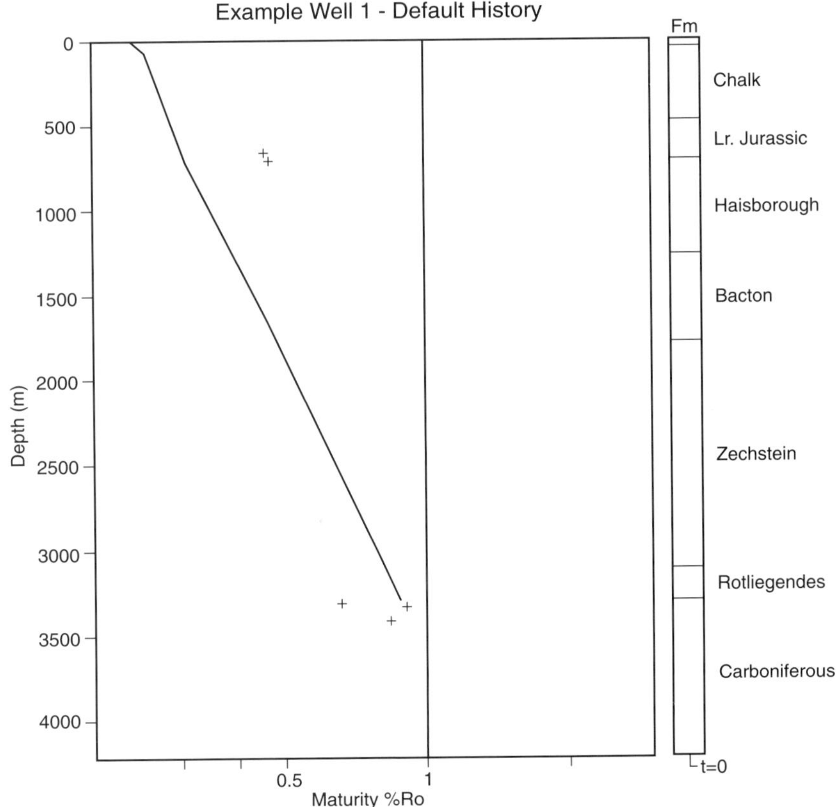

Fig. 13. %Ro versus depth plot — example well 1.

The descriptions above show that no one technique can be used to provide definitive values of uplift. In each well all available data must be used to constrain uplift with values from each technique weighted depending on the stratigraphic position of the sample and the area of the basin being studied. Particularly important is the integration of vitrinite reflectance data with apatite fission track analysis (Fig. 12) to define more clearly the palaeotemperature gradient.

Examples

Three examples are shown to illustrate how apparently conflicting datasets can be reconciled. The wells are taken from the UK Quadrants 43 and 44 (Fig. 1) and all are from areas of salt movement.

Example well 1

Example well 1 shows a case in southern Quad 43 where apatite fission track analysis and vitrinite reflectance data are available. Apatite fission track analysis samples were taken from the Bacton Group, and vitrinite reflectance samples from the Lower Jurassic (above the Zechstein salt) and Westphalian sections.
The %Ro – depth plot (Fig. 13) shows measured reflectance data and a predicted vitrinite reflectance curve assuming continued burial 'default thermal history'. It is apparent from this plot that this burial scenario gives a poor fit to the Jurassic %Ro values, but a reasonable fit to the Carboniferous %Ro data. Some uplift is suggested by the stratigraphy as the Cretaceous is almost at the surface. Two scenarios would explain this dataset: either the thermal gradient was much lower at the time of maximum temperature or the Jurassic section has experienced greater uplift relative to the Carboniferous.

The apatite fission track analysis data from the Bacton (Fig. 14) also indicate higher palaeotemperatures in the past. A traditional interpretation of the apatite fission track analysis and vitrinite reflectance data, assuming uplift at the Laramide event, resulted in an uplift estimate of 2500 m. This necessitated the use of a low palaeogeothermal gradient of 19.7 °C km^{-1} compared to the present day value of 33.7 °C km^{-1}. A second, more likely

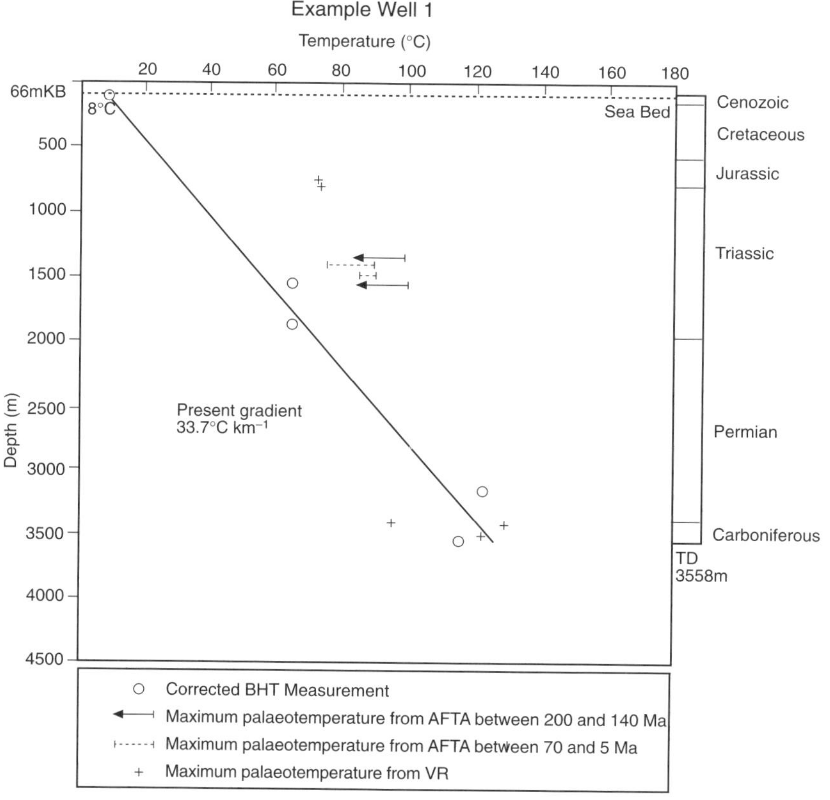

Fig. 14. Bacton apatite fission track analysis — example well 1.

scenario is that the Bacton and Jurassic samples experienced much greater uplift relative to the Carboniferous due to the emplacement of a salt dome. This observation is supported by a seismic line through the well. A tectonic (basement) uplift value of 810 m was derived from the %Ro data and a salt thickening of some 1000 m is predicted. A burial history plot for the well is shown as Fig.15.

The two scenarios obviously have very different implications, not only for the maximum depth of burial suffered by the Carboniferous source and reservoir section and the Rotliegendes reservoir, but also for the prediction of maturity at maximum depth of burial (due to the different geothermal gradients that are required).

Example well 2

Example 2 is a well in southern Quad 44 which is drilled in a low between salt diapirs. It is considered to have maintained a fairly constant salt thickness from deposition. Uplift data were available from apatite fission track analysis, vitrinite reflectance and sonic velocity techniques.

Maturities predicted from the apatite fission track derived 'default burial history' (Fig. 16) over-estimate the %Ro values measured in the Carboniferous. In this well, the geothermal gradient calculated from present day temperatures (36.8 °C km^{-1}) has been over-estimated due to poorly constrained temperature data. Based on areal trends a more realistic gradient is in the order of 32 °C km^{-1}. Incorporating the lower gradient gives a better fit to observed %Ro values in the Carboniferous and a small degree of uplift is suggested.

Apatite fission track analysis data (Fig. 17) are consistent with a burial history that incorporates a small degree of uplift at the Laramide unconformity and sonic velocity techniques also support this interpretation. Uplift in this case is regional tectonic uplift that affects both the pre-salt and post-salt sections. (Fig. 18).

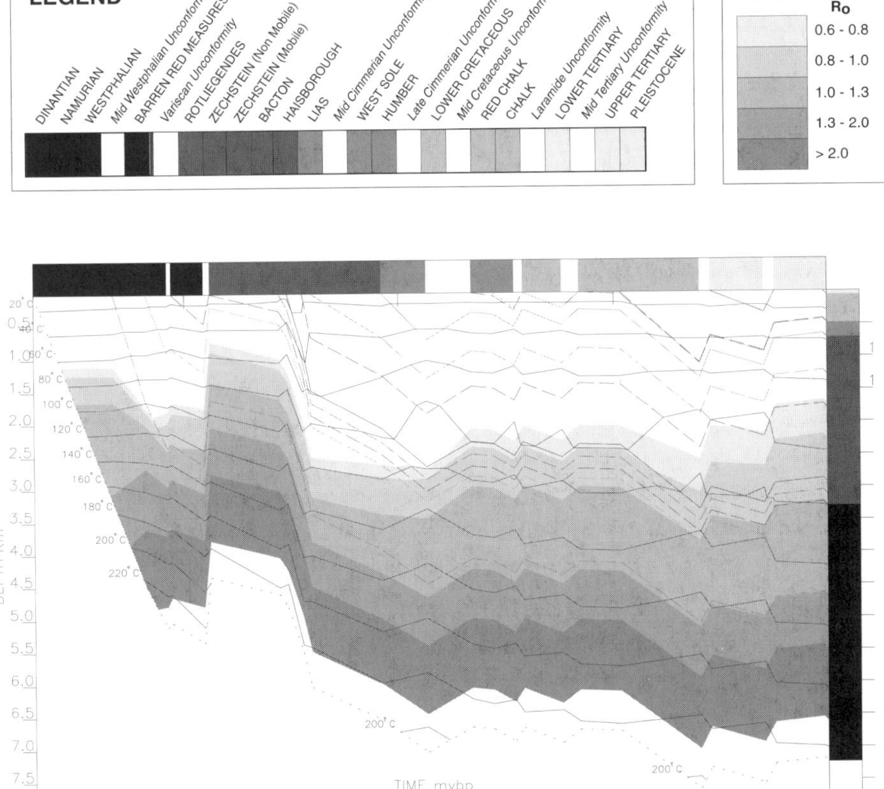

Fig. 15. Burial history plot — example well 1.

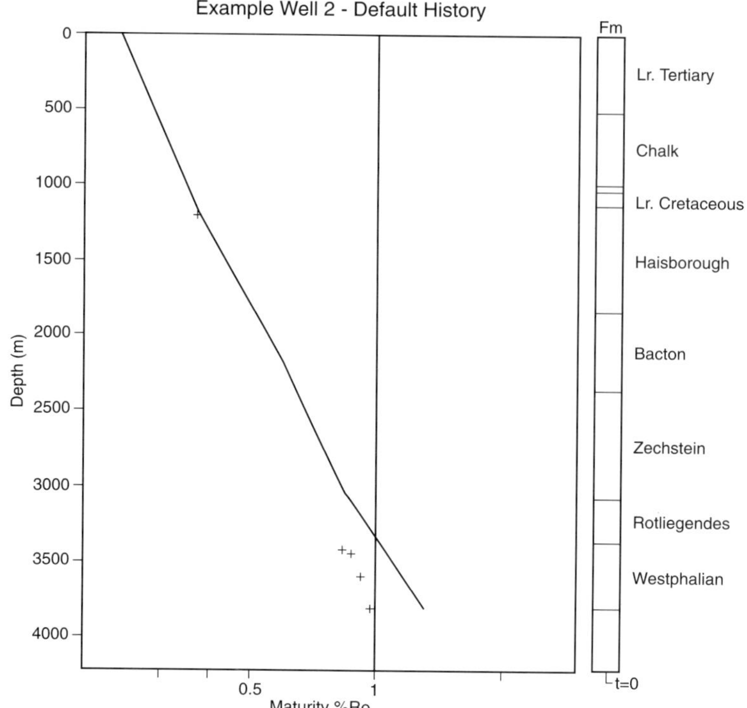

Fig. 16. %Ro versus depth plot — example well 2.

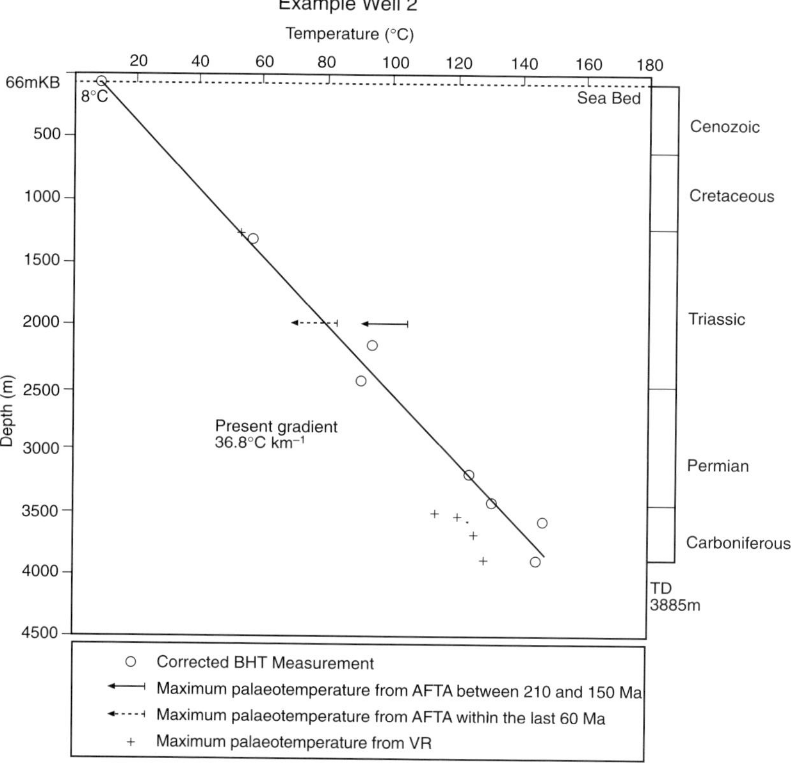

Fig. 17. Apatite fission track analysis — example well 2.

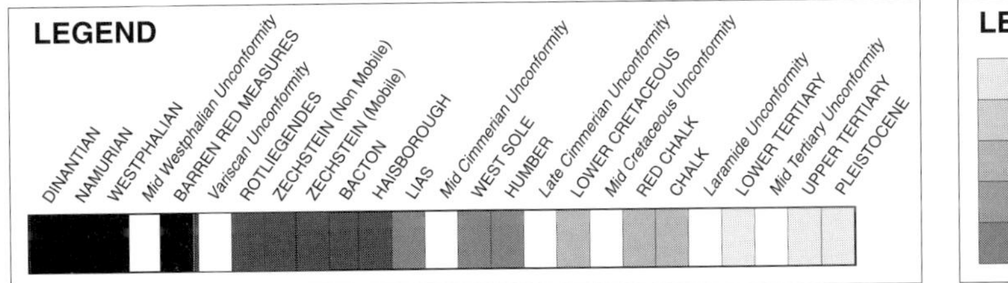

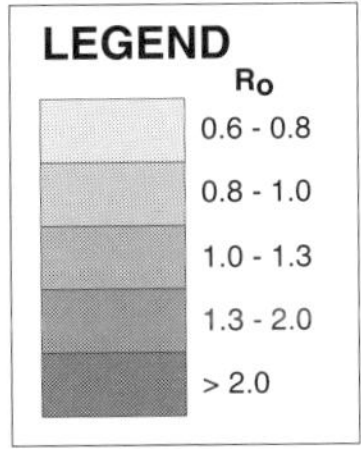

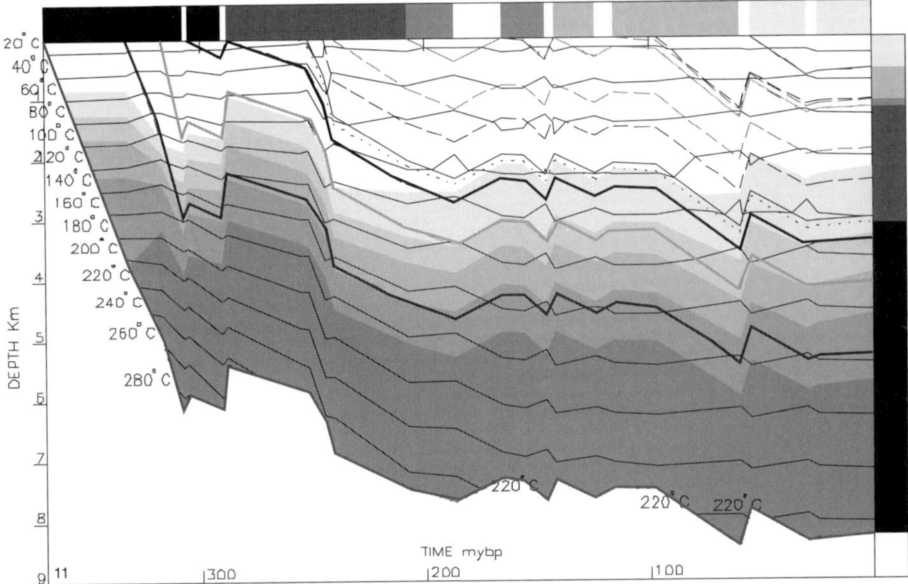

Fig. 18. Burial history plot — example well 2.

Example well 3

Example well 3 is located in northern Quad 43, in an area where Zechstein salt has thickened through time. Due to its position on the margin of the Southern Gas Basin, the well is one of the few where temperatures have always been low enough in the Rotliegendes to permit an apatite fission track analysis sample to be taken at this level.

A comparison of measured vitrinite reflectance values from the Dinantian and predicted values based on the present day thermal regime suggests that the Carboniferous section is at or near its maximum depth of burial at the present day (Fig. 19). The one apatite fission track analysis sample from the Carboniferous suggests maximum temperatures were slightly higher in the past (Fig. 20), although this is consistent with uplift in the order of 200 m assuming a thermal gradient comparable to the present day value.

In contrast, sonic velocity determinations of uplift from the post-salt section are significantly greater, suggesting uplift in the order of 700 m. It is clear that this value may be appropriate for the uplift suffered by the post-Zechstein section, but is not appropriate for the Rotliegendes and Carboniferous. Again the uplift of the post-salt section is a combination of tectonic ≈200 m) and salt-related (≈500 m) uplift. A burial history plot for the well is shown as Fig. 21.

Conclusions

The quantification of uplift in the Southern North Sea is an integral part of thermal history reconstruction. We have shown that no one technique can be relied upon to give absolute values of uplift in all cases. Instead, a combination of the different techniques; apatite fission track analysis, vitrinite reflectance and sonic velocities must be used and integrated with stratigraphic

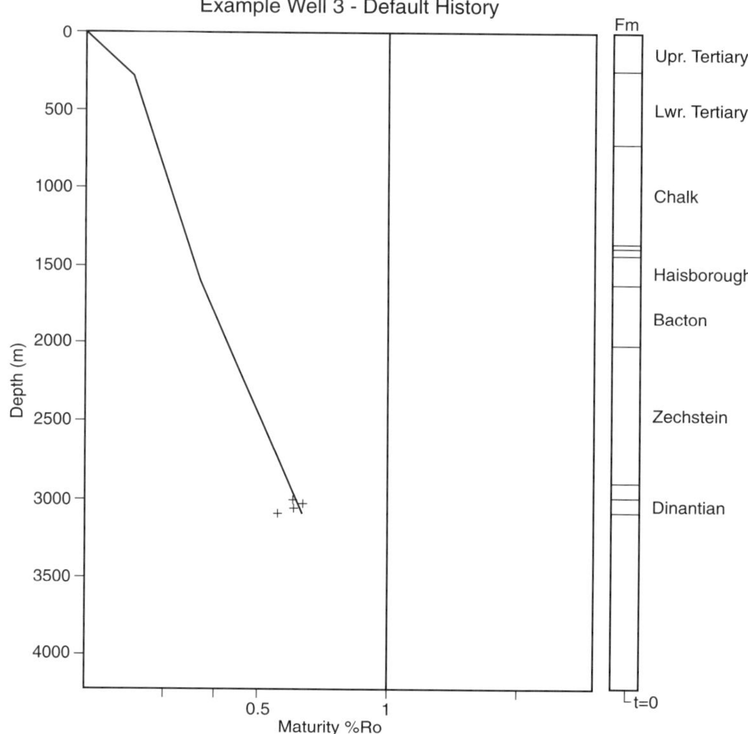

Fig. 19. %Ro versus depth plot — example well 3.

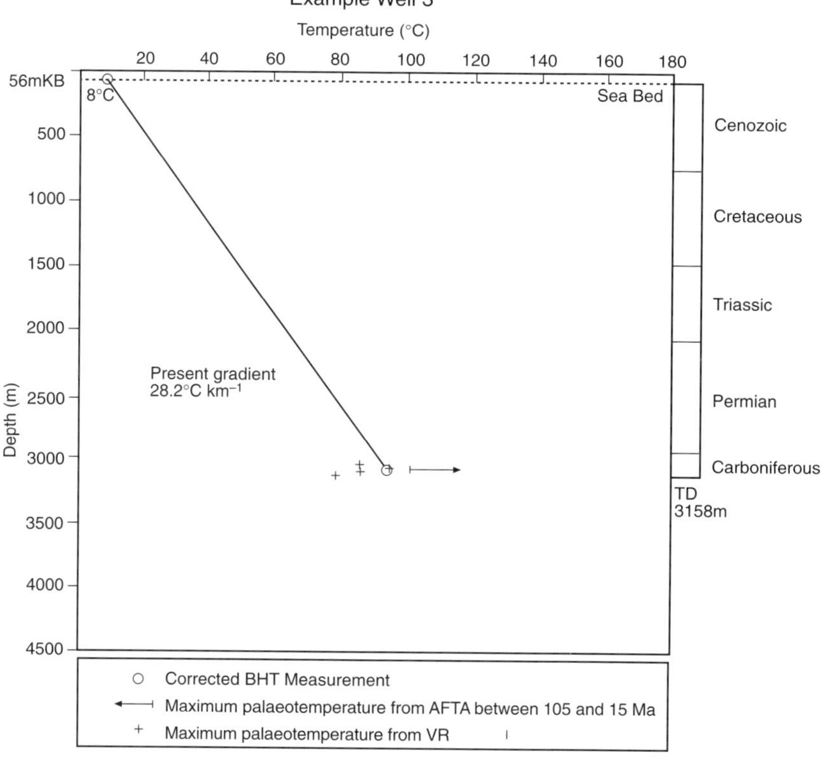

Fig. 20. Apatite fission track analysis — example well 3.

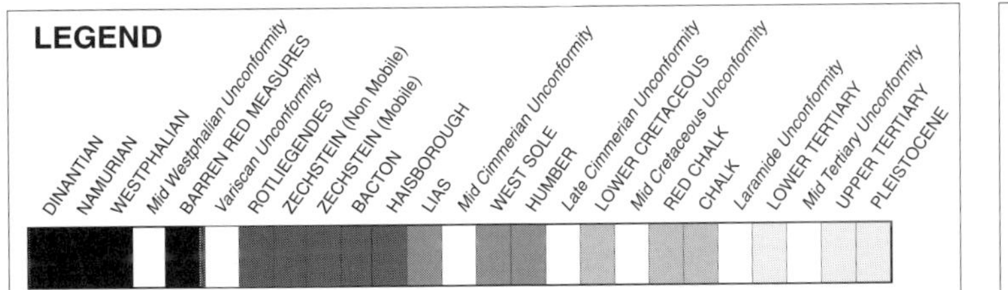

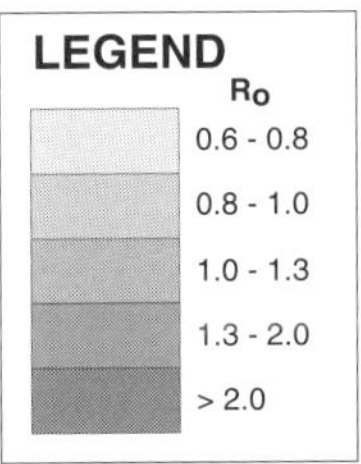

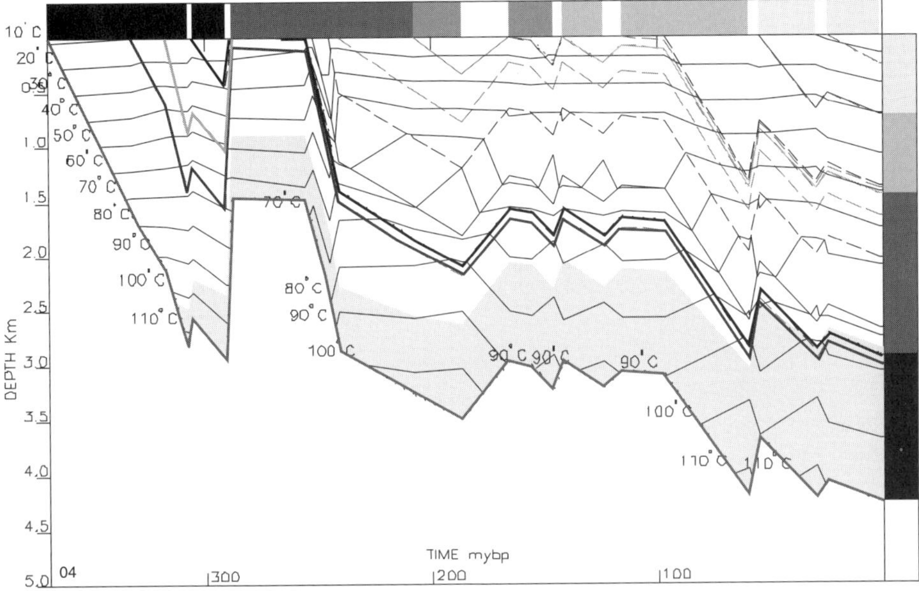

Fig. 21. Burial history plot — example well 3.

analysis and seismic data. Careful thought must be used in formulating sampling programmes to ensure samples and data are acquired from horizons that will give meaningful results. Of particular importance is the need to recognize that differential uplift will occur in areas of salt movement with the post-salt sections experiencing uplift patterns that are dissimilar to those of the pre-salt section. Stratigraphic analysis and seismic data should be used to ensure that uplift determinations obtained from wells are geologically realistic in terms of the structure of the basin.

The seismic section illustrated comes from a non-exclusive survey jointly owned by GeoQuest and Geco-Prakla (UK) Limited. Permission to use this data is gratefully appreciated.

Apatite fission track analysis (AFTA) is a trademark of Geotrack International.

References

BARNARD, P. C. & COOPER, B. S. 1983. A review of geochemical data related to the Northwest European gas province. *In*: BROOKS, J. (ed.) *Petroleum geochemistry and exploration of Europe*. Geological Society, London, Special Publications, **12,** 19–23.

BULAT, J. & STOKER, S. J. 1987. Uplift determination from interval velocity studies, UK Southern North Sea. *In*: BROOKS, J. & GLENNIE, K. (eds) *Petroleum Geology of North West Europe*. Graham and Trotman, London, 293–305.

BURNHAM, A. K. & SWEENEY, J. J. 1989. A chemical kinetic model of vitrinite maturation and reflectance. *Geochimica et Cosmochimica Acta,* **53**, 2649–2657.

CAMERON, T. D. J. ET AL. 1992. *The geology of the southern North Sea*. United Kingdom Offshore Regional Report. London HMSO.

COPE, M. J. 1986. An Interpretation of Vitrinite Reflectance Data from the Southern North Sea. *In*: BROOKS, J., GOFF, J. C. & VAN HOORNE, B. (eds) *Habitat of Palaeozoic gas in N.W. Europe*. Geological Society, London, Special Publications, **23**, 85–98.

GLENNIE, K. W. & BOEGNER, P. L. E. 1981. Sole Pit inversion tectonics. *In*: IILLING, L. V. & HOBSON, G. D. (eds) *Petroleum Geology of the Continental Shelf of North-west Europe*. The Institute of Petroleum, London, 110–120.

GREEN, P. F. 1989. Thermal and tectonic history of the East Midlands shelf (onshore UK) and surrounding regions assessed by apatite fission track analysis. *Journal of the Geological Society, London*, **146**, 753–773.

——, DUDDY, I. R., BRAY, R. J. & LEWIS, C. L. E. 1993. Elevated palaeotemperatures prior to early Tertiary cooling throughout the UK region: implications for hydrocarbon generation. *In*: PARKER, J. R (ed.) *Petroleum Geology of Northwest Europe: Proceedings of the 4th Conference*. Geological Society, London, 1067–1074.

HEMINGWAY, J. E. & RIDDLER, G. P. 1982. Basin inversion in North Yorkshire. *Transactions of the Institute of Mining and Metallurgy*, **91**, 175–186.

HILLIS, R. R. 1993. Tertiary erosion magnitudes in the East Midlands Shelf, onshore UK. *Journal of the Geological Society, London*, **150**, 1047–1050.

KENT, P. E. 1980. Subsidence and uplift in East Yorkshire and Lincolnshire: a double inversion. *Proceedings of the Yorkshire Geological Society*, **4**, 505–524.

LEEDER, M. R. & HARDMAN, M. 1990. Carboniferous of the southern North Sea Basin and controls on hydrocarbon prospectivity. *In*: HARDMAN, R. F. P. & BROOKS, J. (eds) *Tectonic Events Responsible for Britain's Oil and Gas Reserves*. Geological Society, London, Special Publications, **55**, 97–106.

MARIE, J. P. P. 1975. Rotliegendes stratigraphy and diagenesis. *In*: WOODLAND, A. W. (ed.) *Petroleum and the Continental Shelf of North-west Europe, Volume 1, Geology*. Applied Science Publishers, 205–211.

OXBURGH, E. R. & ANDREWS-SPEED, C. P. 1981. Temperature, thermal gradients and heat flow in the south west North Sea. *In*: IILLING, L. V. & HOBSON, G. D. (eds) *Petroleum Geology of the Continental Shelf of North-west Europe*, Institute of Petroleum, London, 114–151.

ZIEGLER, P. A. 1982. *Geological Atlas of Western and Central Europe*, Shell, Masschaapptt b.v.

Analysis of erosion events and palaeogeothermal gradients in the North Alpine Foreland Basin of Switzerland

R. SCHEGG[1] & W. LEU[2]

[1] *Department of Geology and Palaeontology, University of Geneva, 13, rue des Maraîchers, CH-1211 Geneva 4, Switzerland (Present address: Petroconsultants SA, Chemin de la Mairie 24, 1258 Perly, Switzerland)*

[2] *Geoform, Geological Consulting and Studies Ltd, Anton Graff-Str. 6, CH-8401 Winterthur, Switzerland*

Abstract: For basin modelling purposes, uplift and erosion analysis is as important as subsidence analysis. Within the North Alpine Foreland Basin of Switzerland (NAFB) three major interregional unconformities separate the sedimentary succession: the late Palaeozoic, the base Tertiary and the base Quaternary unconformity. The amount of missing section depends on the palaeogeographic basin position and on different inversion mechanisms (thermal uplift, intraplate stress, tectonic uplift or isostatic rebound movements). Several techniques exist to determine the magnitude and timing of maximum palaeotemperatures and palaeogradients, facilitating direct estimation of missing section. In our study coalification and fission track data are analysed and compared with recently published models concerning erosion in the NAFB. The results indicate that each inversion episode is accompanied by significant erosion (up to several km). The following range of removed section values were reconstructed:

(i) late Palaeozoic unconformity: 1000–1200 m;
(ii) base Tertiary unconformity: 800–1800 m;
(iii) base Quaternary unconformity: 1500–3000 m in the SW, up to 700 m in the NE and 4100–4400 m in parts close to the Alpine front.

The reconstruction techniques provide only rough estimates as each approach is based on certain assumptions and calibration uncertainties. Also, the obtained values are higher than those expected from simple evaluation of the preserved stratigraphic record. Calculated palaeogeothermal gradients are very high (80–90 °C km^{-1}) during the late Palaeozoic, slightly higher than normal during the late Cretaceous (30–50 °C km^{-1}) and lower than normal (15–30 °C km^{-1}) during the Cenozoic.

When analysing the present-day stratigraphic record for the purpose of geohistory reconstructions, the geoscientist is in most cases constrained to make some reasonable assumptions on the amount of missing section in unconformities. The preserved rocks reflect only a part of the geological history. The driving mechanisms which control the sedimentary budget (e.g. deposition or erosion) are a complex function of tectonic processes initiating subsidence and uplift, sea-level changes, rock properties, palaeotopography and palaeoclimate.

The increasing popularity of numerical basin modelling tools in hydrogeological and petroleum exploration studies requires a continuous quantification of the burial and uplift history. Only reasonable estimates on the missing thicknesses and lithologies, including their petrophysical properties, allow a meaningful reconstruction of the evolution of compaction trends, fluid flow regimes and thermal gradients or heat flow. A thorough understanding of the pore pressure history, the sequence of diagenetic events and hydrocarbon migration paths is only possible if geologically reasonable assumptions for the time gap represented in major unconformities are made. A successful evaluation of hydrocarbon plays may have to consider potential source rock intervals that were eroded at a later stage or missing sections for the structural reconstruction of cross-sections. The impact of erosion estimates on the results of basin models may be severe, starting already in conventional 1-D analysis (Fig. 1). Different erosion estimates modify the porosity, temperature, maturation and generation history of potential source rocks.

Within the North Alpine Foreland Basin of Switzerland (NAFB) three major interregional unconformities separate the sedimentary succession: the Late Palaeozoic, the base Tertiary and the

SCHEGG, R. & LEU, W. 1998. Analysis of erosion events and palaeogeothermal gradients in the North Alpine Foreland Basin of Switzerland. *In*: DÜPPENBECKER, S. J. & ILIFFE, J. E. (eds) *Basin Modelling: Practice and Progress*. Geological Society, London, Special Publications, **141**, 137–155.

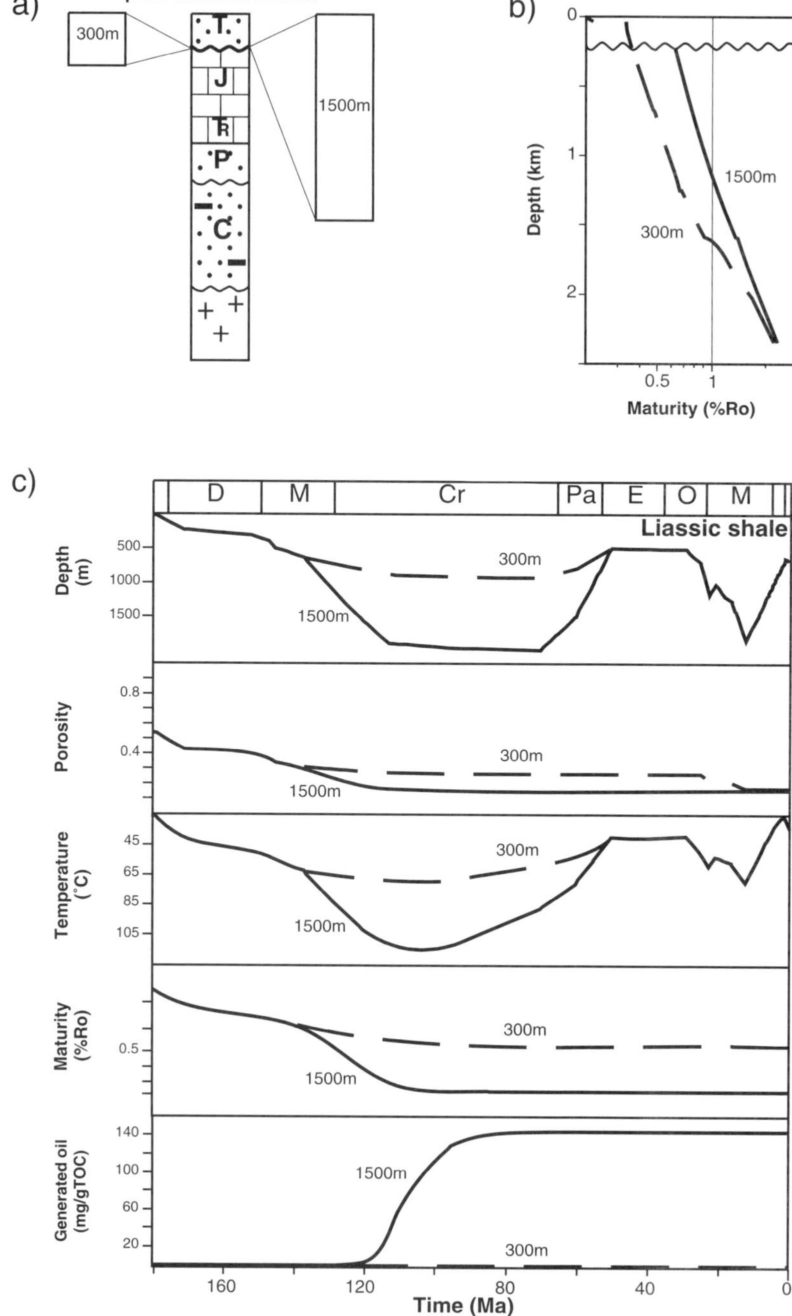

Fig. 1. 1-D basin model for well Weiach with (a) two erosion scenarios of 300 or 1500 m of Jurassic/Cretaceous missing section (all other variables are kept constant), (b) resulting present-day maturity profile; (c) the corresponding burial, porosity, temperature, maturity and oil generation history for a Liassic shale.

base Quaternary unconformity. They represent major inversion and erosion events. The purpose of this paper is a review of postulated burial and erosion models for these three main unconformities in the NAFB (Lemcke 1974; Monnier 1982; Kempter 1987; Brink *et al.* 1992; Kälin *et al.* 1992). In addition to these interpretations, we applied several independent methods on new data from Swiss wells to reconstruct the missing overburden and palaeogeothermal gradients. We want to highlight the uncertainties in such reconstructions and evaluate their geological implications.

Geological setting

The study area covers the Swiss part of the North Alpine Foreland Basin (NAFB). The Swiss Molasse basin is limited to the northwest by the Jura Mountains and to the southeast by the Alps (Fig. 2). It represents a peripheral foreland basin overlying thickened lithosphere (Pfiffner 1986). Its sedimentary fill consists of a southeastward expanding wedge, up to 5000 m thick, of predominantly Tertiary clastics (Homewood *et al.* 1986; 1989). The Molasse Basin (s.s.) started to form in early Oligocene and its sedimentary infill can be subdivided into four lithostratigraphic groups: Lower Marine Molasse (UMM, Rupelian–Chattian), Lower Freshwater Molasse (USM, Rupelian–Burdigalian?), Upper Marine Molasse (OMM, Burdigalian–Langhian?) and Upper Freshwater Molasse (OSM, Langhian–Serravalian). The Molasse rests unconformably on truncated Mesozoic shelf sediments ranging in thickness between 1 and 2 km (Fig. 3). Mesozoic carbonates, shales and clastic rocks overlie a Hercynian basement complex and more locally, deep Permo–Carboniferous grabens (Müller *et al.* 1984; Laubscher 1987; Diebold *et al.* 1991; Gorin *et al.* 1993).

Within the sedimentary succession of the NAFB, three major interregional unconformities are recognized (Fig. 4):

(i) the late Palaeozoic unconformity (Autunian or Saalian);
(ii) the base Tertiary unconformity (late Jurassic–Eocene);
(iii) the base Quaternary unconformity (late Miocene–Pliocene).

The amount of missing section and the related time gap depends on the palaeogeographic position within the basin and on the inversion mechanisms (thermal uplift, intraplate stress, tectonic regime or isostatic rebound).

Methods for estimating amounts of missing section

Comprehensive reviews of commonly applied methods for missing section reconstructions have

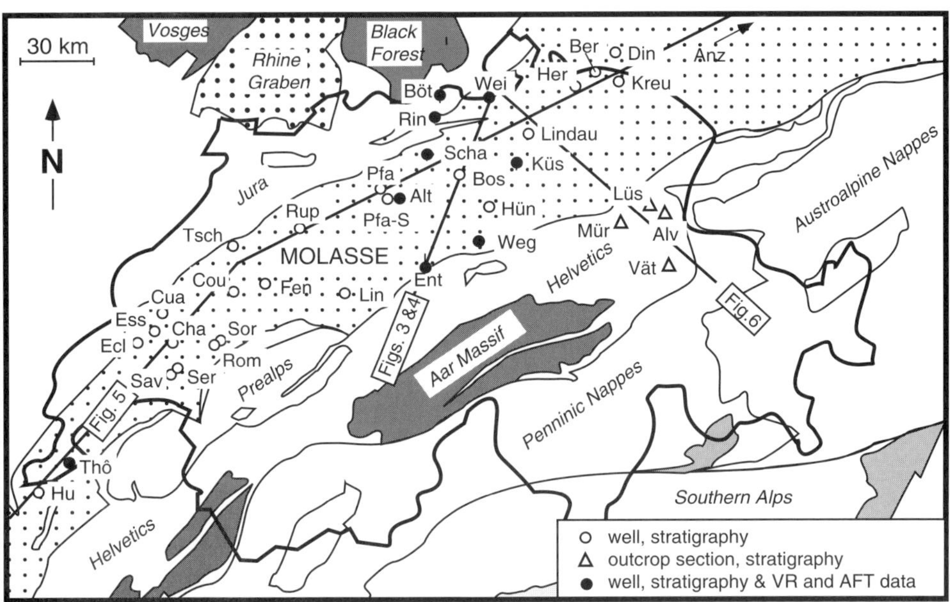

Fig. 2. Simplified tectonic map of Switzerland showing data base of the study and profile lines of Figs 3, 4, 5 and 6. Well abbreviations are explained in Table 1.

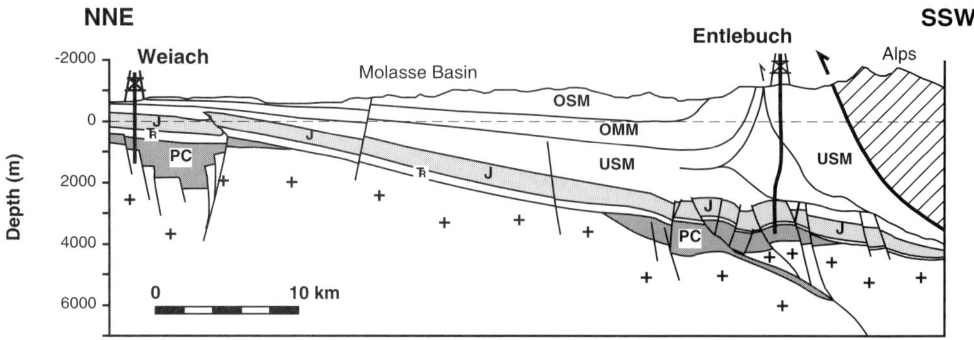

Fig. 3. Simplified geological NE–SW profile through the NAFB modified from Vollmayr & Wendt (1987). See Fig. 2 for location.

been summarized recently by Armagnac *et al.* (1989), Duddy *et al.* (1991) and Løseth *et al.* (1992).

The conventional method for the reconstruction of missing section in unconformities is based on a direct evaluation of the preserved stratigraphic record. Geographical isopach trends of preserved deposits below the erosional unconformity on a basin-wide scale may help to define depocenters and associated maximum thicknesses (≈maximum erosion amount) and to distinguish regional thickness trends (e.g. primary thickness decrease/increase) from erosional thickness reduction. However, one should be aware that the preserved strata may have suffered substantial compaction since the erosion event (reburial of unconformity). Observed thicknesses have therefore to be restored to their pre-erosional situation (decompaction).

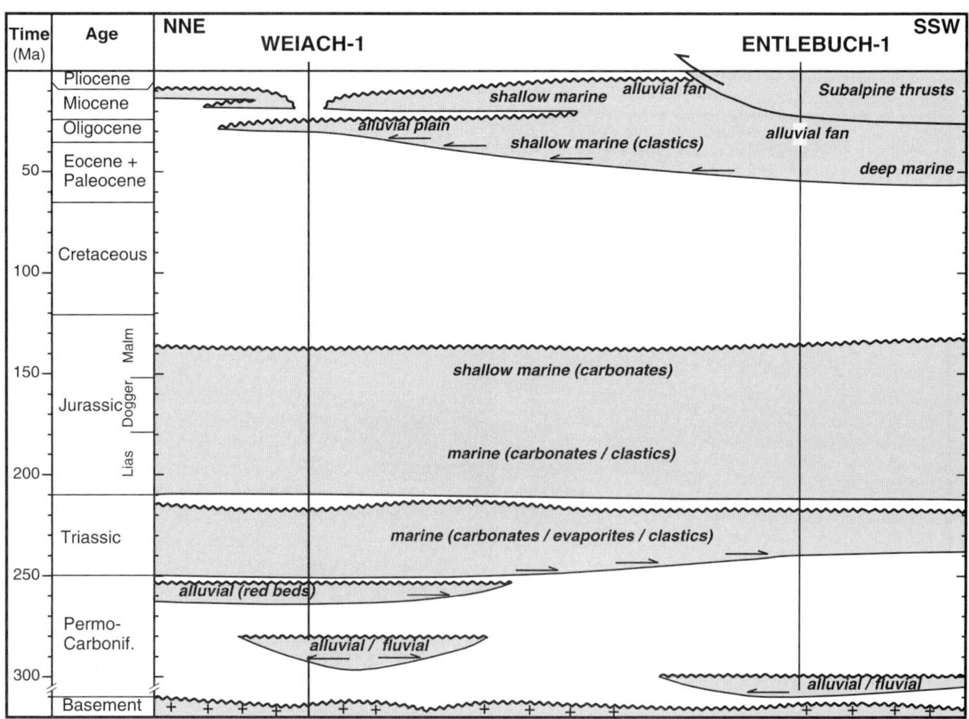

Fig. 4. Chronostratigraphic NE–SW profile showing major sedimentary gaps in the NAFB.

Table 1. *Data base of the study*

well / outcrop* section	abbreviation	TD / thickness	formation at base	VR data	AFT data	references for stratigraphy	references for thermal indicators
Altishofen	Alt	2200	Triassic	4 T		Kopp (1952)	unpublished VR data (Schegg)
Alvier*	Alv	2500	Triassic			Funk (1985)	
Anzing 3	Anz	3300	basement	14 T		Teichmüller & Teichmüller (1975)	Teichmüller & Teichmüller (1975)
Berlingen	Ber	2300	Permian			Büchi *et al.* (1965)	
Boswil	Bos	1800	Malm			Lemcke *et al.* (1968)	Monnier (1982)
Böttstein	Böt	1500	basement	8 M	7 M	NAGRA (1985)	NAGRA (1985), Hurford (1993)
Chapelle	Cha	1500	Early Cretaceous			Lemcke (1959)	Monnier (1982)
Courtion	Cou	3100	Triassic			Fischer & Luterbacher (1963)	Monnier (1982)
Cuarny	Cua	2200	Dogger			Althaus (1947)	
Dingelsdorf	Din	2700	basement			Kämpfe (1984)	
Eclépens	Ecl	2200	Triassic			Vollmayr (1983)	
Entlebuch	Ent	5300	Permo-Carboniferous	11 M		Vollmayr & Wendt (1987)	unpublished VR data (BEW, SWISSPETROL)
Essertines	Ess	2900	Triassic			Büchi *et al.* (1965)	Monnier (1982)
Fendringen	Fen	2000	Lower Cretaceous			Maurer (1983)	
Herdern	Her	2200	basement			Vollmayr (1983)	
Humilly 2	Hu	3100	Carboniferous			Gorin *et al.* (1993)	
Hünenberg	Hün	3300	Malm			Lemcke *et al.* (1968)	Monnier (1982)
Kreuzlingen	Kreu	2600	Permo-Carboniferous			Büchi *et al.* (1965)	Monnier (1982)
Küsnacht	Küs	2700	Malm	4 T		Büchi *et al.* (1961)	Rybach & Bodmer (1980), Monnier (1982)
Lindau	Lindau	2400	basement			Büchi *et al.* (1965)	
Linden	Lin	5400	Triassic			Vollmayr (1983)	Monnier (1982)
Lüsis*	Lüs	1400	Lias			Funk (1985)	
Mürtschen*	Mür	1200	Triassic			Funk (1985)	
Pfaffnau Süd-1	Pfa-S	1200	Malm			Lemcke *et al.* (1968)	Monnier (1982)
Pfaffnau-1	Pfa	1800	basement			Büchi *et al.* (1965)	Monnier (1982)
Riniken	Rin	1800	Permian	11 M	1 M	Matter *et al.* (1987)	Matter *et al.* (1987), Hurford (1993)
Romanens	Rom	4000	Triassic			Vollmayr (1983)	
Ruppoldsried	Rup	1000	Malm			Schlanke *et al.* (1978)	Monnier (1982)
Savigny	Sav	2500	Lower Cretaceous			Lemcke (1963)	Monnier (1982)
Schafisheim	Scha	2000	basement	6 M	5 M	Matter *et al.* (1988a)	Matter *et al.* (1988a), Hurford (1993)
Servion	Ser	1400	Miocene			Althaus (1947)	
Sorens	Sor	3300	Lower Cretaceous			Favini (1970)	Monnier (1982)
Thônex	Thô	2700	Malm	9 T		Jenny *et al.* (1995)	Jenny *et al.* (1995)
Tschugg	Tsch	700	Malm			Schlanke *et al.* (1978), Vollmayr (1983)	Monnier (1982)
Vättis*	Vät	1200	Triassic			Funk (1985)	
Weggis	Weg	2300	Oligocene	5 T		Greber *et al.* (1994)	Schegg (1994)
Weiach	Wei	2500	basement	1 T, 7 M, 17 P	7 M	Matter *et al.* (1988b)	Matter *et al.* (1988b), Hurford (1993)

T = number of data from the Tertiary, M = number of data from the Mesozoic, P = number of data from the Palaeozoic.

Several indirect methods are based on the analysis of vitrinite reflectance (VR), clay mineral transformation, sonic velocity or porosity trends in wells. A regression line through the data points is projected to typical surface values or shifted to a so-called normal depth trend to determine the thickness of removed sediments (e.g. Magara 1976; Dow 1977; Majorowicz *et al.* 1990). The intercept of the logarithmic coalification gradient is commonly assumed to represent an estimate of the eroded thickness (Dow 1977). We used a zero coalification of 0.25 %Rr (Majorowicz *et al.* 1990). In our study, results from the extrapolation of coalification trends are referred to the abbreviation Log%Ro (Tables 2 and 3).

Refined procedures make use of the time/temperature-dependence and associated kinetic models for vitrinite reflectance (EASY%Ro, Sweeney & Burnham 1990) and fission track (Green *et al.* 1989; Duddy *et al.* 1991; Willett 1992), and of palaeothermometers such as homogenization temperatures of fluid inclusions (Barker & Pawlewicz 1986; 1994; Barker & Goldstein 1990).

Estimates of maximum palaeotemperatures determined by the mentioned methods over a range of depths in a vertical sequence provides the possibility of determining the palaeogeothermal gradient immediately prior to the onset of cooling from those maximum palaeotemperatures. The projection of such reconstructed palaeogeothermal gradients to reasonable surface temperatures enables the estimation of removed section.

The following techniques have been applied in this study to determine palaeotemperatures:

(i) EASY%Ro: kinetic modelling of vitrinite reflectance data with EASY%Ro (Sweeney & Burnham 1990);
(ii) AFT: kinetic modelling of apatite fission track data (e.g. Willett 1992);
(iii) B&P '94: correlation between peak temperature and vitrinite reflectance (Barker & Pawlewicz 1994).

Where available, VR and apatite fission track (AFT) data have been used in combination (EASY&AFT) for the composite reconstruction of the geothermal gradient beyond the limit of apatite fission track analysis (i.e. >110 °C). In our study, results from the extrapolation of geothermal gradients are referred to by the abbreviations EASY%Ro, AFT, B&P '94 and EASY&AFT, respectively (Tables 2, 3 and 4).

The kinetic modelling procedure has been carried out for each individual data point (VR or AFT) by assuming a simplified burial and temper-

Table 2. *Estimated erosion amounts for the base Quaternary unconformity*

Well	Log%Ro	+/-	EASY%Ro	+/-	B&P '94	+/-	Lemcke (1974)	Monnier (1982)	Brink *et al.* (1992)
Altishofen	4000	250	6300	200	5200	200	—	—	1500
Anzing 3	0	250	625	200	650	250	630	—	0
Berlingen	—	—	—	—	—	—	670	—	400
Boswil	—	—	—	—	—	—	—	<700	800
Chapelle	—	—	—	—	—	—	—	1300	2000
Courtion	—	—	—	—	—	—	1600	1690	1750
Essertines	—	—	—	—	—	—	2500	<2200	2600
Hünenberg	—	—	—	—	—	—	—	550	100
Kreuzlingen	—	—	—	—	—	—	—	<1300	200
Küsnacht	3800	500	9000	400	4600	400	—	500	300
Lindau	—	—	—	—	—	—	560	—	500
Linden	—	—	—	—	—	—	—	1100	0
Pfaffnau Süd-1	—	—	—	—	—	—	—	<1300	1600
Pfaffnau 1	—	—	—	—	—	—	1620	<1800	1700
Ruppoldsried	—	—	—	—	—	—	—	<1600	2100
Savigny	—	—	—	—	—	—	—	1550	1700
Sorens	—	—	—	—	—	—	—	1100	500
Thônex	2000	250	1750	250	2300	250	—	—	3000
Tschugg	—	—	—	—	—	—	—	<2000	2300
Weggis	4100	1500	4200	500	4400	450	—	—	0

All values in metres. Log%Ro: extrapolation of coalification trends; Easy%Ro: extrapolation of palaeogeothermal gradient, palaeotemperatures calculated combining VR data with the kinetic model of Sweeney & Burnham (1990); B&P '94: extrapolation of palaeogeothermal gradient, palaeotemperatures calculated combining VR data with the method of Barker & Pawlewicz (1994); Lemcke (1974): erosion estimates from subsidence analysis; Monnier (1982): erosion estimates from clay mineral transformation analysis; Brink *et al.* (1992): erosion estimates from seismic interval analysis. Erosion estimates by Monnier (1982) have been increased by 500 m as discussed in the text.

Table 3. *Estimated erosion amounts for the base Tertiary unconformity*

Well	Log%Ro	+/-	EASY%Ro	+/-	AFT	+/-	EASY&AFT	+/-	B&P '94	+/-	Brink *et al.* (1992)
Böttstein	1250	120	1100	100	1150	170	1700	220	1250	130	2500
Entlebuch	1000	150	1800	150	—	—	—	—	1000	200	0
Riniken	1500	380	1500	450	(1 sample)	—	1600	380	1500	380	2300
Schafisheim	1350	500	1320	400	1200	120	800	200	1350	450	2000
Weiach	1200	150	1300	150	800	75	1300	200	1150	150	1300

All values in metres. Log%Ro, Easy%Ro, B&P '94 and Brink *et al.* (1992) refer to different approaches for erosion esimates (for further explanations see Table 2 and text). AFT: extrapolation of palaeogeothermal gradient, palaeotemperatures calculated combining fission track data with the kinetic model of Willett (1992); EASY&AFT: extrapolation of palaeogeothermal gradient, palaeotemperatures from VR and apatite fission track data.

ature gradient history. The resulting temperature path was modified until a satisfactory fit between modelled and measured vitrinite or fission track length/age distribution was achieved. Basic input data and assumptions concerning burial and temperature history in the NAFB can be found in Naef *et al.* (1985); Kempter (1987); Wildi *et al.* (1989); Moss (1992); Schegg (1992; 1993; 1994) and Todorov *et al.* (1993).

Results

An overview of available well data (stratigraphy, thermal indicators) used for this study is given in Table 1 and results for individual wells are summarized in Tables 2, 3 and 4.

Present-day stratigraphic thicknesses: regional trends from stratigraphic investigations

In the following, two schematic profiles (SW–NE in strike-direction, Fig. 5; NW–SE in dip-direction, Fig. 6) based on well data or outcrop sections are presented. This illustrates the geographical variations of stratigraphic thicknesses of preserved deposits below the base Quaternary and the base Tertiary unconformity.

Mesozoic deposits. Mesozoic deposits below the Tertiary clastics of the NAFB (Fig. 5) consist of upper Cretaceous (mainly in the Helvetic realm, local Cenomanian in the western part of the

Table 4. *Estimated palaeogeothermal gradients*

Well	Present geothermal gradient	Palaeogeothermal gradient EASY%Ro	AFT	EASY&AFT	B&P '94
Base Quaternary unconformity: Palaeogeothermal gradient (≈10–20 Ma)					
Altishofen	30	9	—	—	9
Anzing 3	23	19	—	—	20
Küsnacht	27	8	—	—	12
Thônex	31	27	—	—	25
Weggis	30	20	—	—	16
Base Tertiary unconformity: Palaeogeothermal gradient (≈70–100 Ma)					
Böttstein	46	69	52	41	56
Entlebuch-1	28	58	—	—	88
Riniken	49	23*	(1 sample)	22*	23*
Schafisheim	42	28	35	42	27
Weiach	47	38	46	38	41
Late Palaeozoic unconformity: Palaeogeothermal gradient (≈275 Ma)					
Weiach	47	81	—	—	88

All values in °C km^{-1}. Easy%Ro, AFT, EASY&AFT and B&P '94 refer to different approaches for palaeotemperature estimates cited in the text and in Tables 2 and 3.
* Large scattering in original VR data.

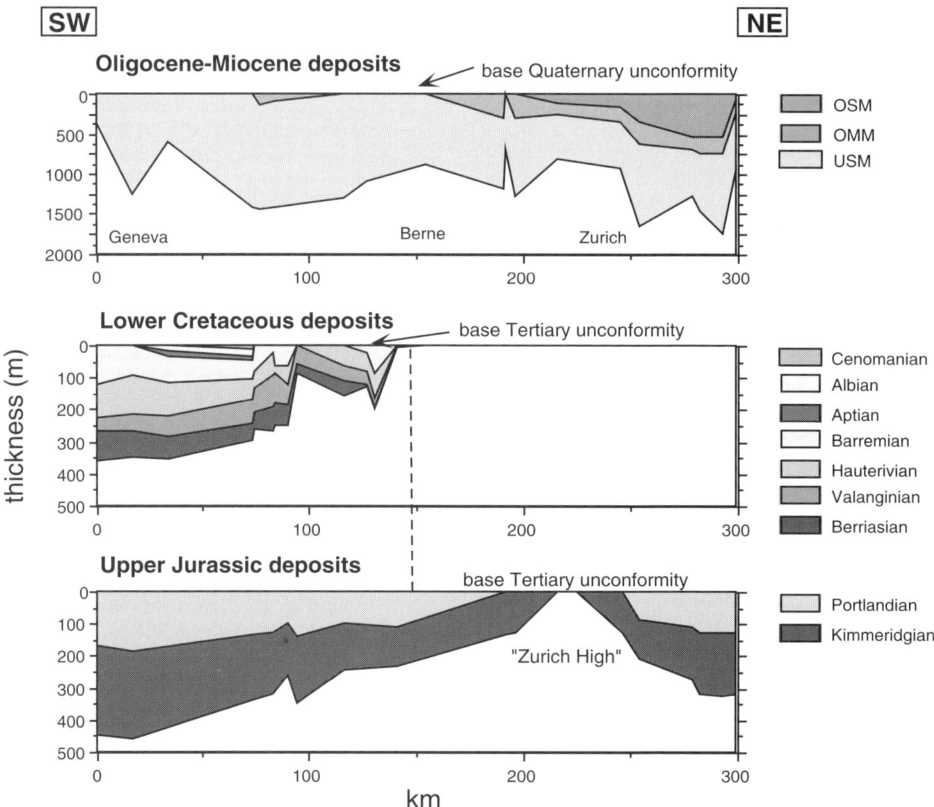

Fig. 5. Schematic SW–NE section (see Fig. 2 for location of the profile) in the NAFB showing present-day thickness variation of upper Jurassic, Cretaceous and Tertiary deposits. OSM = Upper Freshwater Molasse, OMM = Upper Marine Molasse, USM = Lower Freshwater Molasse.

NAFB), lower Cretaceous (Helvetic realm and western part of the NAFB) and upper Jurassic (eastern part of the NAFB) sediments. The section in dip-direction (Fig. 6) indicates a pinch-out of lower Cretaceous and an increase in thickness of upper Jurassic deposits from the Helvetic realm towards the Swiss Plateau. A primary decrease in thickness of lower Cretaceous deposits towards the NE can also be observed in the section in strike-direction, even when this reduction is amplified by an erosional truncation towards the same direction. Maximum thicknesses of lower Cretaceous sediments below the Molasse Basin are less than 350 m. These observations suggest that the area of the NAFB was a region with little or no sediment accumulation throughout the Early Cretaceous.

The same is probably also true for late Cretaceous times: the preserved thickness of the most complete (up to the Maastrichtian) upper Cretaceous section (Alvier, Fig. 6) in the eastern part is only 300 m. According to Villars (1991), maximum thicknesses of preserved Upper Cretaceous sediments in the French Subalpine Chains (lateral equivalent of the Helvetic zone) are about 500 m and decrease towards the external parts.

Concerning the upper Jurassic deposits of Eastern Switzerland, thickness reduction over the 'Zurich High' (broad northwest-trending anticlinal structure, Bachmann *et al.* 1987) is clearly an erosional feature (Figs 5 and 7). Maximum thicknesses of Portlandian and Kimmeridgian formations below the NAFB are between 400–500 m (western part of the NAFB, Fig. 5). The SW–NE profile, however, indicates a slight primary decrease in thickness of Kimmeridgian rocks from the SW to the NE. Therefore, the maximum amount of erosion of Jurassic rocks above the 'Zurich High' is probably smaller than 400 m.

In summary, stratigraphic observations indicate that the post-Mesozoic erosion cut deeper in the eastern part of the NAFB than in the western part.

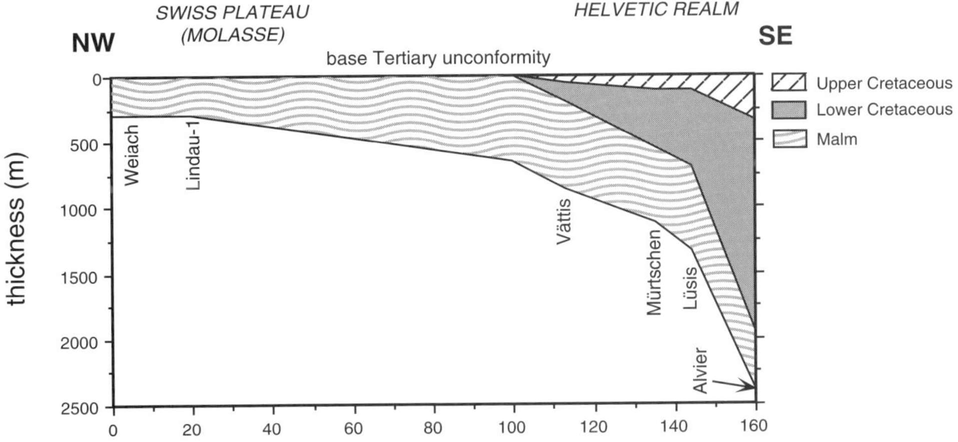

Fig. 6. Schematic NW–SE section (see Fig. 2 for location of the profile) in the NAFB showing present-day thickness variation of upper Jurassic and Cretaceous deposits. Position of stratigraphic sections in the Helvetic realm represents the palaeogeographic position prior to tectonic deformation (Wildi *et al.* 1989, fig. 4).

We estimate that the maximum amount of erosion over the 'Zurich High' is about 600 m (400 m upper Jurassic and 200 m Cretaceous deposits). Consideration of the compaction during Tertiary reburial increases the erosion estimates. Preliminary modelling results using BasinMod® (basin modelling software, Platte River Associates 1993) suggest that the primary depositional thickness of the removed Cretaceous section has to be increased, depending on the basin position of the studied well, by up to 40 %. Therefore, the missing section over the 'Zurich High' could be as high as

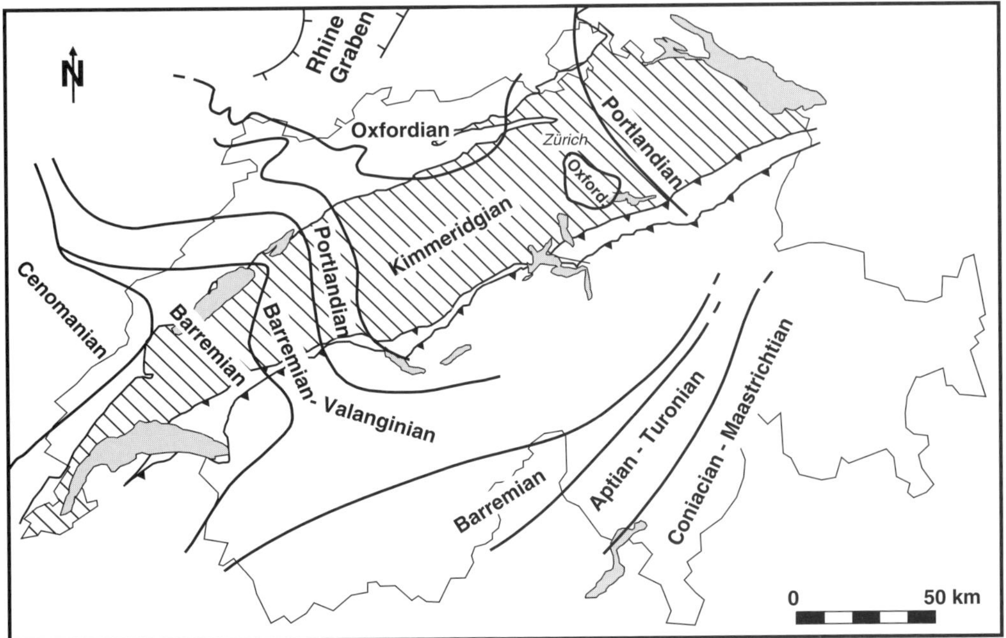

Fig. 7. Base Tertiary subcrop modified after Büchi & Schlanke (1977) and Trümpy (1980). Note that the Alpine part has been restored palinspastically.

800–900 m. Post-Mesozoic erosion in the western part of the Swiss Plateau concerns the upper part of the lower and all of the upper Cretaceous deposits. Even if there is still a debate on whether the Cretaceous transgressions have reached the area of the northeastern NAFB, we are quite confident about modest overall Cretaceous sediment thicknesses.

Tertiary deposits. The section in strike-direction (Fig. 5, uppermost profile) shows the preserved sediments below the base Quaternary unconformity. Note, however, that this section is based mainly on hydrocarbon exploration wells which are situated generally on anticlinal structures. Therefore, additional erosion in former glacial valleys (Pugin & Wildi 1995) will not be accounted for. As the position of this schematic profile is in a more proximal position in the southwestern part of the section than in the northeastern part, Lower Freshwater Molasse (USM) deposits appear to be thicker in this area. This section along the strike of the NAFB clearly demonstrates the reversal of the uplift pattern when compared to the post-Mesozoic history. Preserved thicknesses of the youngest formations (Upper Marine Molasse, OMM; Upper Freshwater Molasse, OSM), even in rather distal positions, may reach values of about 500 m in the eastern NAFB. Maximum preserved thicknesses of OSM deposits in proximal gravel fans of eastern Switzerland are much higher (≈1500 m). Towards the southwest an erosional pinch-out of OSM and OMM can be observed (Fig. 8). Stratigraphic evidence indicates, therefore, that the post-Molasse erosion in the southwestern part of the NAFB may easily attain 2000 m (when including the upper part of the USM, the OMM and the OSM) in the more proximal parts (see Figs 3 and 5).

Extrapolation of coalification trends

Results from evaluations of best-fit regression lines on vitrinite reflectance data (Log%Ro) are presented in Tables 2 and 3.

For the late Palaeozoic unconformity, only the data from the well Weiach are available. We reconstruct 1000 m of missing section.

VR data from five wells in the central and eastern part of the NAFB enabled the estimation of post-Mesozoic erosion (Table 3). Results range from 1000 m (Entlebuch) to 1500 m (Riniken) and are generally higher than those indicated by stratigraphic reconstructions (see previous chapter).

For the post-Molasse erosion, VR data from the wells Altishofen, Thônex, Weggis, Küsnacht and Anzing could be analysed (Table 2). The amount of calculated erosion for the westernmost well Thônex is about 2000 m. A much higher value (4100 m) is obtained for the well Weggis, situated in the Subalpine Molasse of Central Switzerland. Results

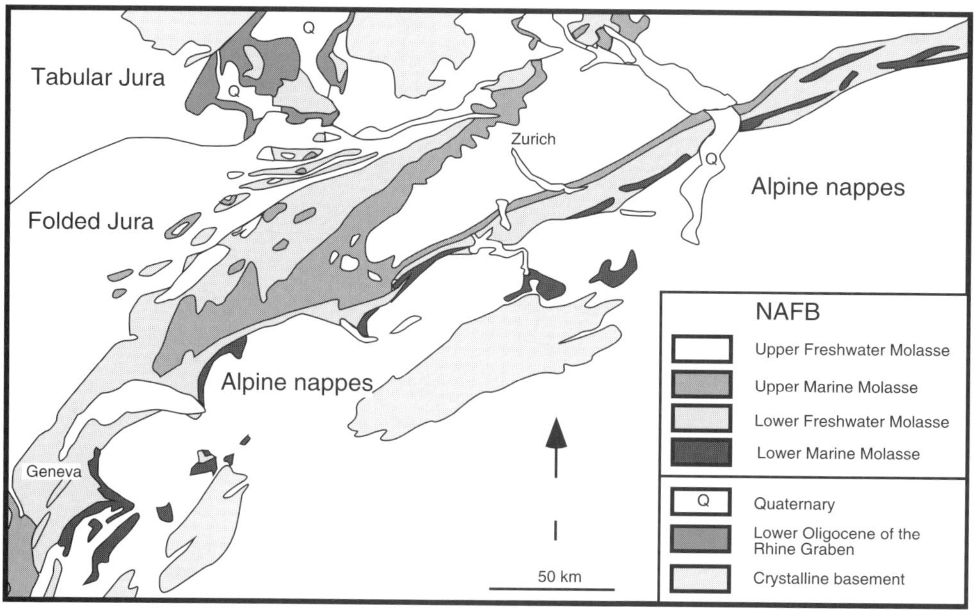

Fig. 8. Simplified geological map of Switzerland showing the effect of the westwards increasing post-Molasse erosion (modified after Keller *et al.* 1990).

for wells Altishofen and Küsnacht indicate that between 3800 and 4000 m of OSM deposits have been eroded. For the easternmost well (Anzing), the extrapolation to a surface coalification value of 0.25 %Rr results in no erosion.

Extrapolation of palaeogeothermal gradients

Results from the evaluation of best-fit regression lines on modelled maximum palaeotemperatures are presented in Tables 2, 3 and 4.

For the late Palaeozoic unconformity in the well Weiach, the estimated missing section is similar to the one obtained from extrapolation of the coalification gradient (1200 m). Inferred palaeogeothermal gradients are very high (81–88 °C km^{-1}).

In Central and Eastern Switzerland, the projection of palaeogeothermal gradients results in post-Jurassic missing sections of 1100–1320 m (EASY%Ro), 800–1200 m (AFT), 800–1700 m (EASY&AFT) and 1150–1350 m (B&P '94) for the wells Böttstein, Schafisheim and Weiach. Corresponding palaeogeothermal gradients are between 28–69 °C km^{-1} (EASY%Ro), 35–52 °C km^{-1} (AFT), 38–42 °C km^{-1} (EASY&AFT) and 27–56 °C km^{-1} (B&P '94). Results from Entlebuch and Riniken have to be viewed with care due to strong mechanical and thermal overprinting, respectively and a large scattering of the original VR data. Estimates of post-Jurassic missing sections are higher than those indicated by stratigraphic reconstructions.

It is important to note that basically it is not possible to distinguish with VR data alone in the Mesozoic section between late Cretaceous and/or Oligocene–Miocene erosion or cooling phases. Only measured apatite age data (Hurford 1993) constrain the timing of maximum temperature to the late Cretaceous.

Estimates for the post-Molasse erosion are similar to those obtained from the extrapolation of coalification gradients. The wells Altishofen and Küsnacht again show very high erosion amounts (4600–9000 m). Resulting palaeogeothermal gradients for the wells Anzing, Thônex and Weggis are lower than normal and range between 19–27 °C km^{-1} (EASY%Ro) and 16–25 °C km^{-1} (B&P '94) and confirm results from Teichmüller & Teichmüller (1986) from the German Molasse Basin. The possible reasons for the very low palaeogeothermal gradients in the wells Altishofen and Küsnacht (8–12 °C km^{-1}) will be discussed below.

Discussion

Our results indicate that the studied inversion episodes in the NAFB are accompanied by significant erosion (up to several km). Such amounts cannot only be accounted for by mechanisms such as base-level lowering or eustatic sea-level fall. Deep-seated tectonic processes must be responsible (see below).

Relevance and uncertainties of results

The large scattering of results (differences of up to several hundred metres) between the different approaches used in this study expresses the uncertainties associated with these techniques. With good data sets an accuracy of no more than several hundred metres may be attained.

Inherent problems of methods which determine the amount of erosion through geological or geochemical parameters that vary regularly with depth and do not significantly change during exhumation have been discussed in greater detail by Duddy *et al.* (1991) and Løseth *et al.* (1992). These methods depend largely on the choice of a typical surface value or the evaluation of 'typical' depth trends but do not provide any information about timing of maximum burial. Concerning the extrapolation of coalification trends, Majorowicz *et al.* (1990) stated that a zero coalification of 0.25 %Rr is a more reasonable assumption than either the 0.15 %Rr employed by England & Bustin (1986) or that of 0.18 %Rr used by Hacquebard (1977), if the reflectance of peat is considered. When logarithmic reflectance gradients are approximated by straight lines, using a least squares method, care must be exercised when the depth interval covered by measured samples is small compared to the length of the extrapolation (Majorowicz *et al.* 1990).

It has to be stressed, that all methods using geothermal gradients are based on the assumption that the rough timing of the palaeotemperature event is known for the determination of a reasonable palaeosurface temperature (palaeoclimate). Only the combined use with fission track data provides some information on absolute timing of the thermal event. An additional simplification is the linear procedure to determine the palaeotemperature gradient. This assumption is only valid in cases where lateral fluid flow and heat transfer can be excluded for the time period of the missing section and when the lithologies of the removed sections are similar to the preserved lithologies (Duddy *et al.* 1994). For buried unconformities additional uncertainties arise from possible younger thermal and mechanical overprinting. According to Katz *et al.* (1988), as reburial of an unconformity progresses, the original offset in VR values at the unconformity decreases, until no statistically significant differences exist in VR values across the stratigraphic boundary. Reburial and further compaction increases the attained pre-erosional coalification and palaeotemperature gradients due to a progressive thickness reduction of the whole profile and results in systematic under-estimation of the missing section.

Intra-Palaeozoic erosion

Amount of erosion. Kempter (1987) calculated directly from VR-data a stratigraphic gap of 1000 m for the late Palaeozoic unconformity in Northern Switzerland. This is in good agreement with our reconstruction of 1000–1200 m. However, all these results have to be interpreted with caution, because the measured vitrinite reflectance values from the Palaeozoic section may have recorded to some extent later maturation phases (reburial and renewed heating during Cretaceous and Oligocene–Miocene), which will result in an over-estimation of missing section.

Inversion mechanisms and thermal regime. Calculated palaeogeothermal gradients are very high (Weiach, 81–88 °C km^{-1}) for the late Palaeozoic and confirm the post-kinematic plutonism and hydrothermal activity at that time in the Central European Variscan orogeny. Kempter (1987), using TTI-modelling, postulated for the same period palaeogeothermal gradients of up to 100 °C km^{-1}.

The Permo–Carboniferous evolution of the study area (for a detailed review see Diebold *et al.* 1991) is controlled by the final phase of the Central European Variscan orogeny which was accompanied by doming and extension of the Variscan chain, postkinematic plutonism, hydrothermal activity and formation of sedimentary troughs and graben. The development of these SW–NE trending pull-apart basins has been related to the formation of a broad dextral transform belt between the Appalachians and the Urals (Arthaud & Matte 1977; Laubscher 1986; 1987; Ziegler 1990). The Saalian tectonic activities mark the end of the Variscan orogenic cycle and are accompanied and followed by an important phase of erosion and peneplanation of the Variscan mountain range giving rise to this Late Palaeozoic unconformity.

Post-Mesozoic erosion

Amount of erosion. Based on porosity and interval velocity analysis in middle Jurassic rocks, Brink *et al.* (1992) concluded that missing post-Jurassic deposits (upper Cretaceous and/or Miocene) reach values of up to 3000 m in southwestern NAFB of Switzerland. Calculated amounts decrease towards the Alpine front and towards eastern Switzerland (Tables 2 and 3). As these techniques are only applied to Jurassic rocks it is not possible to distinguish between upper Cretaceous and/or Miocene missing overburden. It should also be kept in mind that porosity and interval velocity changes may be influenced by burial as well as by diagenetic processes (e.g. cement precipitation). Regional stratigraphical observations in the southwestern part of the NAFB (small thickness of Cretaceous sediments, see discussion above) indicate that the high post-Jurassic erosion estimates by Brink *et al.* (1992) are most probably due to a deep Tertiary burial and subsequent uplift.

The base Tertiary unconformity with a maximum chronostratigraphic gap of Middle Jurassic to Middle Eocene (Herb 1988; Allen *et al.* 1991) underlies the whole NAFB, separating eroded Tethyan platform sediments from progressively younger Tertiary sediments as it is traced northwards into the foreland (Sinclair *et al.* 1991). In the Swiss NAFB and the Jura, Cretaceous sediments are only found in the western part of the country (Lemcke 1974; 1981; Wildi *et al.* 1989), whereas in the eastern part, in the area of the Zurich High (extending from Northeastern Switzerland to the southern Rhine valley, Bachmann & Müller 1991), Mesozoic strata were eroded down to the Oxfordian (Fig. 7). Marine Cenomanian, Turonian and Senonian deposits are mainly found in the French part of the Jura (Guillaume 1966; Trümpy 1980). But locally, Cenomanian deposits also occur in the Vaud and Neuchâtel Jura of Switzerland and in the wells of the Vaud Molasse Basin. Maastrichtian limestones, preserved as boulders in a karst pocket near Bienne (≈10 km NE of the well Tschugg, Fig. 2), are the only relicts left by a late Cretaceous sea covering the Jura (Häfeli 1964).

Important amounts of upper Cretaceous deposits have been observed at the eastern end of the Molasse basin, southwest of the Bohemian Massif (Bachmann *et al.* 1987, Lemcke 1988). There, up to 600–800 m thick Albian to Campanian sediments are preserved beneath the Tertiary clastics (Bachmann & Müller 1991). The Franconian Platform to the northwest became uplifted and exposed during the Early Cretaceous times (Pfeffer 1986). This is illustrated by the erosion and karstification of the upper Jurassic carbonates which show an erosional pre-late Cretaceous relief of a few hundred meters (Schröder 1968; 1987; Pfeffer 1986). In the late Cretaceous (Cenomanian–Santonian), the karst relief was totally buried by fluvial and marine quartz sands. With the Campanian a new uplift and karstification cycle started in this area giving rise to the erosion of most of the upper Cretaceous sediments (erosional remnants in karst pockets).

For this major unconformity in the NAFB, a substantial discrepancy results between estimated erosion amounts based on regional geological observations (up to 600 m, becoming 900 m when compaction is allowed for) and erosion amounts calculated with thermal indicator data (800 to 1800 m, Table 3). These differences remain even when the inherent uncertainties of the methods used are considered and when it is kept in mind that the studied wells are concentrated in a rather small area and external basin position (eastern part of the NAFB). Assuming that these calculated and systematically larger erosion amounts are real would imply that the regional geological evaluation of the preserved stratigraphy misleads the interpretation of the original basin geometry. This would also have consequences for the palaeogeographic reconstructions of the study area and would support the regional interpretations by Ziegler (1990).

On the other hand it may be argued that the basic assumption of a linear projection of the palaeogeothermal or coalification gradient to typical surface values is not justified due to aquifer flow or low thermal conductivities in the eroded lithologies (Duddy *et al.* 1994). In both cases higher palaeogeothermal gradients in the removed section and consequently lower erosion estimates would result. Geological observations (Brink *et al.* 1992, Bachmann & Müller 1991) indicate that Cretaceous deposits of the NAFB are generally more shaley (lower thermal conductivity) than the underlying upper Jurassic carbonates. Our post-Mesozoic erosion amounts may, therefore, be over-estimated.

The hypothesis of a thermal perturbation by fluid flow could only be explained by a potential aquifer of late Jurassic or Cretaceous age. Fractured and possibly karstified carbonates of the Malm and Muschelkalk have been identified by Rybach (1992) as high permeability water conduits. Whether or not these aquifers were already active during the Cretaceous remains an open question. If karstification played an important role, it must have been formed very early. Subaerial exposure of Upper Jurassic carbonates are documented at least for the area of the future German Molasse basin (Lemcke 1987; Pfeffer 1986).

Inversion mechanisms and thermal regime. Allen *et al.* (1991) and Sinclair *et al.* (1991) interpreted the break in sedimentation as a first sign of the arrival of the thrust wedge onto the southern margin of Europe (forebulge unconformity). Bachmann *et al.* (1987) invoked a combination of latest Cretaceous–earliest Tertiary inversion tectonics (Campanian–mid Palaeocene) and the Mid-Palaeocene eustatic lowstand in sea-level. The intra-plate deformations are generally related to stresses that were exerted by the Alpine and Pyrenean orogenic events on the continental forelands of these foldbelts (Ziegler 1987; 1992). Resulting deformations are particularly intense along the southwestern margins of the Bohemian Massif and of the Landshut–Neuötting High, and probably also in the anticlinal Zurich High (Bachmann *et al.* 1987). The uplift of large areas in the external parts of the future Molasse basin resulted in a northwestward truncation of Cretaceous and upper Jurassic strata (Bachmann & Müller 1991).

The evolving mantle diapir below the Rhenish massif (Illies 1974) and volcanic activity in the area of the future Southern Rhine graben (Hüttner 1991) from middle Cretaceous onwards indicates an elevated thermal regime over wide areas of Central Europe. The reconstructed geothermal gradients for this period (30–50 °C km^{-1}, Table 4) are high and are confirmed by fission track data (Hurford 1993). Although this may explain a karstified system with hot water circulation, the palaeo-flow directions and their aerial extent still remain to be identified.

Post-Molasse erosion

Amount of erosion. Lemcke (1974), based on a subsidence analysis of different wells in the NAFB, estimated a missing Tertiary section of 500–600 m in the eastern part of the Swiss Molasse going up to 2500 m in the western part of the basin (Table 2).

Using the clay mineral evolution from 15 wells, Monnier (1982) has evaluated the degree of post-Molasse erosion in the NAFB with reference to the transition zone of smectite to mixed-layer clays. His values correspond to minimum amounts because already for the Küsnacht well (reference level) a minimum erosion amount of 500 m seems to be a reasonable figure in a regional geological context (Pavoni 1957; Schaer 1992). According to Monnier (1982), in wells of the external parts of the NAFB (Essertines, Tschugg, Ruppoldried, Pfaffnau-1 and Pfafnau Süd-1, Boswil), maximum burial of Molasse deposits is not more than 2000 m because the transition zone is never attained. Taking into account the present-day depth of rocks at the base of the Tertiary section, maximum possible erosion amounts vary between 700 m (Boswil) in the east and 2200 m (Essertines) in the west (Table 2). In wells situated in more internal parts of the NAFB, the transition zone has been attained. Monnier (1982) calculated minimum missing sections ranging from 1000–1200 m in the western part to 0 m in the eastern part. Adding an erosion amount of 500 m for the well Küsnacht (Schaer 1992), a range from 1500–1700 m to 500 m results (Table 2). His results are in line with those of Laubscher (1974) and Lemcke (1974).

Sonic and density logs from different wells in the Swiss Molasse have been evaluated by Kälin *et al.* (1992) for porosity–depth trends. According to their compaction trend, late to post-Tertiary erosional gaps in the central Swiss Molasse basin range from about 4 km in the north to as much as 8.5 km for the overthrusted Molasse in the south. Similar to the results of Brink *et al.* (1992) these rather large missing sections may also represent maximum values because of the diagenetic cementation and porosity reduction effect (see above).

The continental Upper Freshwater Molasse (OSM) represents the final filling stage of the NAFB during the middle and late Miocene. Afterwards the Molasse Basin was subjected to erosion giving rise to the base Quaternary unconformity. This erosional stage is well expressed in the present-day outcrop map where from east to west, erosion has progressively cut deeper (Fig. 8). In the western part of the Swiss Molasse Basin, the OSM is not preserved, whereas in the central parts of the gravel fans in eastern Switzerland the OSM measures over 1500 m. Serravalian lake sediments in a syncline of the western Jura (Le Locle, Kälin 1993) are the only signs of a former OSM equivalent sedimentary cover in an external and northwestern position.

The SW–NE profile (Fig. 9) summarizing the results from the different approaches shows the decrease of the post-Molasse erosion in a northeasterly direction (from 1500–3000 m in the SW to 0–700 m in the NE), even when the scattering of individual results is large. Higher erosion amounts towards the Alpine front are indicated by results from the well Weggis (4100–4400 m), situated in the Subalpine Molasse of Central Switzerland. Schegg (1994) explained this important additional overburden by the overthrusting of a thick hanging wall sequence.

There is, however, evidence that our results may not only be influenced by burial. If overburden thickness is the only controlling factor, we should observe a decrease in maturity values of outcrop samples towards the NE. Schegg (1992) showed that this is not the case and presented indications that the palaeogeothermal regime of the NAFB could be influenced by the thermal effects of transient fluid flow. Results for the wells Altishofen and Küsnacht indicate that 3800–9000 m of OSM deposits have been eroded. This is not geologically reasonable. These high figures could be the result of bad data (only four VR values for each well) or could be due to a perturbation of the palaeotemperature profile by lateral transfer of heat by fluids as hypothesized in Schegg (1992). According to Duddy *et al.* (1994), estimating the magnitude of uplift and erosion using VR data in fluid flow dominated regimes is difficult and cannot be unequivocally transferred to the removed section assuming a constant heat flow. They concluded that steady-state fluid flow in the past may be

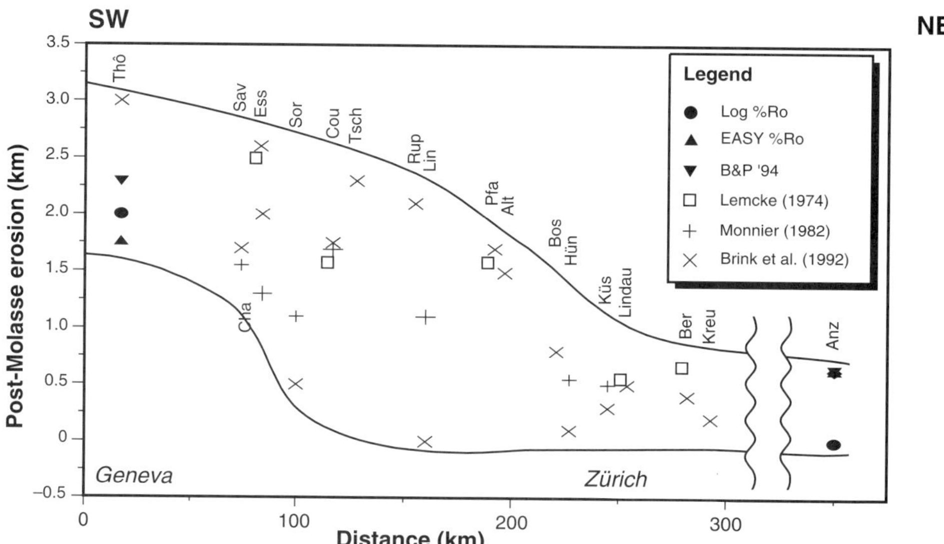

Fig. 9. Comparison of post-Molasse erosion estimates for different approaches in a SW–NE profile. Note that the geological unreasonable erosion estimates (3800–9000 m based on VR data) from the wells Altishofen and Küsnacht have not been included (for explanations see text).

recognized by very low to negative palaeogeothermal gradients below an aquifer. The low resulting apparent palaeogeothermal gradients (8–18 °C km^{-1}) for the wells Altishofen and Küsnacht (Table 4) support such an interpretation. Further, the thermal and stratigraphical data could be reconciled if the palaeogeothermal gradient at the time of maximum palaeotemperatures was non-linear, with a high geothermal gradient in the shallow eroded section. According to Duddy *et al.* (1994), elevated geothermal gradients can be expected above an aquifer where hot fluids flow. In the Tertiary section of the NAFB, the Upper Marine Molasse (OMM) sandstones have been identified as a regional aquifer (Rybach 1992). OMM deposits in the Küsnacht well are at shallow depths (545–1065 m) and a high palaeogeothermal gradient in the eroded section can, therefore, not be excluded.

Inversion mechanisms and thermal regime. All our data and reconstructions based on the Tertiary samples point towards low thermal gradients (15–30 °C km^{-1}) for the Cenozoic and probably indicate some influence by transient fluid flow. Relatively high present-day geothermal gradients (e.g. 48 °C km^{-1} in Weiach) are likely to be related to convective heat transfer by circulating fluids in the vicinity of the Palaeozoic graben system (Schärli & Rybach 1991). That these high temperatures persisted only during most recent times is indicated by the apatite fission track results (Hurford 1993).

There are different inversion mechanisms which may have acted either as amplifying or as self-sufficient processes in parts or over the whole basin:

(1) The Jura 'decollement' was linked to the post-early Miocene Adria–Europe plate convergence ('Fernschub', Laubscher 1961). One of the implications of this thin-skin mechanism is that south of the Jura mountains the surface of the Molasse Basin must have been uplifted above a stationary basement because a southward thickening wedge has been pushed towards the north (Laubscher 1974). Laubscher calculated a 'decollement'-induced Miocene uplift of the Swiss Molasse basin which increases from several hundred of meters in the east and north to over 2000 m in the southwest. His results are in line with those of Lemcke (1974), Monnier (1982) and findings from this study.
(2) The foreland bulge, which ordinarily has a structural relief of no more than few hundred metres, should be expected to mark the external flank of the NAFB (Sinclair *et al.* 1991; Laubscher 1992).
(3) According to Laubscher (1992), in post-early Miocene times, the West-European (Rhine-Bresse–Limagne–Rhone) rift system became the locus of 'constructive interference' with the Alpine forebulge. A new stress system developed, rifting ceased and asthenospheric convection set in, creating hot mantle domes (Laubscher 1992). This new geodynamic situation resulted in the uplift of the Rhine dome (Black Forest–Vosges) and the Loire dome.
(4) According to Lyon-Caen & Molnar (1989), the late Cenozoic uplift of the Molasse Basin and the Alps might be a consequence of a diminution of down welling of mantle material beneath the Alps, thus initiating a large scale isostatic rebound movement in the Alps and the NAFB.
(5) Glacial erosion, especially in former glacial valleys, may be an important erosion mechanism (see discussion in Pugin & Wildi 1995).

Conclusions

A combination of different techniques to estimate the amount of missing sections at unconformities, and a well defined geological context, proves to be a powerful method for the analysis of erosion events. Nevertheless, the independent application of such methods to the same dataset from the NAFB results in a large scatter of values (up to 600 m for erosion estimates and up to 30 °C km^{-1} for palaeogeothermal gradients). This reflects the specific problems of each technique (internal calibration, uncertainties in the kinetics, data quality) and the simplified basic assumptions for each approach (palaeosurface temperature, non-linear palaeogeothermal gradient, unknown thermal conductivity of missing lithologies). Additional pitfalls are caused by compaction-related effects (increase of frozen palaeogradient with time due to an overall thickness reduction during reburial) and thermal overprinting after the time of maximum temperatures.

Our results show that each major unconformity within the sedimentary succession of the Swiss NAFB is accompanied by significant erosion and a specific geothermal regime:

post-Molasse erosion:
SW ≈2250 m (Geneva) 15–30 °C km^{-1}
NE ≈350 m (Zurich)

post-Mesozoic erosion:
NE ≈1300 m 30–50 °C km^{-1}

late Palaeozoic erosion:
NE ≈1000 m 80–90 °C km^{-1}

Calculated values for the post-Jurassic erosion in northeastern Switzerland are higher than those extrapolated from preserved thicknesses. They may be explained by higher palaeogeothermal gradients in the missing section prior to erosion due to lower thermal conductivities in the Cretaceous sediments or due to a thermal perturbation by aquifer flow. As a linear gradient is assumed for the whole section, calculated erosion amounts will be maximum estimates.

In assessing the geothermal and hydrocarbon potential of an area like the NAFB the evaluation of all available data for an optimized estimate of missing section at major unconformities is of prime importance. Fluid flow and thermal history reconstructions are substantially influenced by these parameters. Many of the problems (e.g. lateral heat transport due to aquifer flow) discussed in this study could be overcome, or at least be quantified to some extent, by 2-D or 3-D basin models. We think that the most promising basin modelling strategy is to work with different erosion scenarios (e.g. minimum, maximum, preferred) which are based on estimated values and on regional stratigraphic observations.

We are indebted to NAGRA for providing fission track data, to SWISSPETROL and the Swiss Federal Institute of Energy (BEW) for unpublished temperature and vitrinite reflectance data and the Geological Institute of Berne (A. Matter) for providing samples of well Altishofen. For helpful contributions we would like to thank the following: A. Gautschi, J. Remane, J. Uriarte and W. Wildi. This research was funded by the Swiss national foundation for scientific research (project N° 2000-037585.93/1). We owe our thanks to R. Stoneley and an anonymous reviewer for their critical remarks which improved the manuscript.

References

ALLEN, P. A., CRAMPTON, S. L. & SINCLAIR, H. D. 1991. The inception and early evolution of the North Alpine Foreland Basin, Switzerland. *Basin Research*, **3**, 143–163.

ALTHAUS, H. E. 1947. Erdölgeologische Untersuchungen in der Schweiz. — 1. Die ölführende Molasse zwischen Genfer und Neuenburger See. — 2. Über geologische Untersuchungen im Gebiet der ölführenden Molasse im Kanton Genf. *Beiträge zur Geologie der Schweiz*, geotechnische Serie, **26**, 1–43.

ARMAGNAC, C., BUCCI, J., KENDALL, C. G. & LERCHE, I. 1989. Estimating the Thickness of Sediment Removed at an Unconformity using Vitrinite Reflectance Data. *In*: NAESER, N. D. & MCCULLOH, T. H. (eds) *Thermal History of Sedimentary Basins: Methods and Case Histories*. Springer-Verlag, New-York, 217–238.

ARTHAUD, F. & MATTE, P. 1977. Late Paleozoic strike-slip faulting in southern Europe and northern Africa: Result of a right-lateral shear zone between the Appalachians and the Urals. *Bulletin of the Geological Society of America*, **88**, 1305–1320.

BACHMANN, G. H. & MÜLLER, M. 1991. The Molasse basin, Germany: evolution of a classic petroliferous foreland basin. *In*: SPENCER, A. M. (ed.) *Generation, Accumulation and Production of Europe's Hydrocarbons*. Special Publications of the European Association of Petroleum Geoscientists, **1**, 263–276.

——, —— & WEGGEN, K. 1987. Evolution of the Molasse Basin (Germany, Switzerland). *Tectonophysics*, **137**, 77–92.

BARKER, C. E. & GOLDSTEIN, R. H. 1990. Fluid-inclusion technique for determining maximum temperature in calcite and its comparison to the vitrinite reflectance geothermometer. *Geology*, **18**, 1003–1006.

—— & PAWLEWICZ, M. J. 1986. The correlation of vitrinite reflectance with maximum temperature in humic organic matter. *In*: BUNTEBARTH, G. & STEGENA, L. (eds.) *Paleogeothermics*. Lecture Notes in Earth Sciences, **5**, 79–93, Springer-Verlag, Berlin.

—— & —— 1994. Calculation of vitrinite reflectance from thermal histories and peak temperatures. *In*: MUKHOPADHYAY, P. K. & DOW, W. G. (eds) *Vitrinite reflectance as a maturity parameter. Applications and limitations*. American Chemical Society Symposium Series, **570**, 216–229.

BRINK, H.-J., BURRI, P., LUNDE, A. & WINHARD, H. 1992. Hydrocarbon habitat and potential of Swiss and German Molasse Basin: A comparison. *Eclogae Geologicae Helvetiae*, **85**, 715–732.

BÜCHI, U. P. & SCHLANKE, S. 1977. Zur Paläogeographie der Schweizerischen Molasse. *Erdöl-Erdgas-Zeitschrift*, **93**, 57–69.

——., COLOMBI, C., FEHR, W. R., LEMCKE, K., KOEWING, K., HOFMANN, F., FÜCHTBAUER, H. & TRÜMPY, R. 1961. Geologische Ergebnisse der Bohrung Küsnacht 1. *Bulletin der schweizerischen Vereinigung Petroleum-Geologen und -Ingenieure*, **28**, 7–16.

——, LEMCKE, K., WIENER, G. & ZIMDARS, J. 1965. Geologische Ergebnisse der Erdölexploration auf das Mesozoikum im Untergrund des schweizerischen Molassebeckens. *Bulletin der schweizerischen Vereinigung Petroleum-Geologen und -Ingenieure*, **32**, 7–38.

DIEBOLD, P., NAEF, H. & AMMANN, M. 1991. *Zur Tektonik der zentralen Nordschweiz. Interpretation aufgrund regionaler Seimik, Oberflächengeologie und Tiefbohrungen*. Nagra — National Cooperative for the Disposal of Radioactive Waste, technical Report, **90-04**, Wettingen, Switzerland.

DOW, W. G. 1977. Kerogen studies and geological interpretations. *Journal of Geochemical Exploration*, **7**, 79–99.

DUDDY, I. R., GREEN, P. F., BRAY, R. J. & HEGARTY, K. A. 1994. Recognition of the thermal effects of fluid flow in sedimentary basins. *In*: PARNELL, J. (ed.)

Geofluids: Origin, migration and evolution of fluids in sedimentary basins. Geological Society, London, Special Publications, **78**, 325–345.

——, ——, HEGARTY, K. A. & BRAY, R. J. 1991. Reconstruction of thermal history in basin modelling using apatite fission track analysis: what is really possible? *Offshore Australia Conference Proceedings*, **1**, III-49–III-61.

ENGLAND, T. D. J. & BUSTIN, R. M. 1986. Thermal maturation of the western Canadian Sedimentary Basin south of the Red Deer River (I), Alberta Plains. *Bulletin of Canadian Petroleum Geology*, **34**, 71–90.

FAVINI, G. G. 1970. Utilisation d'un schéma markovien dans la détermination de la tendence commune des variables mesurées dans quelques forages de Suisse Romande. *Bulletin der schweizerischen Vereinigung Petroleum-Geologen und -Ingenieure*, **37**, 35–42.

FISCHER, H. & LUTERBACHER, H. 1963. *Das Mesozoikum der Bohrungen Courtion 1 (Kt. Fribourg) und Altishofen 1 (Kt. Luzern)*. Beiträge zur Geologischen Karte der Schweiz, Neue Folge, **115**.

FUNK, H. P. 1985. Mesozoische Subsidenzgeschichte im Helvetischen Schelf der Ostschweiz. *Eclogae Geologicae Helvetiae*, **78**, 249–272.

GORIN, G., SIGNER, C. & AMBERGER, G. 1993. Structural configuration of the western Swiss Molasse Basin as defined by reflection seismic data. *Eclogae Geologicae Helvetiae*, **86**, 693–716.

GREBER, E., GRÜNENFELDER, T., KELLER, B. & WYSS, R. 1994. Die Geothermie-Bohrung Weggis, Kanton Luzern. *Bulletin der schweizerischen Vereinigung Petroleum-Geologen und -Ingenieure*, **61**, 17–43.

GREEN, P. F., DUDDY, I. R., GLEADOW, A. J. W. & LOVERING, J. F. 1989. Apatite fission track analysis as a paleotemperature indicator for hydrocarbon exploration. *In*: NAESER, N. D & MCCULLOH, T. H. (eds.) *Thermal history of sedimentary basins: Methods and case histories*. Springer-Verlag, New York, 181–195.

GUILLAUME, S. 1966. Le Crétacé du Jura français: le Crétacé moyen et supérieur. *Bulletin du Bureau de Recherches Géologiques et Minières*, **6**, 65–141.

HACQUEBARD, P. A. 1977. Rank of coal as an index of organic metamorphism for oil and gas in Alberta. *In*: DEROO, G., POWELL, T. G., TISSOT, B. & MCCROSSAN, R. G. (eds) *The origin and migration of petroleum in the western Canadian sedimentary basin, Alberta — A geochemical and thermal maturation study*. Bulletin Geological Survey of Canada, **262**, 11–22.

HÄFELI, C. 1964. Ein Maestrichtien-Vorkommen nördlich von Biel (Kt. Bern). *Bulletin der schweizerischen Vereinigung Petroleum-Geologen und -Ingenieure*, **30**, 65–68.

HERB, R. 1988. Eozäne Paläogeographie und Paläotektonik des Helvetikums. *Eclogae Geologicae Helvetiae*, **81**, 611–657.

HOMEWOOD, P., ALLEN P. A. & WILLIAMS, G. D. 1986. Dynamics of the Molasse Basin of western Switzerland. *Special Publications of the international Association of Sedimentologists*, **8**, 199–217.

——, RIGASSI, D. & WEIDMANN, M. 1989. Le bassin molassique suisse. *In*: PURSER, B. H. (ed.) *Dynamique et méthodes d'étude des bassins sédimentaires*. Éditions Technip, Paris, 299–314.

HURFORD, A. 1993. *Fission Track Analysis of Apatite from the Nagra Boreholes of Böttstein, Weiach, Schafisheim and Riniken, Northern Switzerland*. Nagra — National Cooperative for the Disposal of Radioactive Waste, internal Report, Wettingen, Switzerland.

HÜTTNER, R. 1991. Bau und Entwicklung des Oberrheingrabens - Ein Überblick mit historischer Rückschau. *Geologisches Jahrbuch*, **48**, 17–42.

ILLIES, J.H. 1974. Intra-Plattentektonik in Mitteleuropa und der Rheintalgraben. *Oberrheinische geologische Abhandlungen*, **23**, 1–24.

JENNY, J., BURRI, J.-P., MURALT, R., PUGIN, A., SCHEGG, R., UNGEMACH, P., VUATAZ, F.-D. & WERNLI, R. 1995. Le forage géothermique de Thônex (Canton de Genève): Aspects stratigraphiques, tectoniques, diagénétiques, géophysiques et hydrologiques. *Eclogae Geologicae Helvetiae*, **88**, 365–396.

KÄLIN, D. 1993. *Stratigraphie und Säugetierfaunen der Oberen Süsswassermolasse der Nordwestschweiz*. Dissertation Eidgenössisch Technische Hochschule (ETH), Zürich, Switzerland, **10152**.

KÄLIN, B., RYBACH, L. & KEMPTER, E. H. K. 1992. Rates of Deposition, Uplift and Erosion in the Swiss Molasse Basin, Estimated from Sonic- and Density-Logs. *Bulletin der schweizerischen Vereinigung Petroleum-Geologen und -Ingenieure*, **58**, 9–22.

KÄMPFE, C. 1984. *Tiefbohrungen in Baden-Württemberg und umgebenden Ländern*. Arbeiten aus dem Institut für Geologie und Paläontologie an der Universität Stuttgart, **80**.

KATZ, B. J., PHEIFFER, R. N. & SCHUNK, D. J. 1988. Interpretation of discontinuous vitrinite reflectance profiles. *American Association of Petroleum Geologists Bulletin*, **72**, 926–931.

KEMPTER, E. H. K. 1987. Fossile Maturität, Paläothermogradienten und Schichtlücken in der Bohrung Weiach im Lichte von Modellberechnungen der thermischen Maturität. *Eclogae Geologicae Helvetiae*, **80**, 543–552.

KELLER, B., BLÄSI, H.-R., PLATT, N. H., MOZLEY, P. S. & MATTER, A. 1990. *Sedimentäre Architektur der distalen Unteren Süsswassermolasse und ihre Beziehung zur Diagenese und den petrophysikalischen Eigenschaften am Beispiel der Bohrungen Langenthal*. Nagra — National Cooperative for the Disposal of Radioactive Waste, technical Report, **90–41**, Wettingen, Switzerland.

KOPP, J. 1952. Die Erdölbohrung Altishofen. *Bulletin der schweizerischen Vereinigung Petroleum-Geologen und -Ingenieure*, **19**, 21–24.

LAUBSCHER, H. P. 1961. Die Fernschubhypothese der Jurafaltung. *Eclogae Geologicae Helvetiae*, **54**, 221–282.

—— 1974. Basement uplift and decollement in the Molasse Basin. *Eclogae Geologicae Helvetiae*, **67**, 531–537.

—— 1986. The eastern Jura: Relations between thin-skinned and basement tectonics, local and regional. *Geologische Rundschau*, **75**, 535–553.

—— 1987. Die tektonische Entwicklung der Nordschweiz. *Eclogae Geologicae Helvetiae*, **80**, 287–303.

—— 1992. Jura kinematics and the Molasse Basin. *Eclogae Geologicae Helvetiae*, **85**, 653–675.

LEMCKE, K. 1959. Das Profil der Bohrung Chapelle 1. *Bulletin der schweizerischen Vereinigung Petroleum-Geologen und -Ingenieure*, **26**, 25–29.

—— 1963. Die Ergebnisse der Bohrung Savigny 1 bei Lausanne. *Bulletin der schweizerischen Vereinigung Petroleum-Geologen und -Ingenieure*, **30**, 4–11.

—— 1974. Vertikalbewegungen des vormesozoischen Sockels im nördlichen Alpenvorland vom Perm bis zur Gegenwart. *Eclogae Geologicae Helvetiae*, **67**, 121–133.

—— 1981. Das heutige geologische Bild des deutschen Alpenvorlandes nach drei Jahrzehnten Öl- und Gasexploration. *Eclogae Geologicae Helvetiae*, **74**, 1–18.

—— 1987. Zur Frage der alten Verkarstung des Malm im Untergrund des deutschen Molassebeckens und an dessen Nordwestrand. *Bulletin der schweizerischen Vereinigung Petroleum-Geologen und -Ingenieure*, **53**, 33–46.

—— 1988. *Geologie von Bayern. I. Das bayerische Alpenvorland vor der Eiszeit. Erdgeschichte-Bau-Bodenschätze.* E. Schweizerbart'sche Verlagsbuchhandlung, Nägele und Obermiller, Stuttgart.

——, BÜCHI, U. P. & WIENER, G. 1968. Einige Ergebnisse der Erdölexploration auf die mittelländische Molasse der Zentralschweiz. *Bulletin der schweizerischen Vereinigung Petroleum-Geologen und -Ingenieure*, **35**, 15–34.

LØSETH, H., LIPPARD, S. J., SÆTTEM, J., FANAVOLL, S., FJERDINGSTAD, V. *ET AL.* 1992. Cenozoic uplift and erosion of the Barents Sea — evidence from the Svalis Dome area. *In*: VORREN, T. O., BERGSAGER, E., DAHL-STAMMES, Ø. A., HOLTER, E., JOHANSEN, B., LIE, E. & LUND, T. B. (eds) *Arctic Geology and Petroleum Potential.* Norwegian Petroleum Society (NPF) Special Publications, **2**, 643–664.

LYON-CAEN, H. & MOLNAR, P. 1989. Constraints on the deep structure and dynamic processes beneath the Alps and adjacent regions from an analysis of gravity anomalies. *Geophysical Journal International*, **99**, 19–32.

MAGARA, K. 1976. Thickness of removed sediments, paleopore pressure, and paleotemperature, southwestern part of Western Canada Basin. *American Association of Petroleum Geologists Bulletin*, **60**, 554–565.

MAJOROWICZ, J. A., JONES, F. W., ERTMAN, M. E., OSADETZ, K. G. & STASIUK, L. D. 1990. Relationship between thermal maturation gradients, geothermal gradients and estimates of the thickness of the eroded foreland section, southern Alberta Plains, Canada. *Marine and Petroleum Geology*, **7**, 138–152.

MATTER, A., PETERS, T., ISENSCHMID, CH., BLÄSI, H.-R. & ZIEGLER, H.-J. 1987. *Sondierbohrung Riniken, Geologie. Text- und Beilagenband.* Nagra - National Cooperative for the Disposal of Radioactive Waste, technical Report, **86-02**, Wettingen, Switzerland.

——, ——, BLÄSI, H.-R., SCHENKER, F. & WEISS, H.-P. 1988*a*. *Sondierbohrung Schafisheim, Geologie. Text- und Beilagenband.* Nagra - National Cooperative for the Disposal of Radioactive Waste, technical Report, **86-03**, Wettingen, Switzerland.

——, ——, ——, MEYER, J., ISCHI, H. & MEYER, CH. 1988*b*. *Sondierbohrung Weiach, Geologie. Text- und Beilagenband.* Nagra - National Cooperative for the Disposal of Radioactive Waste, technical Report, **86-01**, Wettingen, Switzerland.

MAURER, H. 1983. Sedimentpetrographische Ergebnisse der Bohrung Fendringen 1. *Bulletin der schweizerischen Vereinigung Petroleum-Geologen und -Ingenieure*, **49**, 61–68.

MONNIER, F. 1982. Thermal diagenesis in the Swiss molasse basin: implications for oil generation. *Canadian Journal of Earth Sciences*, **19**, 328–342.

MOSS, S. 1992. Organic maturation in the French Subalpine Chains: regional differences in burial history and the size of tectonic loads. *Journal of the Geological Society London*, **149**, 503–515.

MÜLLER, W. H., HUBER, M., ISLER, A. & KLEBOTH, P. 1984. *Erläuterungen zur geologischen Karte der zentralen Nordostschweiz 1:100 000.* Nagra — National Cooperative for the Disposal of Radioactive Waste, technical Report, **84-25**, Wettingen, Switzerland.

NAEF, H., DIEBOLD, P. & SCHLANKE, S. 1985. *Sedimentation und Tektonik im Tertiär der Nordschweiz.* Nagra - National Cooperative for the Disposal of Radioactive Waste, technical Report, **85-14**, Wettingen, Switzerland.

NAGRA 1985. *Sondierbohrung Böttstein - Untersuchungsbericht.* Nagra — National Cooperative for the Disposal of Radioactive Waste, technical Report, **85-01**, Wettingen, Switzerland.

PAVONI, N. 1957. Geologie der Zürcher Molasse zwischen Albis und Pfannenstil. *Vierteljahresschrift der naturforschenden Gesellschaft Zürich*, **102**, 117–315.

PFEFFER, K.-H. 1986. Das Karstgebiet der nördlichen Frankenalb zwischen Pegnitz und Vils. *Zeitschrift für Geomorphologie*, Neue Folge, Supplement-Band, **59**, 67–85.

PFIFFNER, A. O. 1986. *Evolution of the north Alpine foreland basin in the Central Alps.* Special Publications of the International Association of Sedimentologists, **8**, 219–228.

PLATTE RIVER ASSOCIATES 1993. *BasinMod® 1D for Windows™*, Version 4.0, Boulder, CO, USA.

PUGIN, A. & WILDI, W. 1995. Geological and geophysical exploration of alpine and peri-alpine glacial valleys. *Eclogae Geologicae Helvetiae*, **88**, 183–197.

RYBACH, L. 1992. Geothermal potential of the Swiss Molasse Basin. *Eclogae Geologicae Helvetiae*, **85**, 733–744.

—— & BODMER, PH. 1980. Die geothermischen Verhältnisse der Schweizer Geotraverse im Abschnitt Basel-Luzern. *Eclogae Geologicae Helvetiae*, **73**, 501–512.

SCHAER, J.-P. 1992. Tectonic evolution and vertical movement in Western Switzerland. *Eclogae Geologicae Helvetiae*, **85**, 695–699.

SCHÄRLI, U. & RYBACH, L. 1991. Geothermische Detailkartierung der zentralen Nordschweiz (1:100 000). *Beiträge zur Geologie der Schweiz*, Geophysik, **24**.

SCHEGG, R. 1992. Thermal maturity of the Swiss Molasse Basin: Indications for paleogeothermal anomalies? *Eclogae Geologicae Helvetiae*, **85**, 745–764.

—— 1993. *Thermal maturity and history of sediments in the North Alpine Foreland Basin (Switzerland, France)*. Publications du Département de Géologie et Paléontologie, Université de Genève, Switzerland, **15**.

—— 1994. The coalification profile of the well Weggis (Subalpine Molasse, Central Switzerland): Implications for erosion estimates and the paleogeothermal regime in the external part of the Alps. *Bulletin der schweizerischen Vereinigung Petroleum-Geologen und -Ingenieure*, **61**, 57–67.

SCHLANKE, S., HAUBNER, L. & BÜCHI, U. P. 1978. Lithostrathigraphie und Sedimentpetrographie der Molasse in den Bohrungen Tschugg 1 und Ruppoldsried 1 (Berner Seeland). *Eclogae Geologicae Helvetiae*, **71**, 409–425.

SCHRÖDER, B. 1968. Zur Morphogenese im Ostteil der Süddeutschen Scholle. *Geologische Rundschau*, **58**, 10–32.

— 1987. Inversion tectonics along the western margin of the Bohemian Massif. *Tectonophysics*, **137**, 93–100.

SINCLAIR, H. D., COAKLEY, B. J., ALLEN, P. A. & WATTS, A. B. 1991. Simulation of Foreland Basin Stratigraphy Using a Diffusion Model of Mountain Belt Uplift and Erosion: An Example from the Central Alps, Switzerland. *Tectonics*, **10**, 599–620.

SWEENEY, J. J. & BURNHAM, A. K. 1990. Evaluation of a Simple Model of Vitrinite Reflectance Based on Chemical Kinetics. *American Association of Petroleum Geologists Bulletin*, **74**, 1559–1570.

TEICHMÜLLER, M. & TEICHMÜLLER, R. 1975. Inkohlungsuntersuchungen in der Molasse des Alpenvorlandes. *Geologica Bavarica*, **73**, 123–142.

——, R. & ——, M. 1986. Relations between coalification and palaeogeothermics in Variscan and Alpidic foredeeps of western Europe. *In*: BUNTEBARTH, G. & STEGENA, L. (eds) *Paleogeothermics*. Lecture Notes in Earth Sciences, **5**, 53–78, Springer-Verlag, Berlin.

TODOROV, I., SCHEGG, R. & WILDI, W. 1993. Thermal maturity of Mesozoic and Cenozoic sediments in the south of the Rhine Graben and the Eastern Jura (Switzerland). *Eclogae Geologicae Helvetiae*, **86**, 667–692.

TRÜMPY, R. 1980. *Geology of Switzerland: a Guide Book. Part A: an Outline of the Geology of Switzerland.* Schweizerische Geologische Kommission, Wepf & Co. Publishers, Basel-New York.

VILLARS, F. 1991. *Evolution paléogéographique du domaine delphino-helvétique (entre Chartreuse et Morcles) au Crétacé supérieur (Turonien-Maastrichtien): Biostratigraphie, sédimentologie et dynamique sédimentaire sur une rampe carbonatée.* Publications du Département de Géologie et Paléontologie, Université de Genève, Switzerland, **10**.

VOLLMAYR, TH. 1983. Temperaturmessungen in Erdölbohrungen der Schweiz. *Bulletin der schweizerischen Vereinigung Petroleum-Geologen und -Ingenieure*, **49**, 15–27.

—— & WENDT, A. 1987. Die Erdgasbohrung Entlebuch 1, ein Tiefenaufschluss am Alpennordrand. *Bulletin der schweizerischen Vereinigung Petroleum-Geologen und -Ingenieure*, **53**, 67–79.

WILDI, W., FUNK, H., LOUP, B., AMATO, E. & HUGGENBERGER, P. 1989. Mesozoic subsidence history of the European marginal shelves of the alpine Tethys (Helvetic realm, Swiss Plateau and Jura). *Eclogae Geologicae Helvetiae*, **82**, 817–840.

WILLETT, S. D. 1992. Modelling thermal annealing of fission tracks in apatite. *In*: ZENTILLI, M. & REYNOLDS, P.H. (ed.) *Mac short course hand book on low temperature chronology*, **20**, 53–72.

ZIEGLER, P. A. 1987. Late Cretaceous and Cenozoic intra-plate compressional deformations in the Alpine foreland — a geodynamic model. *Tectonophysics*, **137**, 389–420.

—— 1990. *Geological Atlas of Western and Central Europe 1990*. Shell Internationale Petroleum Maatschappij B.V.

—— 1992. European Cenozoic rift system. *Tectonophysics*, **208**, 91–111.

Geochemical modelling in an organic-rich source rock: the Bazhenov Formation

J. A. HEGRE[1], J. L. PITTION[1], J. P. HERBIN[2] & N. V. LOPATIN[3]

[1] *TOTAL, TEP/DE/DPN, Tour Total, 24 cours Michelet, Cedex 47, 92069 Paris La Défense, France*

[2] *IFP, 1–4 av. de Bois-Préau, BP 311, 92506 Rueil Malmaison Cedex, France*

[3] *The Russian Federation National Geosystems Institute, Varshavskoye shosse 8, 113105 Moscow, Russia*

Abstract: Three wells located in the West Siberian basin have a high density of geochemical information (closely spaced samples) from the Upper Jurassic Bazhenov Formation. The nature of the organic matter is homogeneous and is classified as a marine type II (Hydrogen Indices up to 700 mg of HC g^{-1} of TOC). The organic content is very high with TOC values up to 20 %. The source rock potential is also very good (S2 up to 100 kg HC $tonne^{-1}$ of rock). The maturity level in the wells varies from immature (T_{max} around 435 °C) to near the end of the oil window (T_{max} around 455 °C). This situation is somewhat puzzling since the source rock is at the same depth in the three wells (between 2800–2900 m). The IFP 1-D software GENEX, which is a maturity model that integrates subsidence, thermal reconstruction, hydrocarbon generation and expulsion, was used in order to:

(a) test a geological hypothesis, which could reasonably explain this situation;
(b) investigate the accuracy of the model's ability to calibrate with geochemical data;
(c) test the applicability of calibrating the expulsion saturation threshold from S1 and PI data. The very high S1 values with respect to the immaturity are also discussed. Model results indicate that different expulsion saturation thresholds are required in order to calibrate S1 and PI data. This is probably related to an inadequate definition of the formation's porosity.

Basin modelling is now a common exploration tool. Most basin models are controlled by calibrating with temperature or vitrinite reflectance data. However, these models should also be validated by geochemical data. The analysis of organic matter in sedimentary rocks is critical to interpreting its maturity and petroleum generation potential. Data on organic matter are commonly obtained by Rock-Eval pyrolysis (Espitalié *et al.* 1985; Peters 1986). One of the concerns of this paper is to analyse the problems encountered when calibrating a 1-D basin model with Rock-Eval data instead of thermal maturity data. The data utilized are T_{max}, the temperature for which the S2 peak is maximum, expressed in °C and the Hydrogen Index (HI) = S2/TOC, expressed in mg HC g^{-1} TOC. Rock-Eval Production Index (PI) = S1/(S1+S2) and S1 data were calibrated to estimate the expulsion saturation threshold. All modelling was done with GENEX, a commercially available program developed by the Institut Français du Pétrole (IFP) which simulates compaction, thermal histories, maturity, petroleum generation and expulsion (Ungerer 1990).

The 1-D basin modelling study was conducted for three wells, Ravenskaya 175, Salymskaya 157, and Salymskaya 176 (referred to as Raven, Saly 157, and Saly 176, respectively), located in central Western Siberia (Fig. 1). These wells have a high density of geochemical Rock-Eval data from the Upper Jurassic Bazhenov Formation, the main source rock. In each well, the maturity level indicated differs, varying from immature (T_{max} about 435 °C) to near end of the oil window (T_{max} about 455 °C). This situation is somewhat puzzling since the source rock is at approximately the same depth in the three wells, between 2800–2900 m. Saly 157 (E71°09″, N61°12″) and Saly 176 (E70°35″ N61°09″) are in the Salym Skoye field; Raven (E74°12″ N61°54″) is located to the northeast in the Ravenskoye field. Thus, GENEX was used to test a geological hypothesis which could reasonably explain this situation as well as to investigate problems encountered when calibrating a model with Rock-Eval data (two temperatures and no vitrinite reflectance data were available).

HEGRE, J. A., PITTION, J. L., HERBIN, J. P. & LOPATIN, N. V. 1998. Geochemical modelling in an organic-rich source rock: the Bazhenov Formation. *In*: DÜPPENBECKER, S. J. & ILIFFE, J. E. (eds) *Basin Modelling: Practice and Progress*. Geological Society, London, Special Publications, **141**, 157–167.

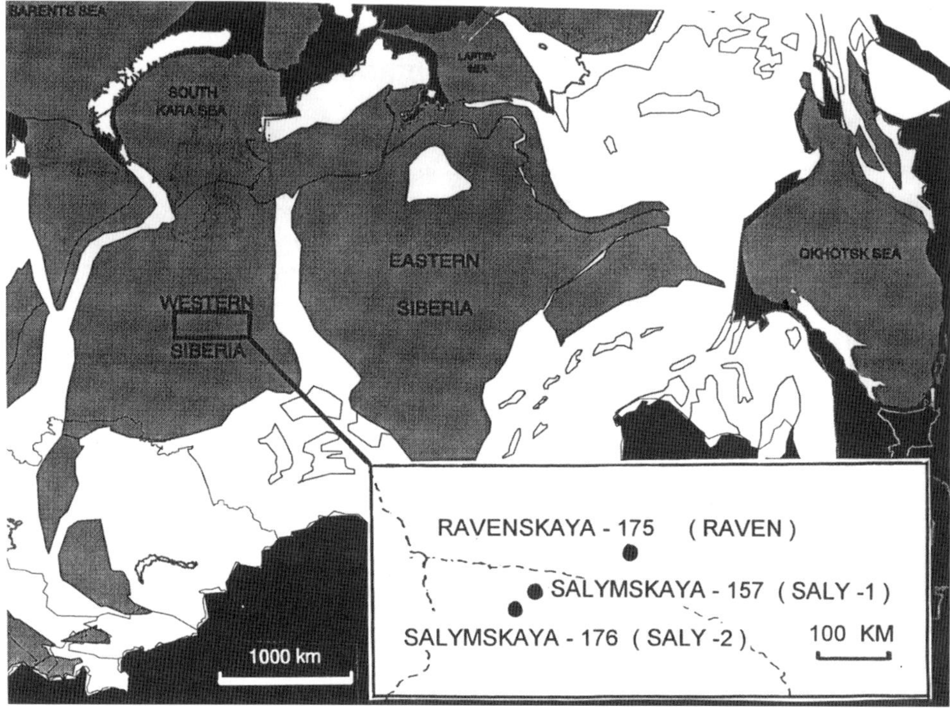

Fig. 1. Location of the three West Siberian wells modelled with 1-D GENEX.

Geological setting

The West Siberian basin is one of the world's largest petroleum provinces. The basin is bounded on the west by the Uralian and Novaya Zemlya uplifts, on the east by the Siberian craton and Taymyr uplift, on the south by the Kazakh and Altay–Sayan uplifts, and on the north by the North Siberian sill. All the geological circumstances required for hydrocarbon generation and accumulation are present: high-quality source rocks, excellent reservoirs, extensive seals, good traps, sufficient temperature for maturation, and absence of significant faulting or erosion to disturb accumulations.

The basement in central Siberia consists mainly of metamorphic rocks and some granitic intrusives of Precambrian and Palaeozoic ages. Much of the Palaeozoic section is also metamorphosed and is considered as basement. This basement is overlain by a thick Mesozoic–Cenozoic section of platform sediments (3–4 km). These sediments were deposited in a broad, shallow sea and have undergone only mild tectonic disturbance since deposition. They consist almost entirely of clastic sediments (sandstones, siltstones and shales), which were deposited in three major transgressive–regressive sedimentary cycles: Triassic–Aptian, Aptian–Oligocene, and Oligocene–Quaternary. Predominantly continental sediments occur at the base of each megacycle grading to largely marine or nearshore sediments at the top. Each cycle is separated by an unconformity (Aleinikov *et al.* 1980). In Early Jurassic time, the West Siberian basin began a steady subsidence, which has continued, with minor interruptions, to the present.

At the top of the Jurassic section is the main source rock, the Bazhenov Formation, which was deposited over much of the basin. The presence of highly organic shale indicates that semi-starved conditions existed in the basin at this time. On seismic sections, the Bazhenov is represented by a maximum flooding surface which is overlain by prograding wedges.

Geochemistry

Rock-Eval data from closely spaced samples in the Bazhenov Formation are available for the three wells. The source rock is at approximately the same depth in each well, between 2850–2900 m in the wells Saly 157 and Saly 176; in Raven, it is 50 m higher, between 2800–2850 m. Even though the

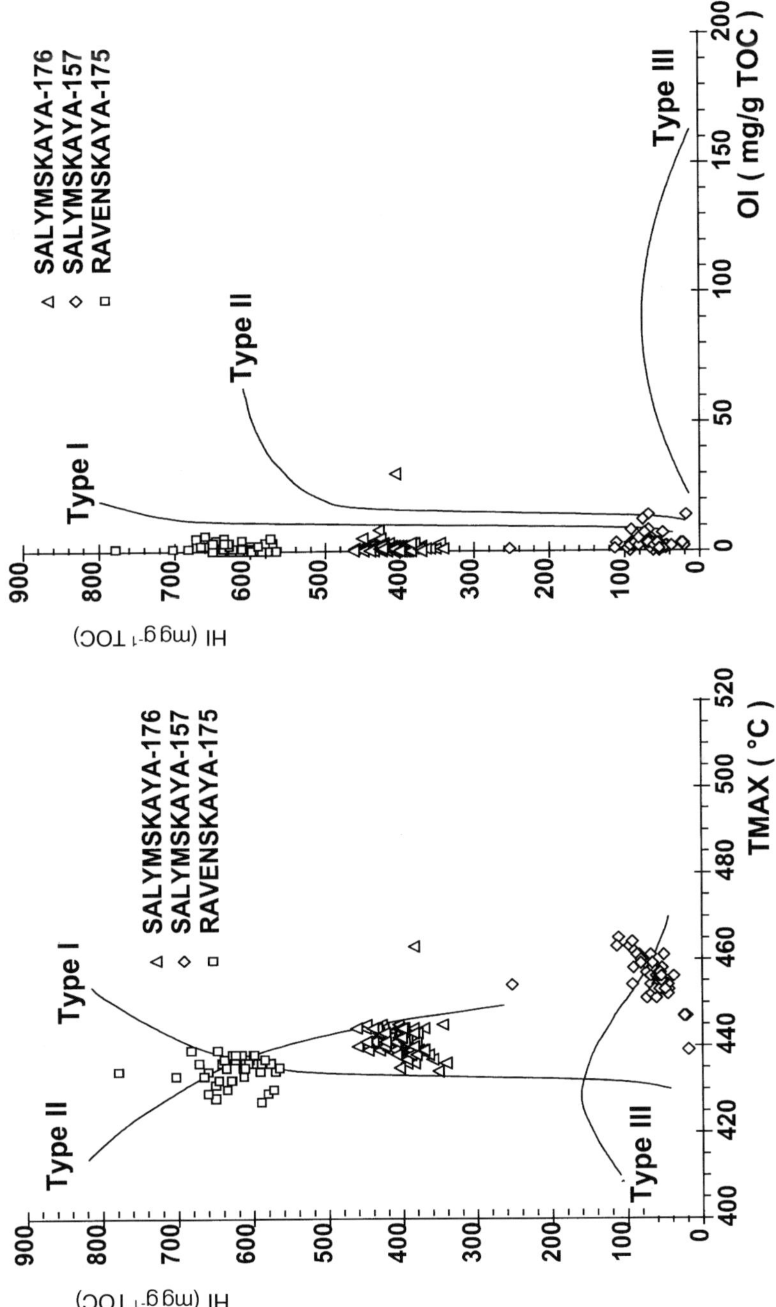

Fig. 2. T_{max} versus HI and OI versus HI plots demonstrate the different levels of maturity for the Bazhenov Formation in three West Siberian wells.

geological environment is similar, different maturity levels are displayed by the geochemical data. The nature of the organic matter is homogeneous and is classified as a typical marine type II kerogen (Hydrogen Index up to 700 mg HC g^{-1} TOC). No data which could indicate possible lithological variations within the Bazhenov Formation were available (well logs, cores, etc.).

The T_{max} versus Hydrogen Index (HI) and Oxygen Index (OI = S3/TOC, expressed in mg CO_2 g^{-1} TOC) versus Hydrogen Index plots clearly show the different levels of maturity in the Bazhenov Formation (Fig. 2). The evolution is seen from immature in Raven (HI = 700 mg HC g^{-1} TOC and T_{max} = 425 °C) to mature in Saly 176 (HI = 420 mg HC g^{-1} TOC, T_{max} = 440 °C) and to almost overmature in Saly 157 (HI = 100 mg HC g^{-1} TOC, T_{max} = 460 °C).

On the TOC versus S2 diagram (Fig. 3), TOC values are almost as high as 30 % in Raven, reflecting the very high content of organic matter. In Saly 176, the TOC ranges from 8–18 %, and in the mature well, Saly 157, the values are between 1 and 16 %. The source rock potential is also very good, S2 up to 180 kg HC $tonne^{-1}$ of rock in the immature well, Raven; most values range from 40–100 kg HC $tonne^{-1}$ of rock. Considering an average thickness of 50 m and S2 of 80 kg HC $tonne^{-1}$ of rock, the Source Potential Index (SPI, the maximum quantity of hydrocarbons that can be generated within a column of source rock under one square metre of surface area, in metric tons of hydrocarbons per square metre) can be estimated to be 10 tonne m^{-2}, which corresponds to a high potential source rock.

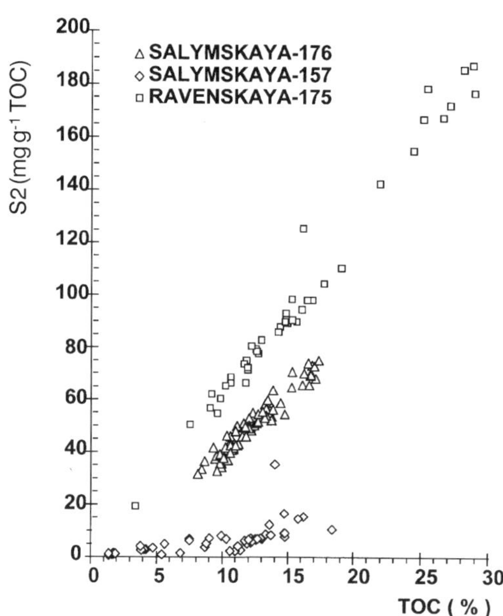

Fig. 3. TOC versus S2 plot.

Proposed hypothesis

As clearly indicated by the geochemical data, the Bazhenov source rock is at dissimilar maturity levels in these wells even though the source rock is at approximately at the same depth. Saly 157 is approximately 25 km from Saly 176 and Raven is about 200 km further to the northeast (Fig. 1). Several geological factors such as erosion, faults, volcanic intrusions, water circulation (hydrothermal, topographic driven), or a change in basement composition could be the cause of this difference in maturity.

According to the geological data available for this region, there is no evidence of any major erosional events, faults or recent volcanic intrusions. Thus, it can be assumed that no major thermal disturbances caused by these geological processes, which would affect the maturity level, occurred from the Jurassic to present day. The burial history curve for the well Saly 157 is shown on Fig. 4. However, a likely cause of this phenomenon is water circulation (either hydrothermally or topographically driven). Since lateral water flow cannot be adequately accounted for in a vertical 1-D model, it was not considered for this study.

Another possibility is a change in the composition of the basement, which would cause the

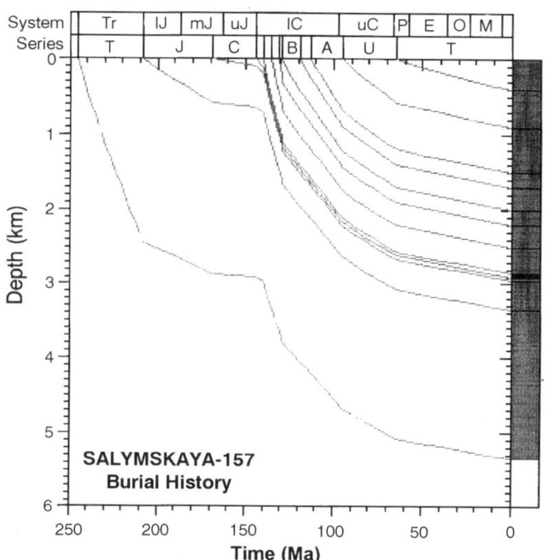

Fig. 4. Burial history graph for the well Saly 157.

radiogenic heat flow produced by the basement to be higher or lower (Roy, *et al.* 1968). For instance, a granitic basement produces a large amount of radiogenic heat (50–130 mW m^{-2}) since it is rich in radioactive elements, whereas a metamorphic basement, being poor in radioactive elements, produces only a small amount of heat (10–30 mW m^{-2}). This type of hypothesis was tested in GENEX. Granitic plutons are known to exist in the vicinity of Saly 157 and 176, which leads to the assumption that the radiogenic heat flow is higher in this area. On the other hand, the field of Ravenskoye (well Raven) is mainly underlain by Precambrian–Palaeozoic volcanic and slightly metamorphic rocks, thus implying that the radiogenic heat flow is lower in this region (Peterson & Clarke 1991).

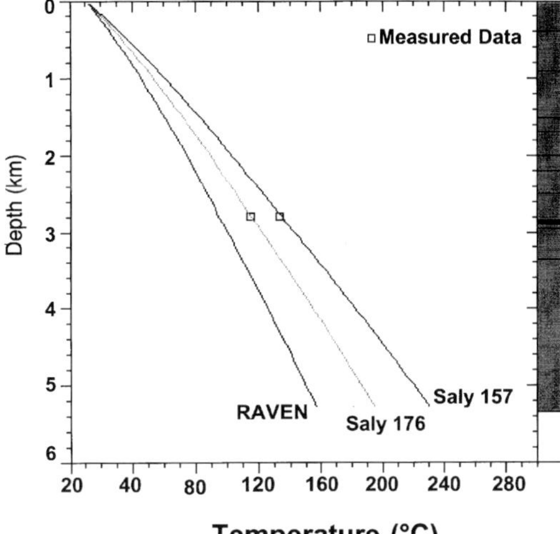

Fig. 5. Temperature versus depth graph for the three wells; constant heat flow = 25 mW m^{-2}, radiogenic heat flow = 40 mW m^{-2} for Saly 157, 30 mW m^{-2} for Saly 176 and 19 mW m^{-2} for Raven.

Thermal maturity modelling

In GENEX, the two main factors used to define the thermal regime are the type of thermal history: constant or variable heat flow which is considered to originate from the bottom of a 40 km thick basement (i.e. the user-defined heat flow starts 40 km beneath the sedimentary column), and the radiogenic heat flow produced within this basement and the sediments. By default, the radiogenic heat flow of the basement is 19 mW m^{-2}, which is applicable to a slightly volcanic–metamorphic Palaeozoic basement.

For this study, a constant heat flow thermal history was chosen because the basin appears to be undergoing normal subsidence and no major erosional events have occurred. In order to account for a possible change in the basement composition, it was assumed that the radiogenic heat flow is different for each well. The constant heat flow thermal history was determined by calibrating with geochemical data (T_{max} and HI). Only two temperatures, one from Saly 157 (134 °C at 2801 m) and the other from Saly 176 (116 °C at 2800 m) were available; these data were plotted on the GENEX temperature graph in order to check the modelled results (Fig. 5). There is no vitrinite reflectance data.

The immature well Raven clearly demonstrates a problem commonly encountered in basin modelling when using geochemical data for calibration purposes. It is very difficult to calibrate an immature well, since the model can not accurately simulate the same results as a Rock-Eval device for the range of T_{max} lower than 435 °C. According to the model, in order to calibrate with measured T_{max} data a constant heat flow of 38 mW m^{-2} is required, whereas a calibration with measured Hydrogen Index data needs a constant heat flow of 22 mW m^{-2} (Fig. 6). This difference in heat flow is quite large and has a major effect on the timing of hydrocarbon generation: nearly no generation for the HI calibration, Transformation Ratio (TR) = 6 % whereas a good Tmax calibration is obtained with TR = 70 % corresponding to an onset of generation at 70 Ma. The Transformation Ratio (TR) is a quantitative parameter that measures the extent to which the hydrocarbon potential has been effectively realized, determined by primary cracking (the amount of hydrocarbons generated by primary cracking to the maximum amount of hydrocarbons that can be generated). In exploration, these types of discrepancies in the timing of generation can induce serious errors when evaluating a play or prospect.

Due to the difficulty in calibrating with T_{max}, more emphasis was put on calibrating with the Hydrogen Indices. The reason for this is that the initial Hydrogen Index of the kinetic parameters can be ascertained from the geochemical data. The thermal history defined for Raven is a basal constant heat flow of 25 mW m^{-2}, and a radiogenic heat flow of 19 mW m^{-2} (for a typical Palaeozoic basement). This results in a good fit with measured Hydrogen Indices and a poor fit with T_{max} (Fig. 7).

For Saly 157 an almost perfect calibration is achieved for both measured T_{max} and Hydrogen Index data (Fig. 8) with a constant heat flow of 25 mW m^{-2} at the base of the crust and a radiogenic heat flow of 40 mW m^{-2}; a high value was used to simulate a granitic intrusion possibly located in the basement beneath this well. In the well Saly 176, a

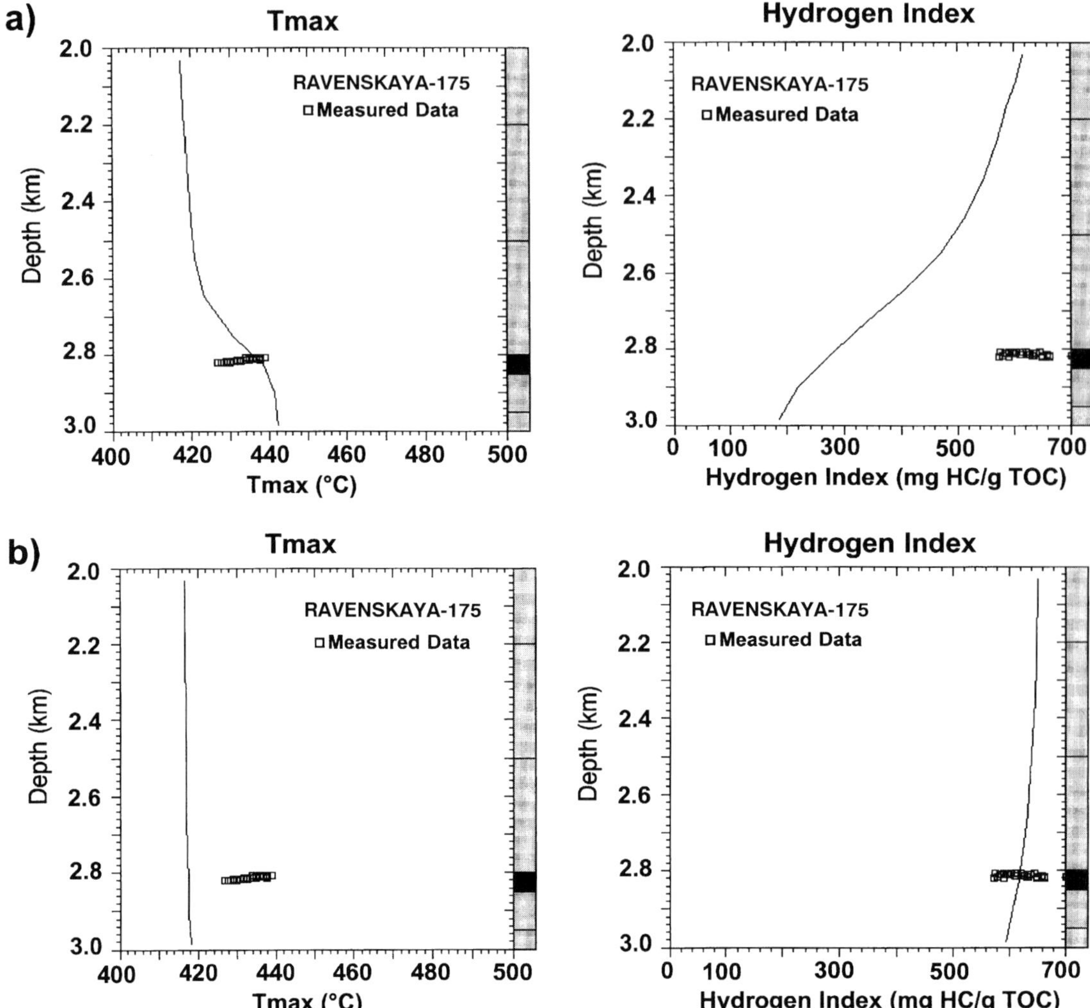

Fig. 6. Sensitivity calibration on Ravenskaya-175: (a) constant heat flow = 38 mW m^{-2}; good T_{max} calibration and poor HI calibration, (b) constant heat flow = 22 mW m^{-2}.

good HI calibration (Fig. 9) is obtained with a constant heat flow of 25 mW m^{-2} and a cooler radiogenic heat flow, 30 mW m^{-2}, implying that a granitic intrusion may be in the vicinity. The calculated temperature curve for this well and Saly 157 do coincide with the temperature values from these well (Fig. 5). The discrepancy between measured and calculated T_{max} values is about 10 °C, which can be considered as an acceptable error range.

Based on the above defined thermal histories, modelling results indicate that the present day transformation ratio for the Bazhenov Formation is 94 % in Saly 157, 55 % in Saly 176 and 11 % in Raven. This appears to be in agreement with the conventional maturity zones given by the geochemical data: Saly 157 is in the oil–gas transition zone, Saly 176 is in the oil zone and Raven is immature.

In conclusion, it is evident that even though the three wells were modelled with the same basal constant heat flow of 25 mW m^{-2}, the amount of radiogenic heat flow generated by the basement has sufficient influence on the sediment's thermal regime to affect the maturity level. The maturity differences observed in these wells could possibly be linked to only a change in basement composition. A majority of geological thermal distur-

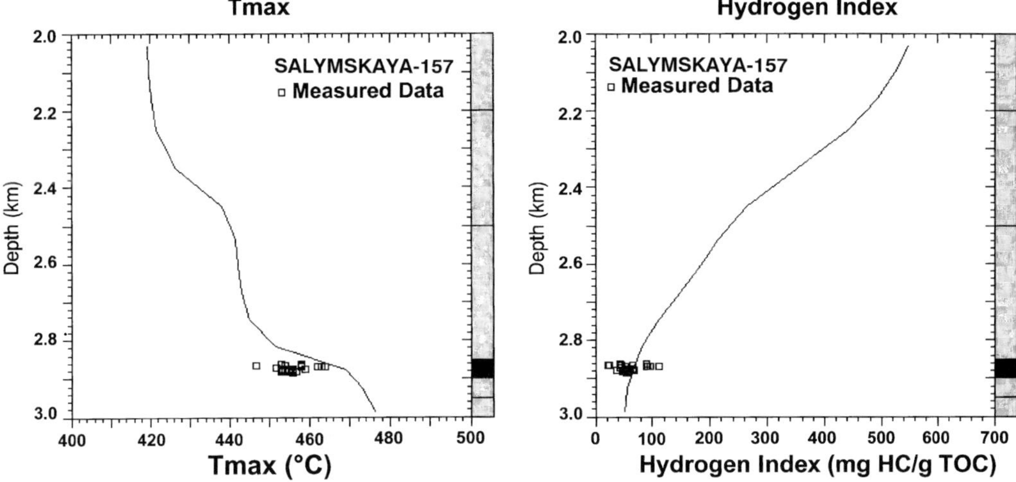

Fig. 7. Geochemical calibration results from the immature well Raven. A constant basal heat flow of 25 mW m^{-2} and a radiogenic heat flow of 19 mW m^{-2} were defined. The HI computed curve falls within the measured data, but the T_{max} curve is lower than the measured values. The Bazhenov Fm is shown in black.

bances originate from the upper basement and not from within the sedimentary column. Therefore, adjusting the basement's radiogenic heat flow according to its composition is more realistic than adjusting a sedimentary column's heat flow to match measured data.

It has been shown that Rock-Eval parameters such as Hydrogen Index and T_{max} can be compared and calibrated with model results, but there are some constraints. In the present case, the Hydrogen Index was a more reliable tool for calibration since the initial Hydrogen Index of the source rock could be estimated. The problems encountered for calibrating T_{max} are related to the difficulty to

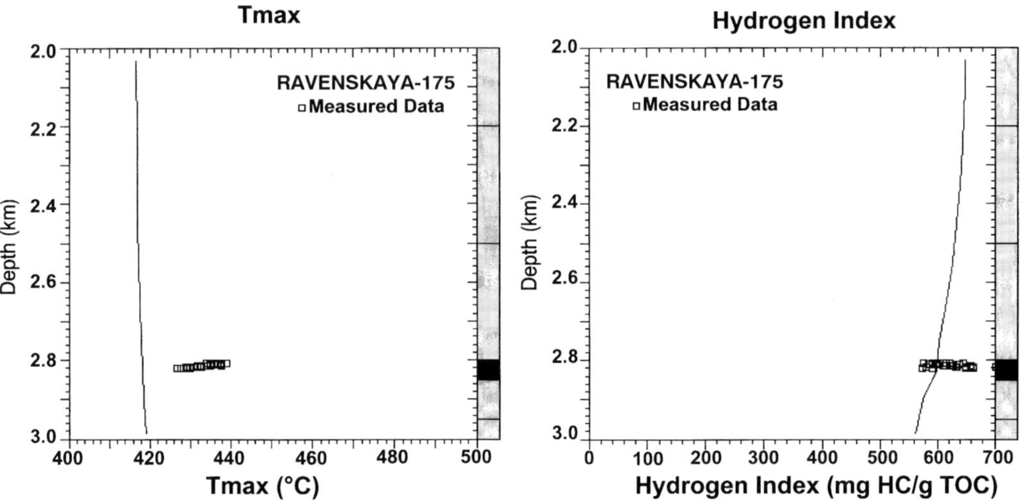

Fig. 8. A good calibration is obtained in the almost overmature well Saly 157 with a constant heat flow of 25 mW m^{-2} and a radiogenic heat flow of 40 mW m^{-2}. The higher radiogenic heat flow is used to test the possibility of a granitic intrusion. Both HI and T_{max} calculated curves intersect measured data. The Bazhenov Fm is shown in black.

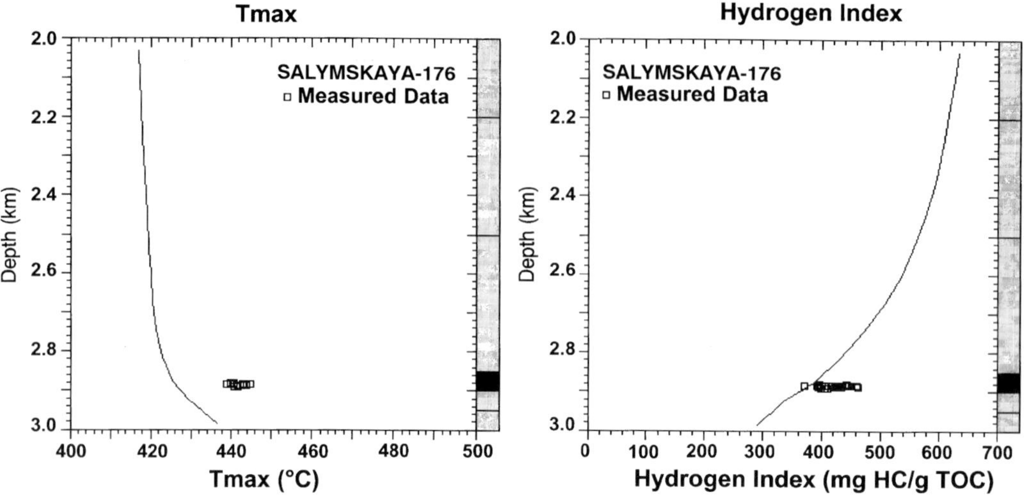

Fig. 9. For Saly 176, the mature well, the HI curve calibrates with measured data, but the T_{max} curve is below values. A constant heat flow of 25 mW m^{-2} and a radiogenic heat flow of 30 mW m^{-2} were used. The Bazhenov Fm is shown in black.

properly simulate the T_{max} evolution mainly in the low maturity ranges. T_{max} prediction requires an accurate estimation of the rates of generation during pyrolysis, which is a rigid requirement for an empirical model.

Source rock expulsion

Models of expulsion use a simplified zero-dimensional treatment, based either on a saturation threshold (Ungerer *et al.* 1988) or on fluid flow equations – relative permeability curves (Braun & Burnham 1989; Düppenbecker & Welte 1991). These models account for the density changes of petroleum when its composition changes from oil to gas and for the porosity decrease as a result of compaction. The model of Düppenbecker & Welte (1991) stresses the influence of pore size distribution and fracturing. Modelling primary migration still presents difficult fundamental problems, but empirical expulsion models have already improved the qualitative and quantitative assessment of the petroleum available for secondary migration. In these models, the physics of expulsion is not sufficiently well known to give predictive values of the saturation threshold or of the relative permeabilities.

A simplified model based on a saturation threshold, coupled with one-dimensional models of backstripping and palaeotemperature reconstruction is used to simulate expulsion in GENEX. For a given maturity level the expulsion process depends on two main factors:

(a) the quantity of the organic matter (TOC);
(b) the saturation threshold.

In modelling, the initial TOC from an immature sample should be used. The final or observed TOC, which can be significantly different from the initial TOC in mature source rocks that have expelled hydrocarbons (Rullkötter *et al.* 1988), is computed by the model. This type of computation allows the determination of the initial TOC via a trial and error process by comparing the modelled results to measured values. Thus, the TOC can be used to control and calibrate the expulsion model, but this method does not account for TOC data scattering. In this case study, it was decided to determine the initial TOC by this method and use the same initial TOC in the three wells. The initial TOC was estimated to be 20 %, which is slightly lower than some of the reported values from the immature well, Raven (Fig. 3). But the final TOC computed by the model corresponds with the mean of the observed values for each well: computed final TOC for Raven (immature) = 19.6 % (mean of measured values = 17.6 %), for Saly 176 (mature) = 14.4 % (mean of measured values = 13.4 %), and for Saly 157 (very mature) = 11.1 % (mean of measured values = 10.7 %).

The remaining unknown is the saturation threshold. According to GENEX, expulsion starts once a certain hydrocarbon's saturation threshold (S_{max}) (by default 30 %) is reached. This means

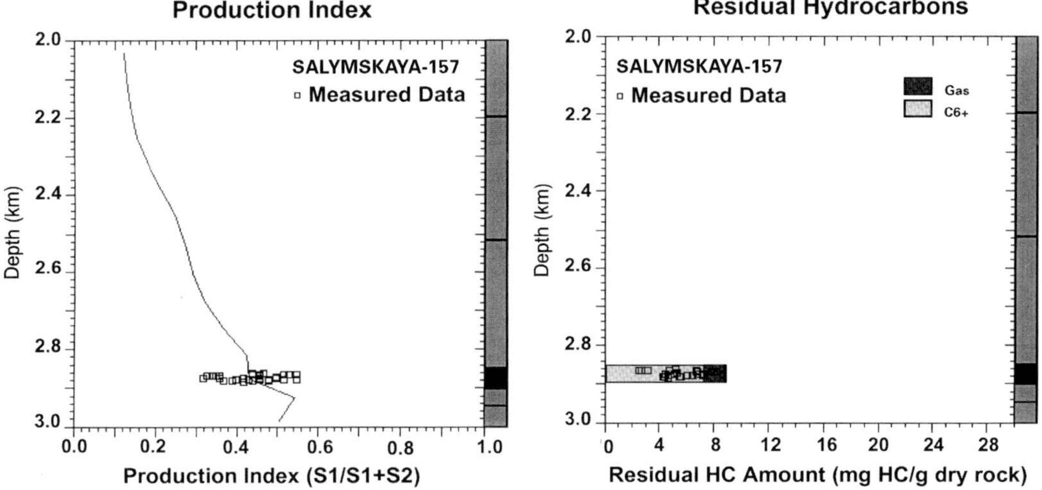

Fig. 10. In order to calibrate with measured PI and S1 data in the almost overmature well Saly 157, a saturation threshold of 50 % is required. The TOC is 20 %. The Bazhenov Fm is shown in black. C6 refers to oil, gas to C5–C1 and coke to residue.

that 30 % of a layer's total pore volume (dependent on porosity not permeability) must be filled with hydrocarbons before expulsion will occur. If the saturation threshold is increased, it will cause expulsion to occur later in time. There is no direct method to measure this threshold. Nevertheless, a calibration of the GENEX calculated residual petroleum and Production Index (PI = S1/(S1 + S2)) with Rock-Eval measured S1 and PI allows a better determination of this threshold (Forbes *et al.* 1991). If the model predicts excess amounts of residual petroleum or higher Production Indices in mature source rocks, the saturation threshold must be decreased, and vice versa.

The main limitation in the use of S1 data for model control is that during the Rock-Eval analysis

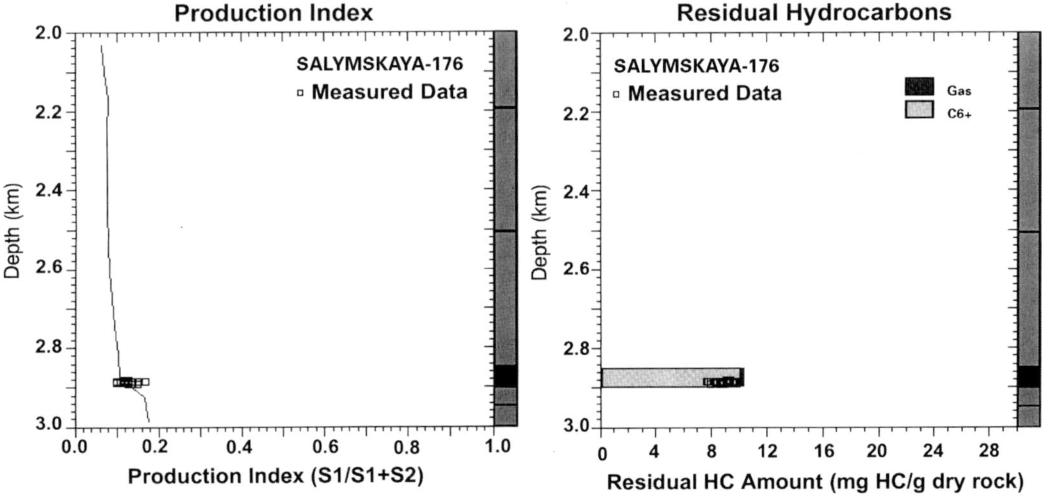

Fig. 11. For Saly 176, the mature well, a saturation threshold of 35 % (TOC = 20 %) calibrates with measured PI and S1 data. The Bazhenov Fm is shown in black. C6 refers to oil, gas to C5–C1 and coke to residue.

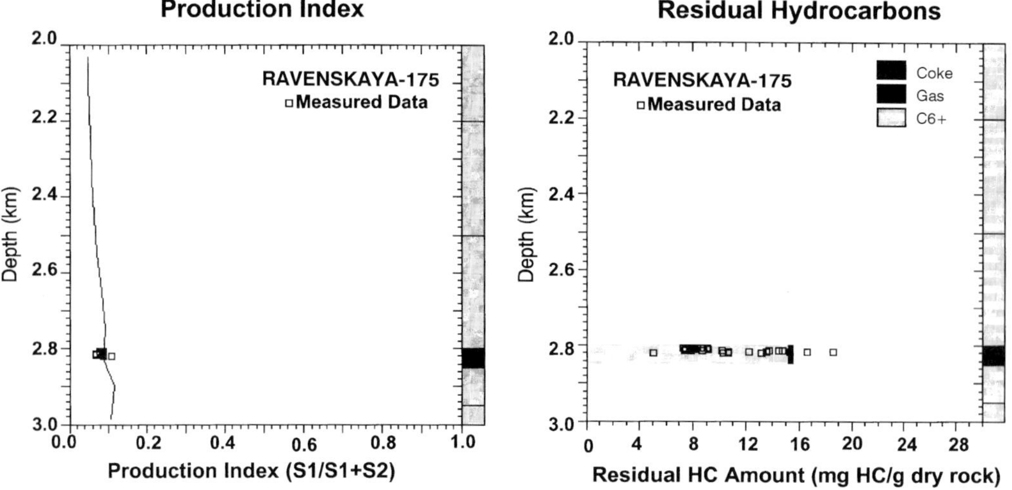

Fig. 12. For Raven 175, in spite of low TR 6 %) a good calibration of the high S1 values are obtained with a saturation threshold of 50 %. C6 refers to oil, gas to C5–C1 and coke to residue.

the light ends are often lost by evaporation prior to analysis and the heavy ends (resins and asphaltenes) are usually not detected (Ungerer 1990). This implies that model results should be corrected for the amounts of light and heavy fractions lost. However, these types of corrections are very difficult since the proportion of losses is variable and poorly documented. Generally, observed Production Index trends are more clearly defined than S1 trends. Therefore, Production Indices may be more reliable than S1 data to control expulsion models.

The applicability of the PI and S1 calibration method was tested in the three Siberian wells. For Saly 157 (very mature), an S_{max} of 50 % seems to calibrate fairly well with both PI and S1 (Fig. 10). Due to the range in values, it is very difficult to determine the best calibration. The data from the well Saly 176 (mature) show less scatter and calibrate with a saturation threshold of 35 % (Fig. 11). It is interesting to note that in spite of the immature stage of Raven (T_{max} about 430 °C, TR = 6 %), the observed S1 are high (7 to 15 kg t^{-1}). Calculated residual petroleum and PI are rather well calibrated with the high S1 values by using a S_{max} of 50 % (Fig. 12).

The difference of saturation thresholds used in the wells could be explained by the fact that the porosity used for shale in GENEX (10.4 %) may not correspond to the porosity of the Bazhenov Formation (no data are available to check this hypothesis). The saturation threshold is dependent on the porosity, which may be in error by several percent. Furthermore, the porosity linked to the organic matter is not taken into account.

A change in porosity appears to be indicated on the S1 versus TOC diagram (Fig. 13). It clearly shows the typical decrease in TOC with maturity as well as the high S1 values (greater than 15 kg t^{-1})

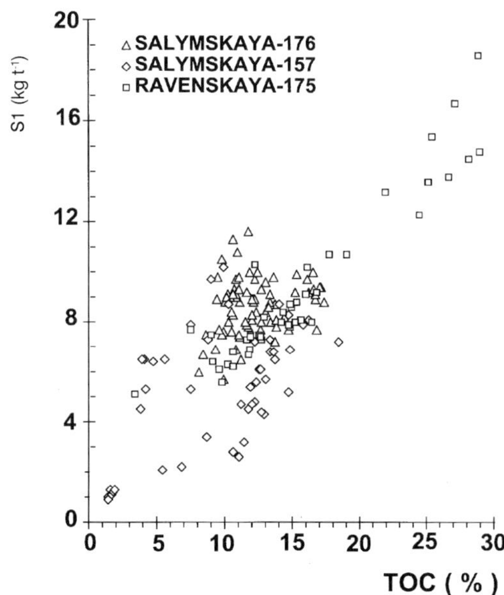

Fig. 13. TOC versus S1.

from the immature well Raven. However, the unusual aspect of this diagram is that the S1 seems to increase with increasing TOC. Normally, this is not the case and S1 should be the same at a given maturity regardless of the TOC since it corresponds to the shale's porosity which is independent of TOC. Thus, this could reflect a change in porosity; a much higher available porosity in the organic matter in addition to the shale's porosity.

Conclusions

(1) It is evident from geochemical data and modelling results that some thermal disturbance is necessary in order to cause the different levels of maturity seen in the Bazhenov Formation. The hypothesis of a change in basement composition (radiogenic heat flow) provided a probable solution, but it does not eliminate other possible causes (e.g. hydrothermal).

(2) This study has shown that the Hydrogen Index was a more reliable calibration tool than T_{max}. The modelled T_{max} values did not always correspond to Rock-Eval values. However, thermal maturity (temperature and vitrinite reflectance) data is required to confirm and support any model calibrated with only geochemical data.

(3) In expulsion models, S1 and PI data can be used to calibrate the saturation threshold. Some discrepancies between measured and computed values are to be expected since the measured S1 does not exactly represent the modelled S1 (i.e. the residual hydrocarbons).

(4) Modelling results from the wells Saly 157 and Saly 176 indicate that different saturation thresholds, 50 % and 35 %, respectively, must be used in order to calibrate with PI and S1 data from the Bazhenov Formation. This phenomena could be related to the difficulty in estimating the correct porosity (including organic porosity) to be used in the model.

References

ALEINIKOV, A. L., BELLANVIN, O. V., BULASHEVICH, YU. P., TAVRIN, I. F., MAKSIMOV, E. M. ET AL. 1980. Dynamics of the Russian and West Siberian Platforms. *In*: BALLY, A. W. *et al.* (eds) *Dynamics of Plate Interiors*, American Geophysical Union and Geological Society of America, 53–71.

BRAUN, R. L. & BURNHAM, A. K. 1989. *A mathematical model of oil generation, degradation and expulsion.* Lawrence Livermore National Laboratory, Rep. UCRL 101708.

DÜPPENBECKER, S. J. & WELTE, D. H. 1991. *Petroleum expulsion from source rocks — insight from geology, geochemistry and computerized numerical modelling.* Proc. of the 13th World Petroleum Congress, Buenos Aires, October 1991 (**3**), Wiley, London.

ESPITALIÉ, J., DEROO, G. & MARQUIS, F. 1985. La pyrolyse Rock-Eval et ses applications. *Revue de l'Institut Français du Pétrole*, **40**, 563–579 & 755–784.

FORBES, P. L., UNGERER, P. M., KUHFUSS, A. B., RIIS, F. & EGGEN, S. 1991. Compositional Modeling of Petroleum Generation and Expulsion: Trial Application to a Local Mass Balance in the Smorbukk Sor Field Haltenbanken Area Norway, *American Association of Petroleum Geologists Bulletin*, **75**, 873–893.

PETERS, K. E. 1986, Guidelines for evaluating petroleum source rock using programmed pyrolysis: *American Association of Petroleum Geologists Bulletin*, **70**, 318–329.

PETERSON, J. A. & CLARKE, J. W. 1991. *Geology and Hydrocarbon Habitat of the West Siberian Basin.* American Association of Petroleum Geologists Studies in Geology No. 32.

ROY, R. F., BLACKWELL, D. D. & BIRCH, F. 1968. Heat generation of plutonic rocks and continental heat flow provinces, *Earth and Planetary Science Letters*, **5**, 1–12.

RULLKÖTTER, J., LEYTHAEUSER, D., HORSFIELD, B., LITTKE, R., MANN ET AL. 1988. Organic matter maturation under the influence of a deep instrusive heat source. A natural experiment of hydrocarbon generation and expulsion from a petroleum source rock (Toarcian Shale, Northern Germany). *Organic Geochemistry*, **13**, 847–856.

UNGERER, P. 1990. State of the art of research in kinetic modelling of oil formation and expulsion. *Organic Geochemistry,* **16**, Nos. 1–3, 1–25.

——, ESPITALIÉ, J., BEHAR, F. & EGGEN, S. 1988. Modélisation mathématique des interactions entre craquage et migration lors de la formation du pétrole et du gaz. *Comptes Rendus de l'Academie des sciences. Série 2, pp.* 927–934.

A multidisciplinary approach to modelling secondary migration: a Central North Sea example

W. A. SYMINGTON[1], K. E. GREEN[2], J. HUANG[2], R. J. POTTORF[2] & L. L. SUMMA[2]

[1] *Esso Exploration and Production UK Ltd., Esso House, Ermyn Way, Leatherhead, Surrey KT22 8UY, UK*

Present address: Exxon Production Research, P. O. Box 2189, Houston, TX 77252, USA

[2] *Exxon Production Research, P. O. Box 2189, Houston, TX 77252, USA*

Abstract: Numerical simulations assist in understanding secondary migration, and can help estimate which traps in a basin may have received a hydrocarbon charge. This is illustrated by a suite of numerical models used to track the secondary migration of hydrocarbons into an Upper Jurassic Fulmar discovery in the Central North Sea. The models include areal flow calculations using Exxon's proprietary reservoir simulator, and cross-section calculations using TEMISPACK, a commercial basin modelling package. While the application of conventional reservoir simulation techniques to geologic time-scale problems is itself a novel approach, the most important development has been in the integration of fluid flow modelling with techniques for dating the timing of hydrocarbon charge. These techniques have been used to constrain and validate the model results. In addition, sensitivity studies permit evaluation of the required permeability for effective lateral migration.

The areal calculations provide information about migration timing and the fill history of several traps. These fields are charged from mature Kimmeridge source rocks, mapped within the local drainage area. Several model runs were used to evaluate the sensitivity of migration timing to carrier bed permeability and capillary pressure characteristics. Calculations indicated that for reservoir quality carrier beds (permeability ≥ 1 mD) secondary migration can be geologically instantaneous, and that even low permeability silts can serve as effective migration pathways.

A cross-sectional model was used to integrate cross-stratal and strata-parallel migration. The model included hydrocarbon expulsion from Kimmeridge source rocks into the Fulmar sandstone and lateral migration up the flank of the structure. Drilling results indicate that leakage into younger Tertiary strata has occurred. The calculation suggests a portion of the top seal breached by 10–20 Ma. Evidence of partial seal loss includes a seismic chimney interpreted above the predicted breached seal, and an appraisal well high on the Jurassic structure which encountered hydrocarbon-stained reservoir rocks with an apparent palaeohydrocarbon contact. Furthermore, hydrocarbons in the overlying Tertiary reservoirs can be geochemically tied to the remaining Jurassic accumulation. Finally, analysis of authigenic cements, particularly the K–Ar ages of fibrous illite, suggests that hydrocarbons occupied the trap until approximately 10 Ma. Analysis of Tertiary hydrocarbon fluid inclusions further indicates that hydrocarbons entered these Tertiary reservoirs approximately 10 Ma. These latter two techniques now enable the modeller to apply temporal constraints to hydrocarbon charge simulation. Such constraints allow 'history matching' analogous to field scale reservoir simulations.

Numerical simulations have assisted in understanding secondary migration into an Upper Jurassic Fulmar discovery in the Central North Sea. The simulations included one-dimensional thermal history models which quantify source maturation and yield, areal flow calculations of migration timing and trap fill history, and cross-sectional flow models integrating cross-stratal and strata-parallel migration. The one-dimensional thermal history models and areal flow calculations were run with Exxon proprietary software for thermal modelling and reservoir simulation. The cross-sectional flow models were run using TEMISPACK, a commercial basin modelling package.

The multi-disciplinary nature of the study is evident in the diverse kinds of observations which support the simulation results. These include:

(a) The results of drilling in the modelled area;
(b) Observations made from 3-D seismic data including the interpretation of a gas chimney;
(c) The analysis of authigenic cements including

SYMINGTON, W. A., GREEN, K. E., HUANG, J., POTTORF, R. J. & SUMMA, L. U. 1998. A multidisciplinary approach to modelling secondary migration: a Central North Sea example. *In*: DÜPPENBECKER, S. J. & ILIFFE, J. E. (eds) *Basin Modelling: Practice and Progress*. Geological Society, London, Special Publications, **141**, 169–185.

both hydrocarbon and aqueous fluid inclusion analysis, and K–Ar ages of diagenetic illite;
(d) The biomarker geochemistry of the reservoired hydrocarbons.

Geological setting

The analysis was conducted on a gas condensate discovery in the Central Graben of the Central North Sea. The geological setting is summarized in Fig. 1. Major faults separating structural provinces are shown on this map. Important elements of the stratigraphy include the Jurassic Fulmar and Triassic Skaggerak reservoirs. The Kimmeridge Clay provides the primary source rock in the area. Younger reservoir intervals include the Tay, Forties, and Sele formations in the Early Tertiary.

The primary reservoir is in the Jurassic Fulmar sandstone. We believe the Kimmeridge source beds expelled hydrocarbons downward into the Fulmar. Hydrocarbons then migrated up-dip under the action of buoyancy. The Kimmeridge clay and Heather shale acted as a regional top seal below which this migration took place. After the primary reservoir had filled, a seal breach over a portion of the field permitted hydrocarbons to escape to younger Tertiary reservoirs.

The field is a structural trap cored by a basement horst block. The trap geometry has been modified as a result of Zechstein salt withdrawal. Salt evacuation along basement faults has created synclines to the north and south of the structure. These deep sub-basins provide the hydrocarbon kitchen areas.

The synclines can be seen in Fig. 2, a drainage analysis map with depth structure contours on the Base Cretaceous. Colours on this map indicate depth. The depth in the northern salt withdrawal syncline reaches 6600 m. The southern salt withdrawal syncline has a maximum depth of 5600 m. The crest of the structure is at 4300 m, and the map area is 50 by 40 km. The top of the Fulmar reservoir is below the base of the Cretaceous, but is not seismically mappable. We have therefore used the base Cretaceous unconformity as a proxy for the structure of the migration surface. The area, subjected to late thermal sag, has been buried uniformly during the late Cretaceous and Tertiary. As a result, the structure of the migration surface has remained unchanged since the onset of hydrocarbon generation, 70–80 Ma. The map displays a drainage analysis of the surface, the black lines marking the individual drainage polygons of each independent structural closure. The posted flow lines are valid assuming hydrocarbons move solely under the action of buoyancy within a continuous carrier bed. They indicate flow into the shaded structural closures.

Although numerous faults are posted on the surface, their throws are interpreted to be less than the thickness of the sealing lithologies overlying the reservoir. As a result these faults should be dip-sealing. Their primary impact on migration is

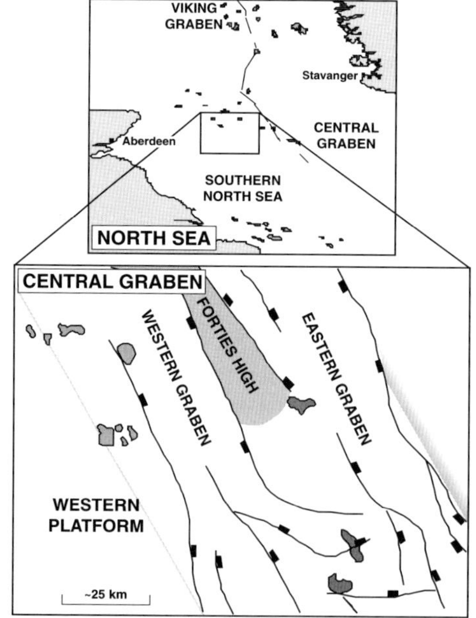

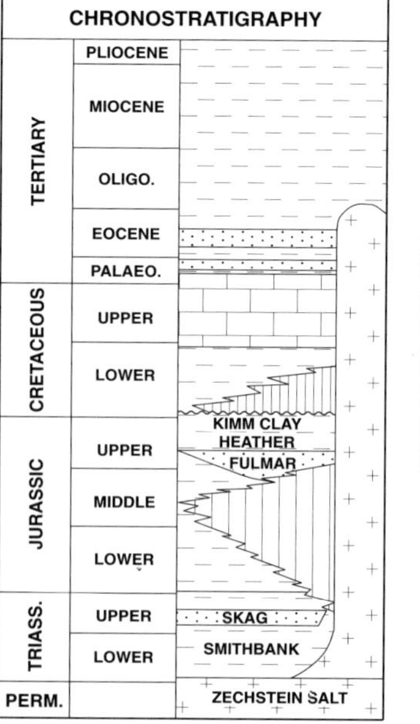

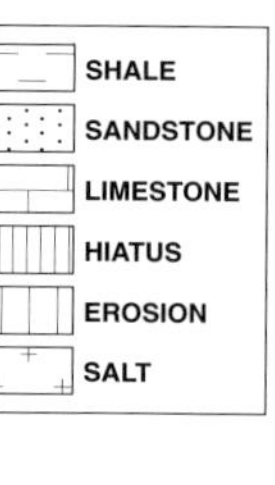

Fig. 1. Geological Setting — Central Graben, Central North Sea.

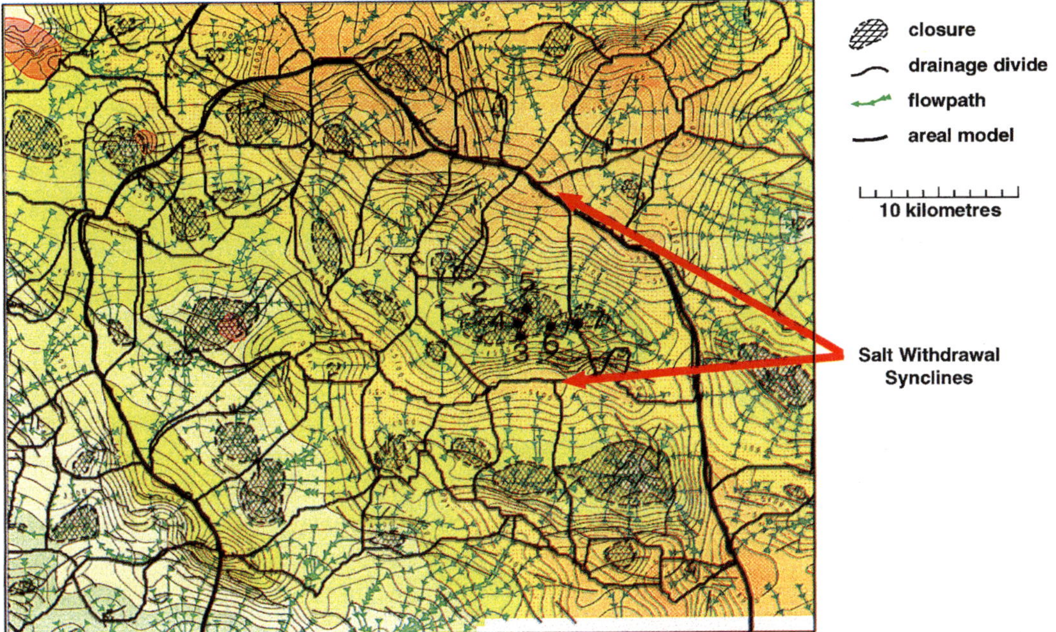

Fig. 2. Base Cretaceous drainage analysis. The black lines mark individual drainage polygons for each independent structural closure (cross-hatched). The posted flow lines are valid assuming hydrocarbons move solely under the action of buoyancy within a continuous carrier bed. They indicate flow into the structural closures. Pertinent wells drilled on the structure are also marked.

therefore to increase the local dip of the carrier bed. It is also likely that they set up a number of smaller traps which must be filled as hydrocarbons migrate into the main accumulation.

The map also shows the pertinent wells drilled on the structure. Wells 3, 5, and 6 encountered gas condensate in the Jurassic Fulmar sands. Wells 4 and 7 discovered condensate in the Tertiary Sele Formation. Wells 1 and 2 were both drilled to the Jurassic but were dry holes. The #2 well encountered hydrocarbon stains which are interpreted as a palaeohydrocarbon accumulation.

The heavy outline is the boundary of a model built for areal flow calculations. The outline conforms with the drainage polygons, and represents a closed migration cell. As a result, hydrocarbons generated outside the model are unlikely to impact migration within the model and vice versa.

A more compete description of the geology and hydrocarbon migration in the Central Graben of the Central North Sea is given by Cayley (1986) and Erratt (1992).

Areal flow modelling

Figure 3 gives a three-dimensional view of the base Cretaceous in the modelled area. The view is from the northeast looking at the deep northern syncline. The field and several nearby structures can be seen. The slope out of the syncline is up to 20 degrees. The white lines draped on the structure are flowpaths which would be followed by hydrocarbons moving under the action of buoyancy.

The black lines are the calculational grid used for areal flow modelling. Conventional reservoir simulators model fluid flow in porous rock by dividing the rock volume of interest into an array of grid blocks. The arrays may be two-dimensional areally, two-dimensional in cross-section, or three-dimensional. The blocks may be of variable sizes. They are generally nearly orthogonal and conform to the structure of the rock stata. Rock properties such as porosity, permeability, and capillary pressure and relative permeability versus saturation are defined on a block-by-block basis. This provides a mechanism for modelling virtually any distribution of rock facies, both areally and stratigraphically. Generally, modelling accuracy is limited by the ability to describe the true distribution of rock facies. The grid cells used in this model are 1 km by 1 km and the Fulmar migration conduit is modelled as a 75 m thick continuous sand.

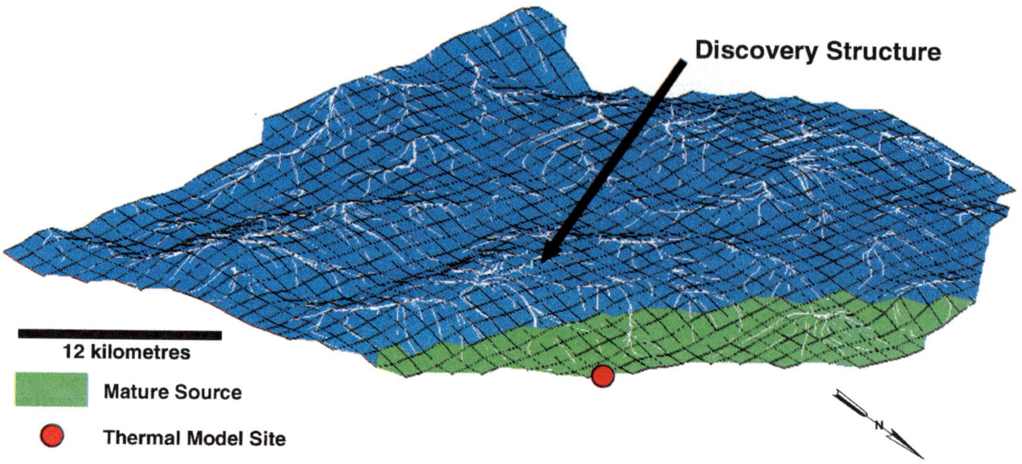

Fig. 3. Three-dimensional view from the northeast of the modelled area.

To incorporate vertical segregation effects within the Fulmar migration conduit, the calculational model employs the concept of a gravity–capillary equilibrium between the migrating hydrocarbons and water. According to this concept, described by Coats (1971), vertically-averaged 'psuedo' relative permeabilities can be related to the vertically averaged water and hydrocarbon saturations. As a result, the model calculates both the average hydrocarbon saturation in each grid block and a thickness of the migrating hydrocarbon layer.

In conventional reservoir simulators, fluids may be introduced and/or removed at model boundaries via 'wells'. In most reservoir simulations the model 'wells' correspond to physical wells. To model secondary migration, 'wells' were used to introduce hydrocarbons into the model at appropriate locations, as they were expelled from the Kimmeridge source into the Fulmar migration conduit. The green shading in Fig. 3 indicates where the source is calculated to have expelled significant hydrocarbons. These calculations were made with a one-dimensional thermal model and measured source rock kinetics. They provide input with respect to the areal fluid flow model.

An example one-dimensional thermal calculation is summarized in Figs 4 and 5. Figure 4 shows the burial depth and temperature of the Kimmeridge source rock since its deposition in the late Jurassic. The calculation is for a site in the northern syncline along a regional seismic line. Its location is marked on Fig. 3. The episode of rapid burial and temperature increase starting at 70 Ma corresponds to the onset of significant hydrocarbon generation.

The maturation history for the site is displayed in Fig. 5. The figure shows a calculated vitrinite reflectance indicating maturity and oil and gas yields in grams per 100 grams of original total organic carbon in the source rock. The oil and gas yields are consistent with the expelled hydrocarbons being a moderate gravity under-saturated oil. We expect this to have been the character of the hydrocarbons at the time of migration, and have used these fluid characteristics in the areal flow calculations. The fact that the reservoired hydrocarbons are currently a gas condensate is explained by a combination of late gas addition, indicated in the gas yield history, and continued maturation and cracking in the trap. Neither of these mechanisms is included in the areal flow calculations.

Thermal calculations were used to develop a relationship between present source depth and the time of onset of hydrocarbon generation. With this relationship and assuming a 20 million year maturation period, model expulsion rates were specified. Hydrocarbons were 'injected' into the grid blocks of the areal flow model using these rates. These rates indicate that expulsion began in the deepest part of the syncline at 80 Ma, and reached a peak value of about 1.5 STB km^{-2} a^{-1}. Expulsion occurred in a band of blocks that effectively moved up structure as the area was buried.

A primary question we chose to address with the model was how quickly hydrocarbons would migrate from the syncline up into the structures. The model was therefore run for a spectrum of permeability levels between 10 mD and 0.001 mD. Rock capillary pressure characteristics were estimated to correspond with each permeability level. A realistic estimate of the actual

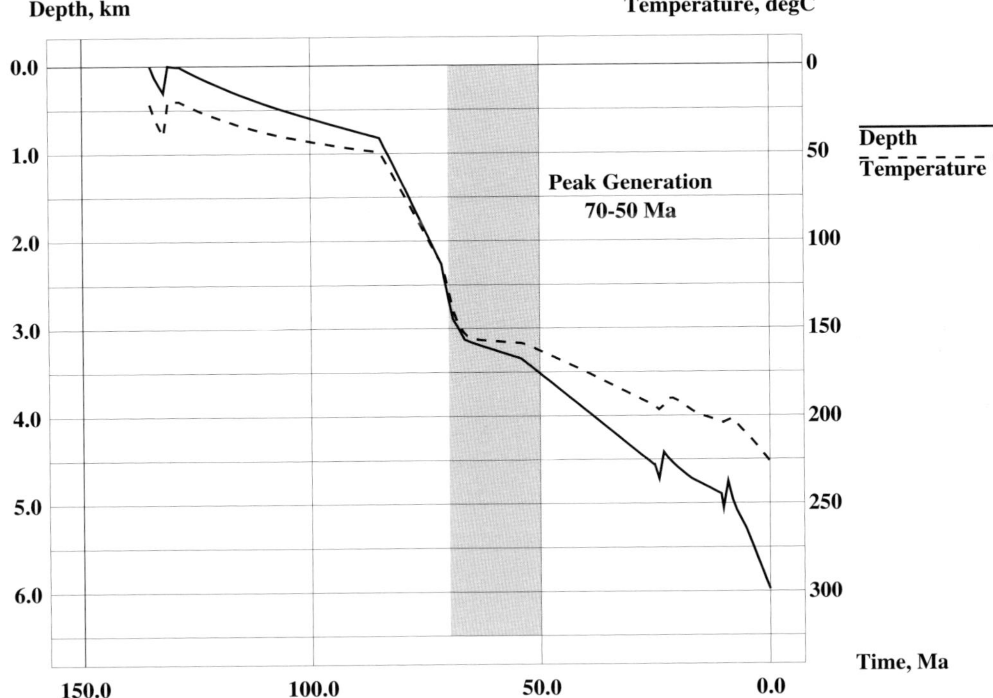

Fig. 4. Source burial depth and temperature.

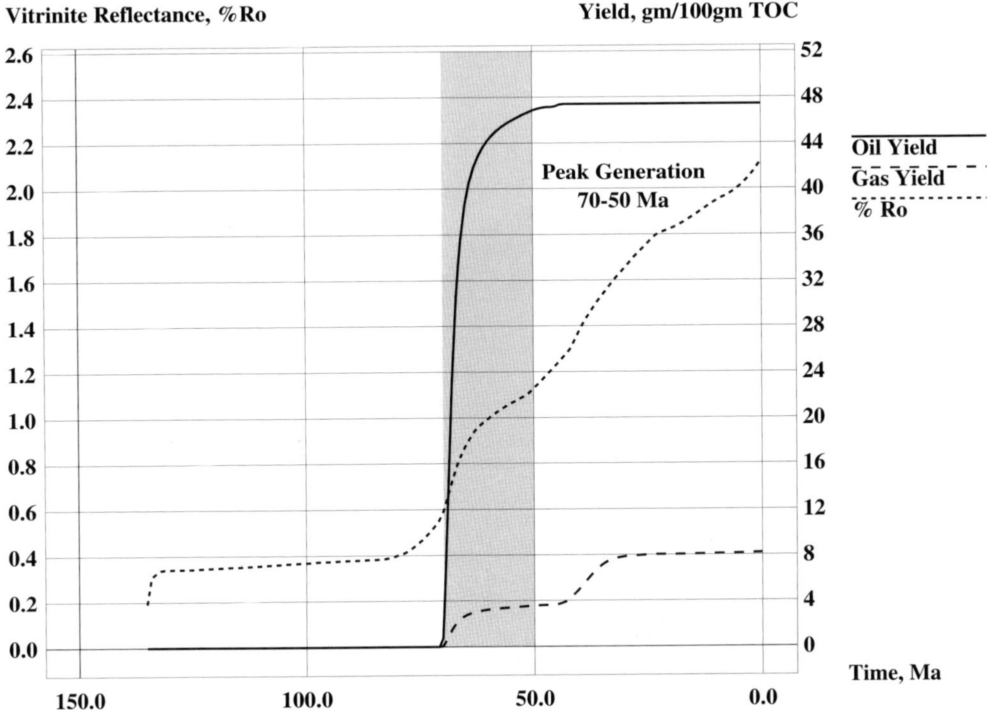

Fig. 5. Source maturity and yield.

migration conduit permeability is probably 1–10 mD.

Figures 6–10 show hydrocarbon distributions for several cases. Each figure is a contour map of calculated oil saturations. Saturations are contoured so that black corresponds to areas that have not been contacted by oil. Blue corresponds to areas that have a very small but finite oil saturation. Higher saturations are indicated by lighter colours.

Figure 6 displays oil saturations in a 10 mD migration conduit at 78 Ma, just 2 million years after oil is first expelled at the deepest point in the syncline. Oil has migrated 10–15 km and is starting to fill traps on its pathway out of the syncline. At this permeability level, the rate of trap filling is limited by the rate of expulsion rather than the time required for migration.

Figure 7 shows the same 10 mD conduit at 40 Ma. Oil has accumulated significantly in several traps, and saturations remain low in portions of the migration pathway that are not part of an accumulation. This indicates only a thin layer of oil is required for migration at this permeability level.

Figure 8 shows the present day hydrocarbon distribution for the 10 mD conduit. It indicates that the major traps in the area should be nearly filled, but that hydrocarbons have not spilled into structures in the western part of the modelled area.

Figure 9 examines the effect of lowering the conduit permeability to 0.1 mD. The figure shows present day calculated oil saturations for this case. At this lower permeability migration does not proceed as rapidly. In addition, there are higher oil saturations in the migration pathway, indicating that a relatively thicker layer of oil is required to migrate through the lower permeability rock. Also, the main hydrocarbon accumulation does not appear quite as full, indicating some of the oil is still in the migration pathway.

Figure 10 shows that at a permeability of 0.001 mD, migration is virtually shut-down. The calculated present-day oil saturations are primarily indicative of where hydrocarbons were sourced rather than where they are migrating to. As a result, a significant oil column is built up in the migration conduit.

Figures 11 and 12 summarize these results. Figure 11 shows the representative thickness of the migrating hydrocarbon layer as a function of permeability. At higher permeabilities significantly less rock volume will be contacted by migrating hydrocarbons, and 1–2 m of high quality reservoir rock is sufficient for effective migration.

As described above, for high permeabilities, trap fill timing is controlled by the rate of expulsion rather than the time required for migration. This

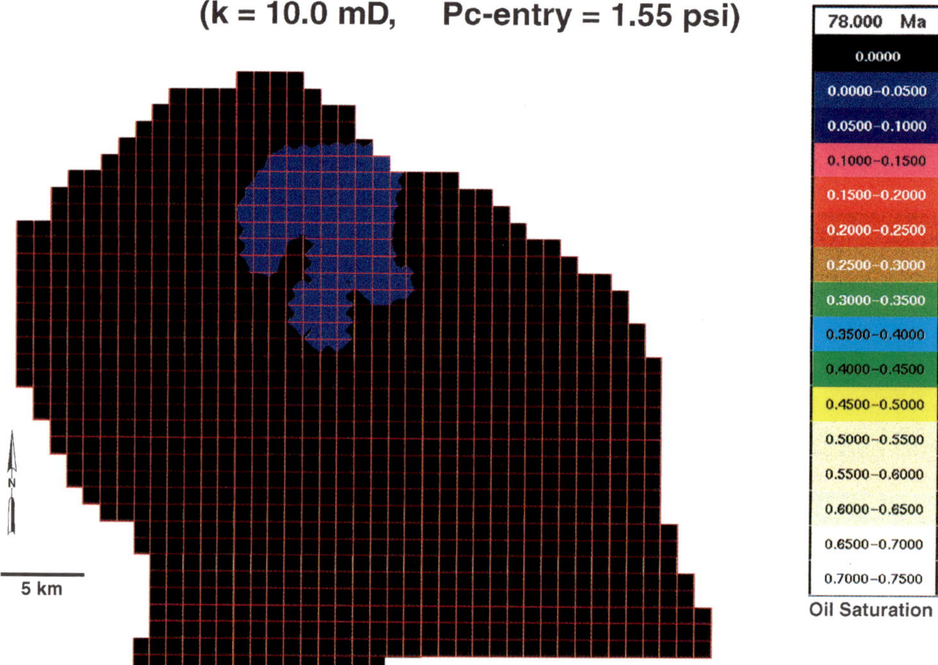

Fig. 6. Hydrocarbon distribution in a 10 mD carrier bed, 2 Ma after onset of generation. Oil has migrated 10–15 km.

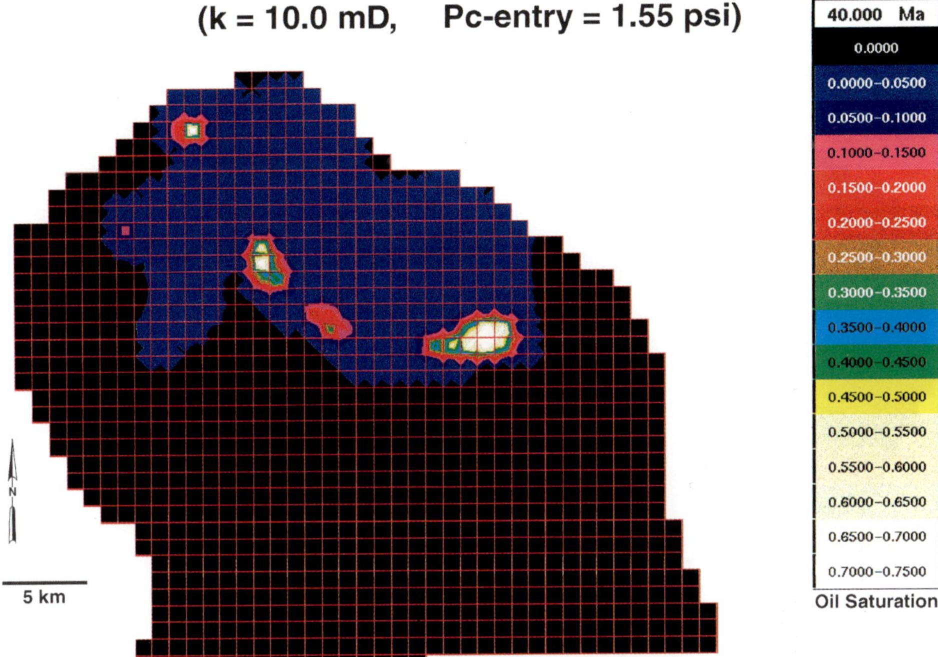

Fig. 7. Hydrocarbon distribution in a 10 mD carrier bed, 40 Ma. Oil has accumulated in several traps, but oil saturations remain low (<1%) in the migration pathway.

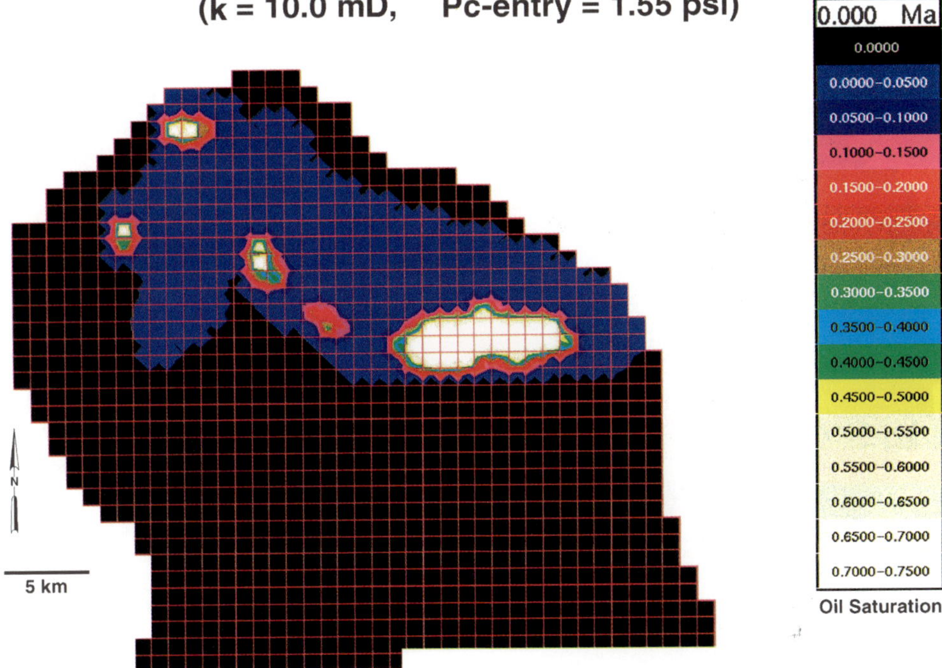

Fig. 8. Hydrocarbon distribution in a 10 mD carrier bed, present day. The major traps in the area are nearly filled.

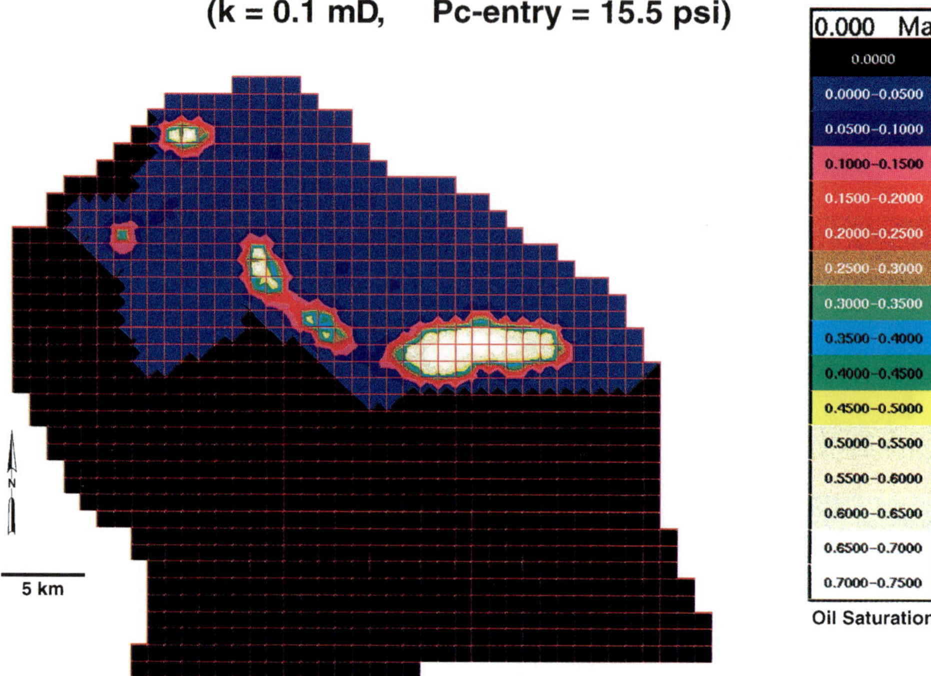

Fig. 9. Hydrocarbon distribution in a 0.1 mD carrier bed, present day. At this permeability, slightly higher oil saturations (>5% in some blocks) are required in the migration conduit, and slightly less oil accumulates in the traps.

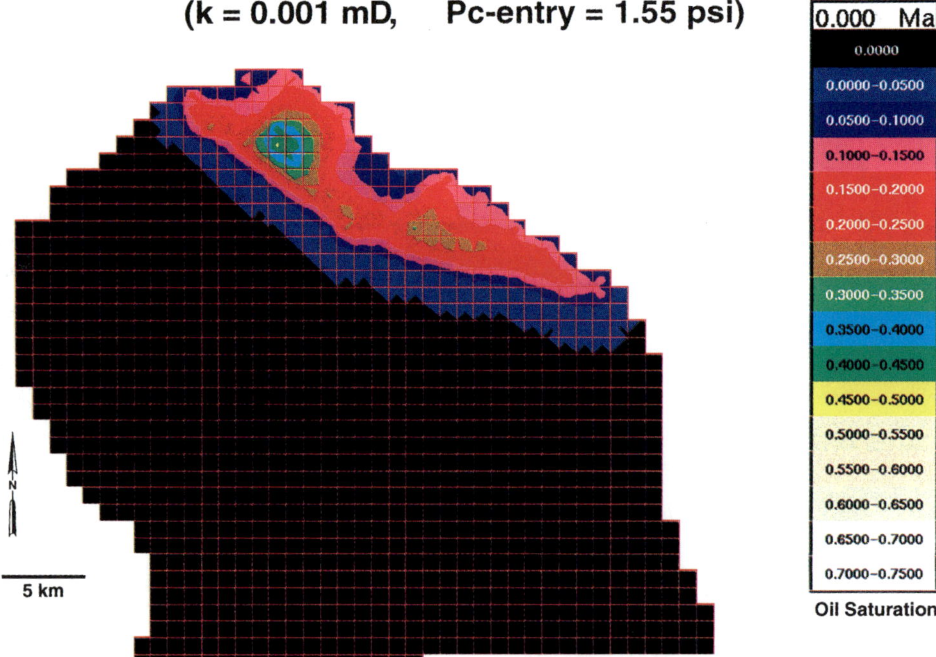

Fig. 10. Hydrocarbon distribution in a 0.001 mD carrier bed, present day. At this permeability, migration is virtually shutdown.

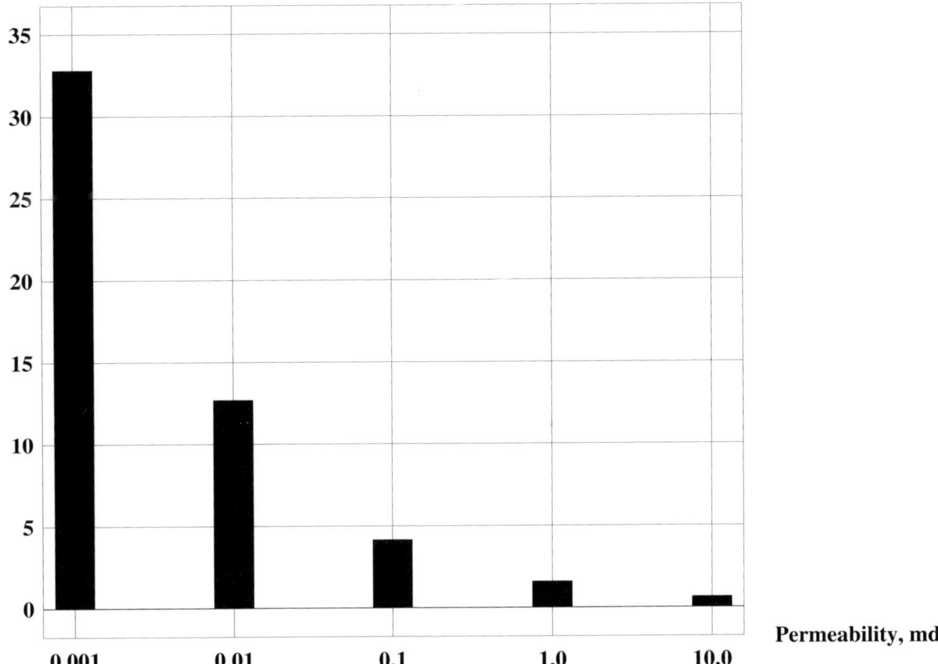

Fig. 11. Migrating hydrocarbon layer thickness layer versus permeability.

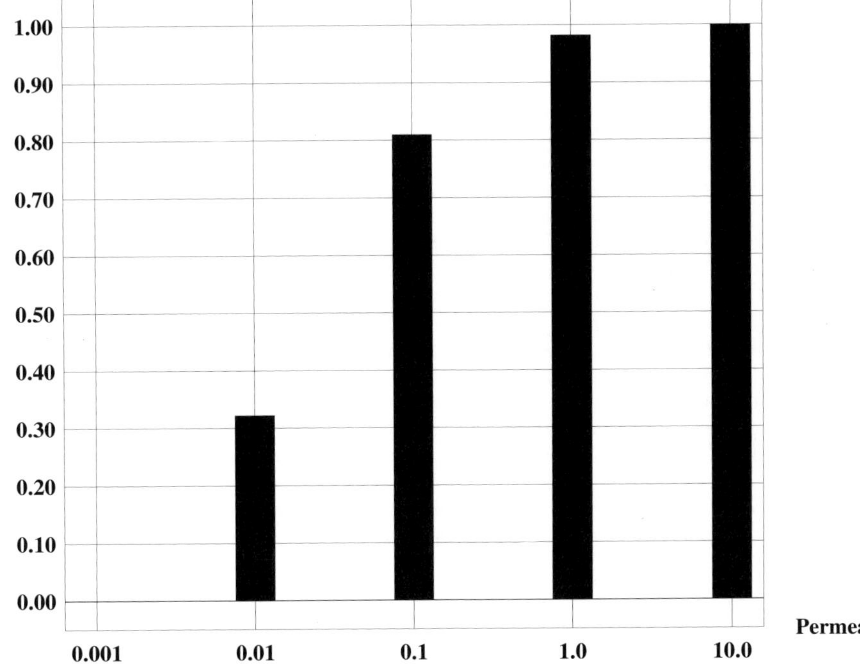

Fig. 12. Simulated migration efficiency versus permeability.

allows the definition of a site-specific migration efficiency which is the volume of hydrocarbons trapped for a given permeability, divided by the amount that would have been trapped with a highly permeable conduit. Figure 12 displays this calculated efficiency as a function of permeability. Naturally, high permeability conduits have efficiencies near 1.0. For lower permeabilities, a significant volume of oil remains in the migration pathway, and the efficiency is less than 1.0. Significantly, migration can be adequate in rocks with permeabilities as low as 0.01 mD. This indicates that some very silty, poor reservoir rocks can be viable migration pathways. As observed earlier though, permeabilities as low as 0.001 mD are not viable.

Cross-sectional flow modelling

Although areal models provide valuable information concerning the timing of dominantly lateral migration, they cannot address cross-stratal migration issues. Therefore, to integrate cross-stratal and strata-parallel migration, a cross-sectional fluid flow model was built. The model was run with TEMISPACK, a commercially available basin modelling package.

In contrast to the areal flow calculations which included only the rock volume of interest for secondary migration, TEMISPACK simulates the deposition and burial of the entire stratigraphic column. The model is composed of a two-dimensional array of grid blocks. The boundaries between layers of grid blocks are chronostratigraphic surfaces. The length of grid blocks horizontally is variable. For each block in the TEMISPACK cross-section, lithology and other rock properties are defined. The rock properties include porosity, permeability, capillary pressure, and thermal conductivity. In addition, compaction characteristics must be defined. All rock properties are permitted to vary with compaction state, which is characterized by the effective overburden stress. The calculations include compaction, overpressuring, fluid flow of water and hydrocarbons, heat flow by conduction and advection, and kinetic calculations for the generation of hydrocarbons. The formulation requires that the model be run iteratively until the presumed depositional layer thicknesses result in a good match between present-day calculated and observed layer thicknesses. A more complete description of the TEMISPACK simulator is given by Ungerer (1990).

The inclusion of compaction and overpressuring of rock strata above and below the Fulmar migration conduit leads to an additional migration drive mechanism not included in the areal flow calculations. Compaction waters can sweep hydrocarbons along as they escape from shales and move laterally through the Fulmar migration conduit. This drive mechanism acts in combination with the buoyancy drive included in the areal flow calculations.

A southwest–northeast cross-section was chosen for the model following a regional seismic line. Figure 13 shows the location of the line, posted on the base Cretaceous structure. The section goes through the western portion of the field and into the northern syncline.

Several factors influenced the choice of this line. The seismic coverage of the area consists of a 3-D survey over the discovery and a set of regional 2-D lines. Because the 3-D survey does not extend into the northern syncline, a cross-section following a regional 2-D line was chosen. The chosen line has the important feature that the overlying shale sequence is relatively thin above the crest of the structure. This is important because cross-stratal migration is most likely to occur where the overlying shale is thinnest. The expected mechanism for this is the lateral transfer of overpressures developed in the syncline up onto the crest of the structure. If the overlying shale is thin enough, these pressures can overcome its sealing capacity enabling cross-stratal migration. It would also be desirable for the cross-section model to cut through the deepest part of the syncline and the absolute crest of the structure. Although the chosen cross-section does not meet this criterion, there is nearly 1.4 km of relief on the Fulmar migration conduit in the cross-section. Therefore, the chosen cross-section should provide a reasonable representation of the mechanisms involved in cross-stratal migration.

Figure 14 shows the stratigraphy and lithologies that were input to the cross-section model. The modelled section starts in the Permian. Of particular note are the Fulmar sand unit and the overlying source shale. For each layer in the model, rock properties consistent with the modelled lithology were chosen. Comparison of modelled to observed overpressures confirmed that the chosen rock properties were reasonable. The cross-section model is 18 km long and nearly 7 km deep at the present time.

The cross-section model results are shown in Figs 15–17. In each, the colours indicate modelled oil saturations, and the green arrows indicate the dominant hydrocarbon flow direction.

Figure 15 shows results calculated for 25.2 Ma. At this point in time, oil is already filling the structure, and hydrocarbons are being swept to the southwest by compaction waters escaping through the thin overlying shale section that seals the western portion of the discovery.

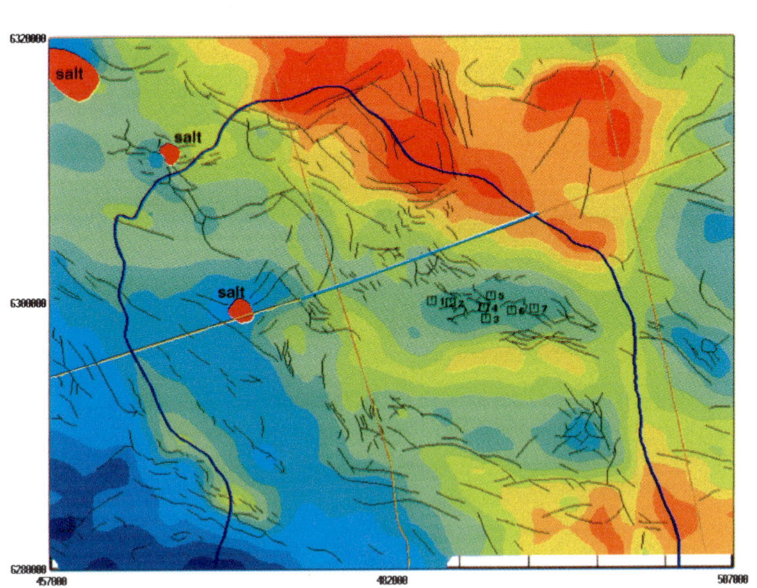

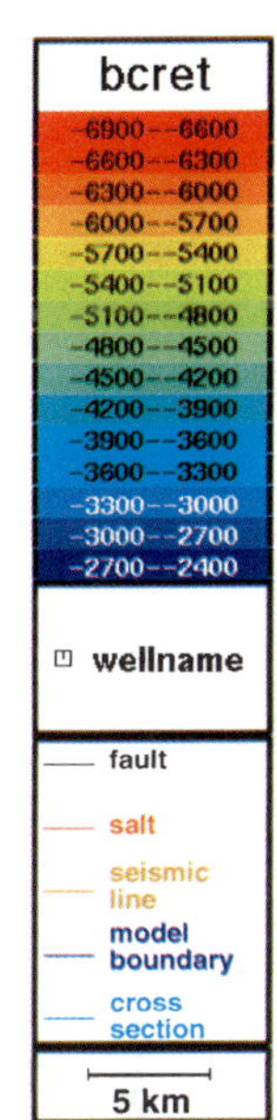

Fig. 13. Cross-section model location, posted on the Base Cretaceous structure map.

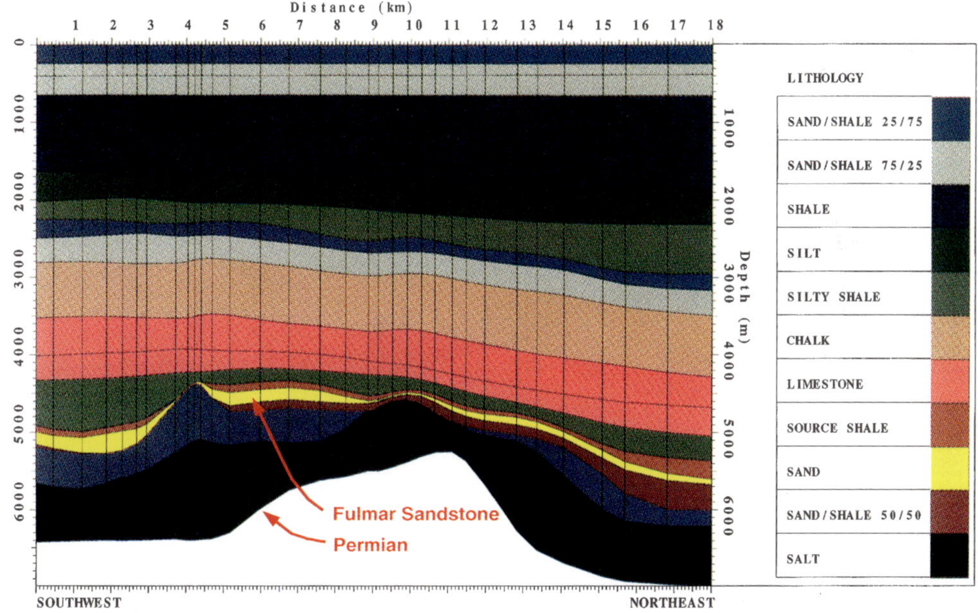

Fig. 14. Cross-section model stratigraphy and lithology.

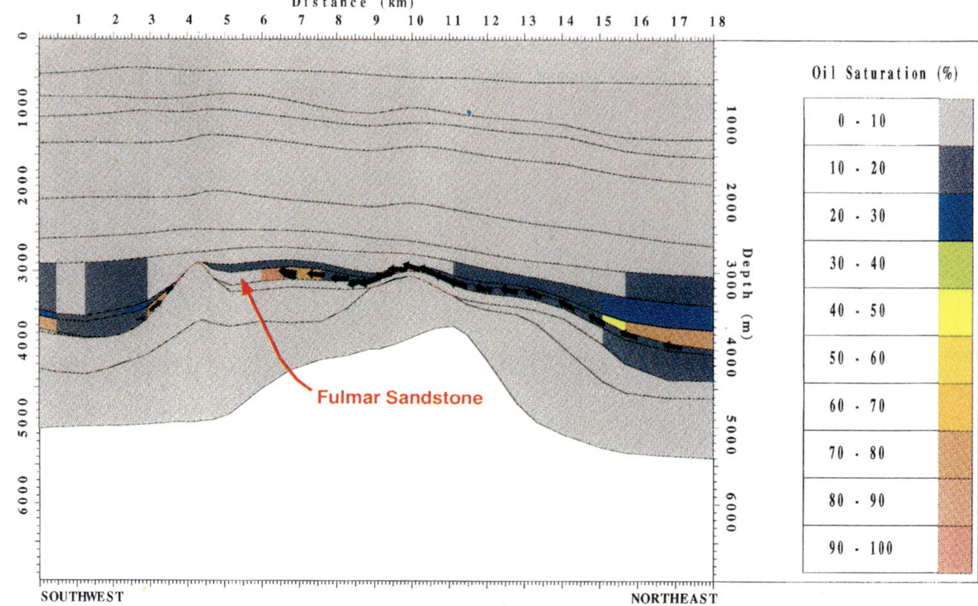

Fig. 15. Hydrocarbon distribution in cross-section model, 25.2 Ma.

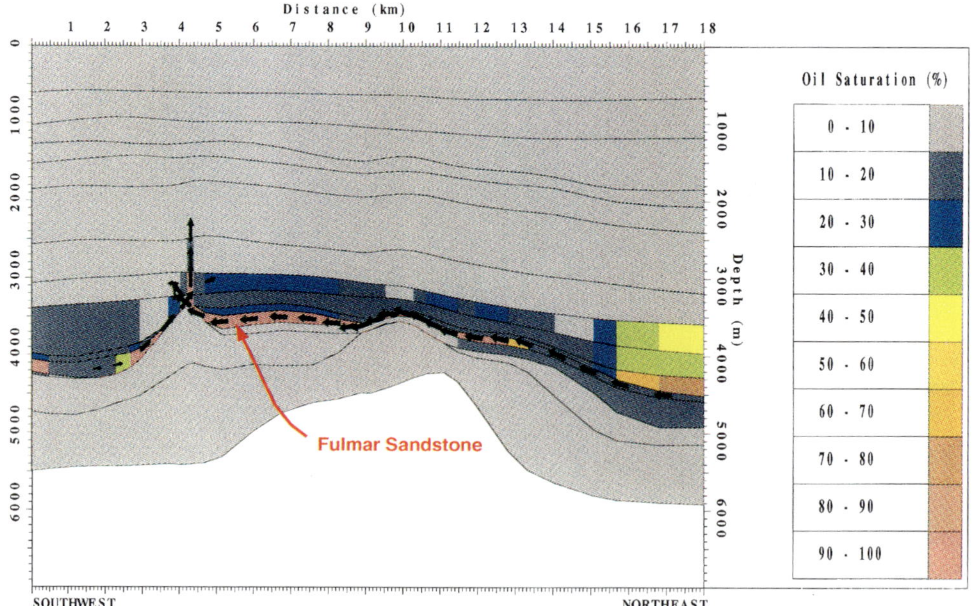

Fig. 16. Hydrocarbon distribution in cross-section model, 10.5 Ma.

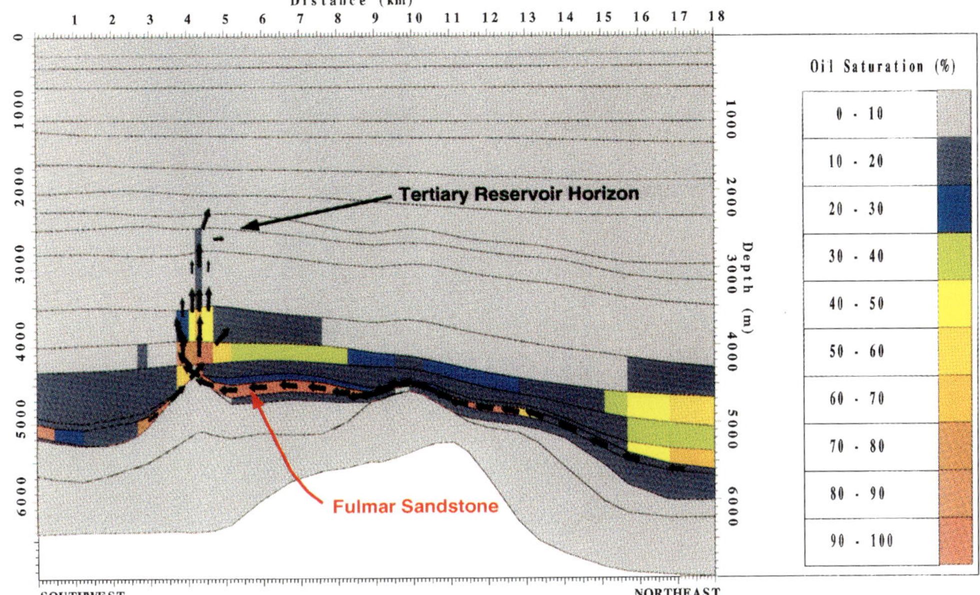

Fig. 17. Hydrocarbon distribution in cross-section model, present day.

At 10.5 Ma, shown in Fig. 16, the Fulmar sand is thoroughly oil saturated, and a capillary seal breach has occurred, allowing hydrocarbons to escape into younger sections.

The calculation for the present day, shown in Fig. 17, shows the process continuing and significant quantities of hydrocarbons leaving the Jurassic and migrating into the Cretaceous and Tertiary.

The escape of hydrocarbons from the Jurassic is a feature of the model which permits direct comparison with geologic observations. In the model, hydrocarbons escape via a capillary seal failure, precipitated by the escape of compaction waters. Since the western portion of the field has not retained a sufficient hydrocarbon column to be controlled by capillary seal, it is likely that the actual seal failure was more catastrophic, but may have been preceded by a capillary breach. As discussed below, the timing of the modelled capillary seal failure agrees well with independent estimates of seal failure timing.

Geological observations

Together, these two flow models depict a scenario in which hydrocarbons migrate rapidly from the syncline into the trap, and then subsequently leak through the capillary seal where the overlying shale sequence is thinnest. This scenario is supported by several observations.

First, a seismic 'gas chimney' is interpreted to exist over the western portion of the discovery. Figure 18 shows the chimney on a SW–NE line from a 3-D seismic survey made over the area. The discontinuous reflectors and incoherent seismic data indicate the area in which hydrocarbons are thought to be migrating. In such seismically disrupted zones, ray paths become distorted due to gas saturation and fracturing in the Cretaceous chalk and other porous strata. We have mapped out the location of this chimney. Its plume-like areal distribution and the location of the example seismic line are shown on Fig. 19.

Several observations from authigenic cements also support this scenario. The cross-section in Fig. 20 shows wells and depths from which cores were taken and analysed. Fluid history analysis from cements can follow several lines of reasoning. Differences in cement paragenesis between hydrocarbon-saturated and water-saturated rocks can offer clues to trap fill timing. Trap fill timing can also be deduced from analysing hydrocarbon and aqueous fluid inclusion homogenization temperatures and comparing these temperatures with calculated thermal history. To work effectively the fluid inclusion technique requires four components:

(a) a conventional or rotary sidewall core sample;
(b) detailed petrography, along with measurements of fluid phase behaviour conducted with a

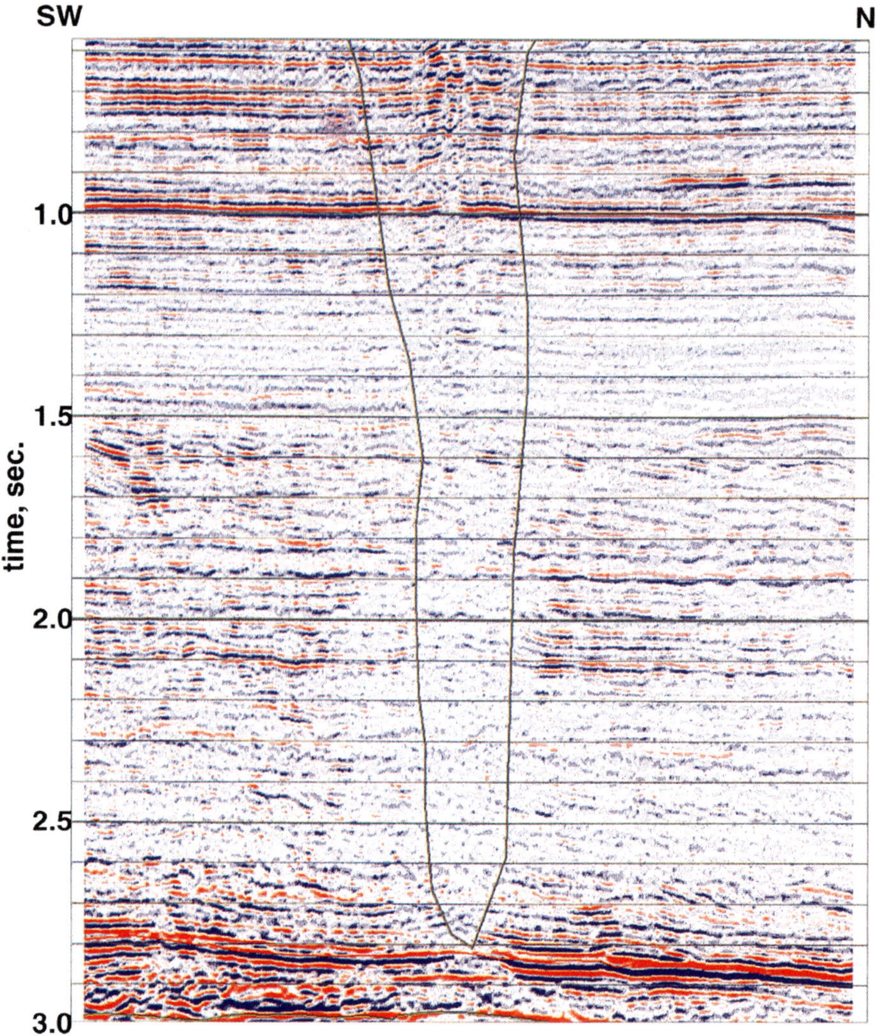

Fig. 18. Seismic chimney over western portion of discovery.

special heating and freezing stage attached to a petrographic microscope;

(c) an estimate of the inclusion fluid compositions and their pressure, volume, temperature behaviour;

(d) an estimate of the temperature–pressure history of the rock from burial history analysis.

A discussion of these techniques is given by Goldstein (1994). Analysis of hydrocarbon fluid inclusions in the #3 and #6 wells, along with cement paragenesis indicate that a significant hydrocarbon accumulation must have existed at the discovery location by 30 Ma.

The timing of seal failure, in the western portion of the field, is constrained by an analysis of the K–Ar ages of diagenetic illite. K–Ar ages from diagenetic illite can be used to estimate the time of hydrocarbon emplacement. The technique requires recognition that diagenetic illite can exist in two phases. First, a mixed layer smectite–illite phase resulting from the smectite–illite transition for which reaction kinetics are known, and, second, a fibrous illite phase. Formation of the fibrous illite phase is impeded by the presence of hydrocarbons, and it is therefore commonly assumed that the age of the finest fraction of diagenetic illite approximates the timing of hydrocarbon emplacement. The

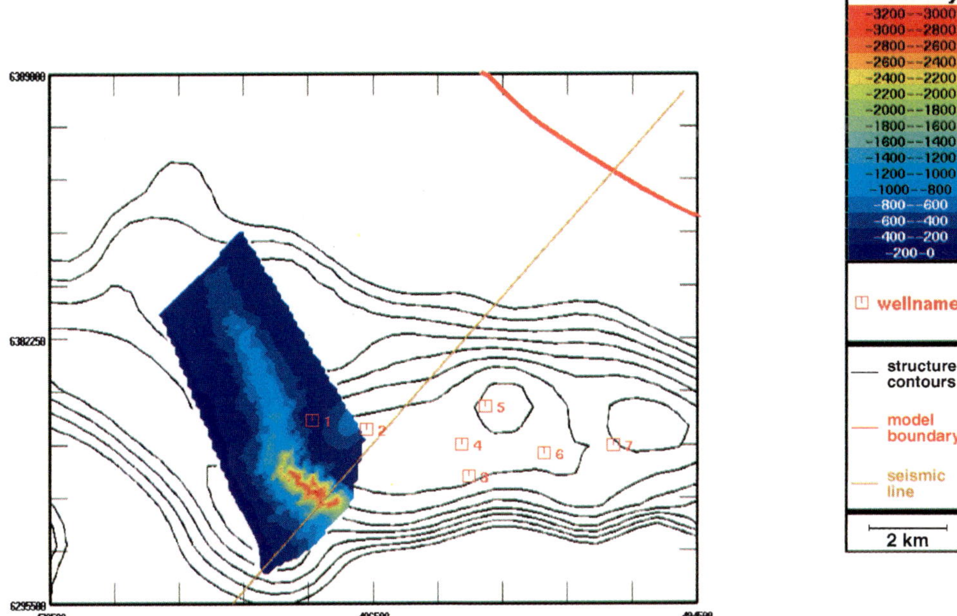

Fig. 19. Areal distribution of seismic chimney.

technique is discussed in detail by Hamilton *et al.* (1992).

In this case the tendency of hydrocarbons to inhibit fibrous illite growth was used to estimate the time of trap failure in the western part of the field. Illite ages were measured on core samples from wells #1, #2, and #6. The ages reflect an increase in the abundance of fibrous illite in well #2 which

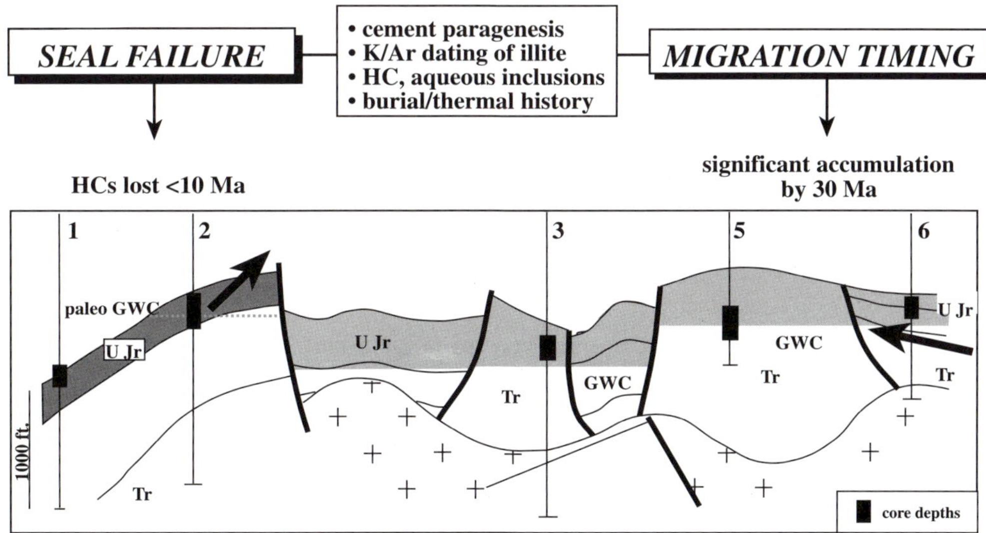

Fig. 20. Wells and cored intervals for interpretation of authigenic cements.

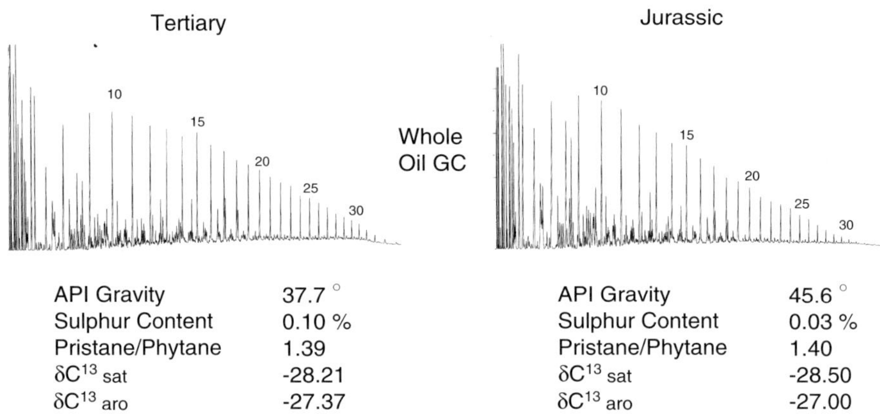

Fig. 21. Geochemistry of Jurassic and Tertiary hydrocarbons.

occurred after the seal was breached in that area and the hydrocarbons escaped. The increase in fibrous illite abundance allows us to estimate that the seal failure occurred more recently than 10 Ma. At this time the formation pressure had likely climbed to approach the fracture gradient.

Finally, hydrocarbon fluid inclusions were also found in a Tertiary reservoir penetrated by the #7 well, east of the cross-section illustrated in Fig. 20. These inclusions indicated that a significant hydrocarbon accumulation was reservoired in the Tertiary as early as 10 Ma, probably coincident with the seal breach. This is also the timing of hydrocarbon charge for a number of other Central North Sea Chalk/Tertiary fields.

To confirm that the hydrocarbons reservoired in the Tertiary originated in the Jurassic, there are a number of geochemical similarities between the two hydrocarbons. Figure 21 shows whole oil gas chromatagrams of both the Tertiary and Jurassic condensates. In addition to the general similarity of the GC traces, the pristane/phytane ratios, and the carbon isotope measurements show close correspondence between these hydrocarbons.

In addition, the Jurassic hydrocarbons are interpreted to be slightly more mature than the Tertiary hydrocarbons, since several biomarkers are absent from the Jurassic condensate. It is likely that hydrocarbons remaining in the Jurassic have continued to mature and crack after the escape of some hydrocarbons to the Tertiary, given the present Jurassic reservoir temperature of 335 °F.

Conclusions

This study leads us to both general observations and site-specific conclusions.

In the case of reservoir-quality migration conduits with permeabilities of 1.0 mD or more, secondary migration can be effectively instantaneous on a geologic time-scale. The volume of rock contacted by hydrocarbons can be quite small. Additionally, low permeability silts, with permeabilities in the range of 0.01 to 1.0 mD, can serve as effective migration conduits. The volume of rock contacted by hydrocarbons in silty carrier beds will be larger. Migration losses can be expected to vary directly with the volume of rock contacted.

More specific to this site, cross-section modelling suggests that capillary seal failure initiated the escape of hydrocarbons from Jurassic to Tertiary reservoirs. This correlation between hydrocarbons reservoired in the Jurassic and Tertiary is supported by the presence of a seismic chimney and geochemical similarities between the two hydrocarbons.

Finally, analysis of authigenic cements supports modelled reservoir fill timing and seal breaching. These data enable models to be 'history matched' to both drilling results and migration timing constraints.

The work described in this paper was undertaken at Exxon Production Research in Houston. The authors would like to thank both Exxon Production Research and Esso Exploration and Production UK Ltd for their support of this work and for permission to publish the results.

References

CAYLEY, G. T. 1987. Hydrocarbon Migration in the Central North Sea. *In*: BROOKS, J. & GLENNIE, K. (eds) *Petroleum Geology of North West Europe*. Graham & Trotman, London, p. 549.

ERRATT, D. 1993. Relationships between Basement Faulting, Salt Withdrawal, and Late Jurassic Rifting, UK Central North Sea. *In*: PARKER, J. R. (ed.) *Petroleum Geology of Northwest Europe: Proceedings of the 4th Conference*. Geological Society, London, pp. 1211–1219.

COATS, K. H., DEMPSEY, J. R. & HENDERSON, J. H. 1971. The Use of Vertical Equilibrium in Two-Dimensional Reservoir Performance, SPEJ (March, 1971) p. 63. Reference on VE pseudo-functions.

GOLDSTEIN, R. H. & REYNOLDS, T. J. 1994. *Systematics of Fluid Inclusions in Diagenetic Minerals*. SEPM Short Course Notes, **31**.

HAMILTON, P. J., GILES, M. R. & AINSWORTH, P. 1992. K–Ar Dating of Illites in Brent Group Reservoirs, A Regional Perspective. *In*: MORTON, A. C., HASZELDINE, R. S., GILES, M. R. & BROWN, S. (eds) *Geology of the Brent Group*, Geological Society, London, Special Publications, **61**, 377–400.

UNGERER, P., BURRUS, J., DOLIGEZ, B., CHÉNET, P. Y. & BESSIS, F. 1990. Basin evaluation by integrated two-dimensional modeling of heat transfer, fluid flow, hydrocarbon generation, and migration. *Bulletin of the American Association of Petroleum Geologists*, **74**, 309–335.

Comparative studies of pre- and post-drilling modelled thermal conductivity and maturity data with post-drilling results: implications for basin modelling and hydrocarbon exploration

T. T. Y. HO[1], R. P. JENSEN[2], S. K. SAHAI[1], R. H. LEADHOLM[3] & O. SENNESETH[4]

[1] *Conoco Inc., P.O. Box 1267, Ponca City, OK 74603, USA*

[2] *Basin Modelling Consultant, P.O. Box 25, Greensboro, VT 05841, USA*

[3] *Conoco Inc., P.O. Box 2197, Houston, TX 77252, USA*

[4] *Conoco Norway Inc., P.O. Box 488, N-4001, Stavanger, Norway*

Abstract: The primary objective of this paper is to demonstrate the successful application of maturation modelling using velocity-based thermal conductivity in the whole cycle of hydrocarbon exploration (frontier, developing and mature) on the Halten Terrace, Offshore Norway. This seismic geochemical method, used as a critical technique for selecting a favourable block in the early stages of exploration in the early 80s, enabled Conoco and its partners to make the first oil discovery and the subsequent discovery of the giant Heidrun Field in the area north of the 62nd parallel.

Additional data (modelled and measured) on the thermal conductivity and Ro values are now available, and provide an excellent opportunity to compare the original (pre-drilling) basin models and related modelled results of the early 80s with 1990s (post-drilling) state-of-the-art data and models. The results of the comparative study indicate that the velocity-based conductivity compares very well with measured data, particularly in an overpressured area where conventional porosity (modelled) based thermal conductivities were inadequate. The pre-drilling predicted values of heat flow, geothermal gradient and vitrinite reflectance (based on Arrhenius equation) and oil window limits also compared favourably with post-drilling measured results. Thus, the method of calculating thermal conductivity from seismic data provides a useful tool to integrate geochemistry and geophysical (seismic) data, to calibrate maturation models, and to enhance the value of geochemistry and basin modelling in hydrocarbon exploration.

Basin modelling has been used for more than fifteen years in oil exploration. However, most of the applications reported in the literature are confined to a post-drilling evaluation of oil/gas prospects that are often of secondary importance once a discovery is made. For this reasons there are still some sceptics among exploration managers in the oil industry about the role and application of basin modelling or related geochemical techniques in a frontier area where the well data required for sophisticated basin modelling are absent or scarce, and where the chance of making a major discovery is greater than in a developed area. Perhaps if we can demonstrate a successful application of geochemistry and its related basin modelling techniques that lead to a major oil discovery, exploration managers and others could be convinced of the usefulness of basin modelling. With that goal in mind, this paper documents a case history of our successful use of seismic data. These data were often the only data available, in maturation modelling during the early phase of the exploration of the Halten Terrace in the Mid-Norway Continental Shelf in the early 1980s.

The Halten Terrace has a simple burial history and geologic setting (Heum *et al.* 1986), therefore, it is an ideal setting to test basin modelling concepts. Another objective of this paper is to compare the pre-drilling modelled data with post-drilling measured and modelled data, and to discuss the significance of comparative studies in both oil exploration and basin modelling. The post-drilled data in this paper includes not only measured maturity levels but also modelled results from publicly available modelling programs such as the Yükler program (Yükler & Welte 1980) and BasinMod. The predicted data used in this comparative study were published by Leadholm *et al.* (1985).

Primarily, however, the objective of this paper is to demonstrate that velocity-based thermal conductivity can be one of the best parameters to calibrate

Ho, T. T. Y., Jensen, R. P., Sahai, S. K., Leadholm, R. H. & Senneseth, O. 1998. Comparative studies of pre- and post-drilling modelled thermal conductivity and maturity data with post-drilling results: implications for basin modelling and hydrocarbon exploration. *In*: Düppenbecker, S. J. & Iliffe, J. E. (eds) *Basin Modelling: Practice and Progress*. Geological Society, London, Special Publications, **141**, 187–208.

temperature for basin modelling. This is important because recent work by several workers (Chapman *et al.* 1984; Waples & Kamata 1993; Wangen & Throndsen 1994; Zwach *et al.* 1994) indicated some problems in the method used to calculate thermal conductivity by conventional basin modelling packages. Waples & Kamata (1993) showed that all of the porosity models used in the existing public domain basin modelling packages incorrectly calculate the porosity that is, in turn, used conventionally to calculate thermal conductivity. A similar problem was pointed out earlier by Chapman *et al.* (1984) in their work on heat flow in the Uinta Basin. In addition, Wangen and Throndsen (1994) show that 1-D modelling of thermal conductivity by conventional modelling is inadequate because it fails to honour the effects of water flow and overpressure. Recently, Jensen & Dorè (1993) and Zwach *et al.* (1994) demonstrated that hydrocarbon saturation, which is neglected in the conventional modelling, also significantly affects the calculated thermal conductivity based on a conventional method. These problems in the calculated thermal conductivity significantly effect basin modelling results because the calculated porosity is used to calculate thermal conductivity. Using these inaccurate thermal conductivities will lead to a significant error in temperature and maturity calculations as thermal conductivity is approximately an exponential function of porosity (see Chapman *et al.* 1984; Brigaud *et al.* 1990).

The importance of accuracy in calculated thermal conductivity is illustrated in the spider diagram (sensitivity graph) shown in Fig. 1. The diagram depicts the sensitivity of modelled Ro values to various input parameters. The parameter values are similar to those found in the study area (see Fig. 1). Fig. 1 shows that the modelled Ro value is very sensitive to any error in thermal conductivity (*k*), in particular when the *k* value is under-estimated. For example, a 20 % decrease in *k* resulted in an increase of more than 30 % in the modelled Ro value; in contrast a 20 % increase in *K* value caused only a 20 % decrease in modelled Ro value. This aspect of parameter sensitivity will be discussed later.

In this study we will concentrate our discussion on 1-D maturation modelling rather than on the sophisticated integrated basin modelling which has become popular in the last five years, and is more applicable to well-explored basins. Why do we restrict our attention only to 1-D maturation modelling in this study? Firstly, in frontier exploration, as a rule, we do not have the luxury of having subsurface geological and geochemical well data (such as fluid properties, kinetic parameters, etc.) for elaborate basin modelling involving reaction kinetics and fluid dynamics such as carried out recently by Ungerer *et al.* (1990) and Forbes *et al.* (1991) in the Smørbukk Sør Field of the Halten Terrace. Often the only data available in frontier areas are suitable only for maturation calculations.

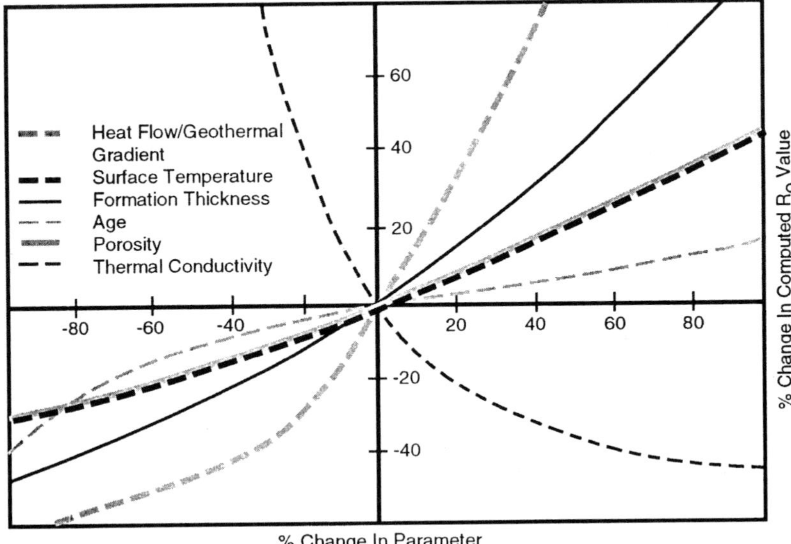

Fig. 1. Spider diagram showing the sensitivity of computed Ro value to various input parameters. The following control values were used in a base case in the modelling to construct the diagram: heat flow of 1.5 HFU, geothermal gradient of 31.25 °c km^{-1}, matrix and water thermal conductivities of 4.84 and 1.45 mcal cm^{-1} sec^{-1} °c^{-1}, respectively, porosity of 18 %, age of 200 Ma, and depth of 6100 m (20 000 ft).

Secondly, maturation modelling is basic to the subsequent modelling of oil generation/preservation, migration and accumulation, because it provides the proper temperature calibration necessary for the subsequent modelling of hydrocarbon generation, migration and entrapment. The importance of temperature was demonstrated recently by Hermanrud *et al.* (1990). They showed that temperature is one of the most sensitive parameters effecting the outcome of integrated basin modelling related to the calculation of possible reserves. A similar conclusion related to hydrocarbon-generation modelling was reached by Cao & Lerche (1990).

History of hydrocarbon exploration in the Halten Terrace

Drilling on the Haltenbanken started in 1980, as one of the first ventures north of the 62nd parallel. This venture was not a commercial success until the third well in 1981 that led to the gas-condensate discovery in the Midgard Field (Ekern 1987). This disappointing initial venture was attributed mainly to immature Jurassic source rocks (Hollander 1982; Ronnevik *et al.* 1983; Campbell and Ormassen 1987). The situation changed after the first oil (condensate) discovery in well Statoil 6407/1-2 (in the present Tyrihans Field). The discovery was made by a consortium of four oil companies of which Conoco was a member. This milestone was very significant because it sparked the oil industry's interest in the area, and because it was the first oil ever discovered north of the 62nd parallel in the Norwegian Sea (Vielvoye 1984; Leadholm *et al.* 1985; Whitley 1992). Since this first oil discovery there have been seven major oil fields found (Larsen & Heum, 1988; fig. 2). From the standpoint of basin modelling, this first oil discovery is significant because it could be one of the very few examples of the early and successful applications of basin modelling techniques in a frontier area.

Conoco used the seismic–geochemical method in our 'crude' basin modelling technique developed in 1982 (Ho & Sahai 1982) to select this prospect (coded as prospect C then) during the 5th Round License Application (see Whitley 1992). This initial maturation modelling was further refined and applied in 1983 to a large scale maturation

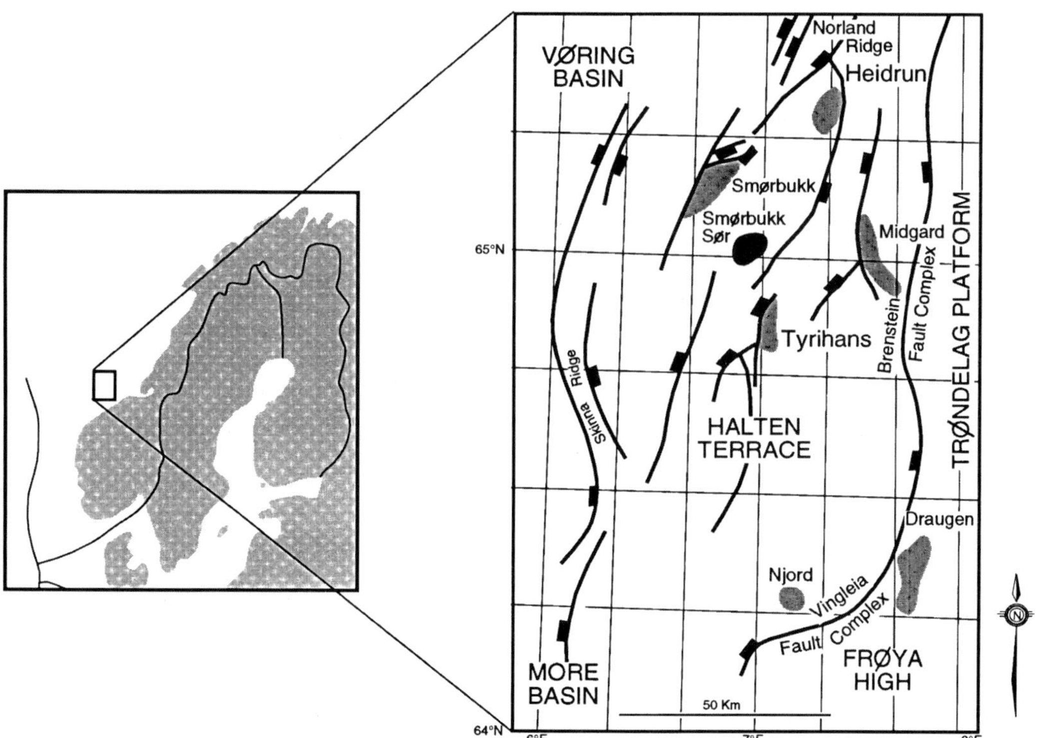

Fig. 2. Geological setting and the distribution of oil/gas fields on the Halten Terrace.

modelling of Offshore Norway during the 8th Round License Application in 1983; the work was published later by Leadholm *et al.* (1985). This expanded work involved mapping of calculated Ro values on time–structure maps. The work led to our selection of Block 6507/7 that resulted in the major discovery of about 750 MMB of recoverable oil in the Heidrun Field (Leadholm *et al.* 1985; Koenig 1986; Whitley 1992), the largest oil field north of the 62nd parallel. For these reasons our discussion in this paper will focus on the work in these blocks to illustrate the points just mentioned.

Brief geological setting and related background information

Detailed descriptions of the structural, stratigraphic and geological history of the Halten Terrace have been published by Bukovics *et al.* (1983), Hollander (1984), Bukovics and Ziegler (1985), Heum *et al.* (1986) and Dorè (1991). A summary of the geological history of the area is available in the paper by Larsen & Heum (1988) and Jensen & Dorè (1993). Of course such comprehensive information on the geological setting was not available at the time the prospects discussed here were evaluated for the lease applications in 1981 to 1983. Only sketchy geological data such as those published by Hollander (1982) and gross stratigraphic information by Ronnevik *et al.* (1983) were available for correlating to a seismic section during the 5th Round. Naturally, more well data were at our disposal (from six wells) during the 8th Round in 1983 for mapping as were Ro values calculated from about 100 seismic data points (Leadholm *et al.* 1985).

The Halten Terrace (see Fig. 2) is located between the 64th and 66th parallel north, and between six and eight degrees east longitude. Physiographically, it is bounded to the west by the Sklinna High (West Haltenbanken High) that makes a gradual transition into Nordland Ridge to the north. On the east side it is separated from Trøndelag Platform by the Vingleia and Bremstein Fault Complexes. The south margin is marked by the deeply subsided Sklinna High that separates the terrace from the deep Møre Basin to the south-west.

The major source rocks (Spekk Formation and Aare Formation of the Båt Group shown in Fig. 3) on the Halten Terrace were deposited mainly during the Jurassic. The sedimentary basin that hosts these source rocks in the Halten Terrace was formed by a regional passive subsidence during Late Triassic to about middle Jurassic time. This slow subsidence was followed by rapid subsidence induced by an east–west rifting during upper Jurassic time and by

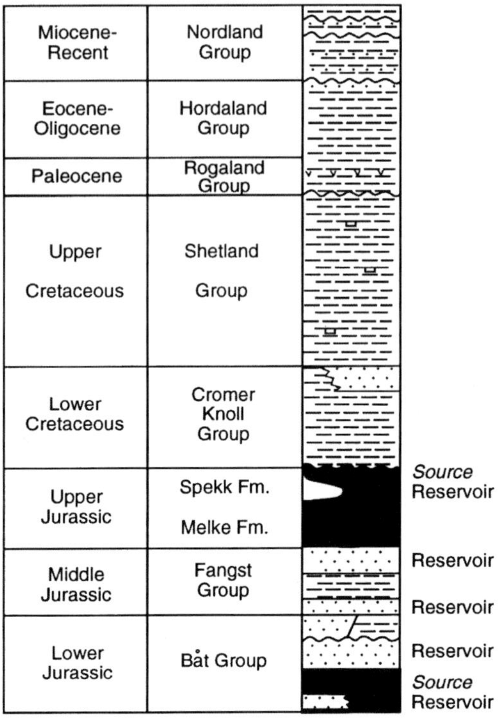

Miocene-Recent	Nordland Group	
Eocene-Oligocene	Hordaland Group	
Paleocene	Rogaland Group	
Upper Cretaceous	Shetland Group	
Lower Cretaceous	Cromer Knoll Group	
Upper Jurassic	Spekk Fm. Melke Fm.	*Source* Reservoir
Middle Jurassic	Fangst Group	Reservoir Reservoir
Lower Jurassic	Båt Group	Reservoir *Source* Reservoir

Fig. 3. Generalized Halten Terrace lithostratigraphy. (After Jensen & Dore 1993).

a north–south right lateral wrenching during Early to middle Cretaceous (Larsen & Heum 1988). The regional subsidence was reactivated during the late Cretaceous and Tertiary. During the Pliocene rapid sedimentation due to westward shelf progradation was induced by the uplift of the Fennoscandian shield. This resulted in deposition of up to 1000 m of clastic sediments (Heum *et al.* 1986; Larsen & Heum 1988). The sedimentation history is graphically represented in Fig. 4. The figure shows the burial history of the potential source rocks, the Jurassic Spekk Formation and Aare (or Åre) Formation (Båt Group) in well Statoil 6407/1-2 located in the Tyrihans Field.

Method

Thermal conductivity and related thermal parameters

The basis of our early basin modelling on the Halten Terrace was the derivation of thermal conductivity from seismic interval velocity (Ho & Sahai 1982; Leadholm *et al.* 1985). This method is based on the empirical relationship between

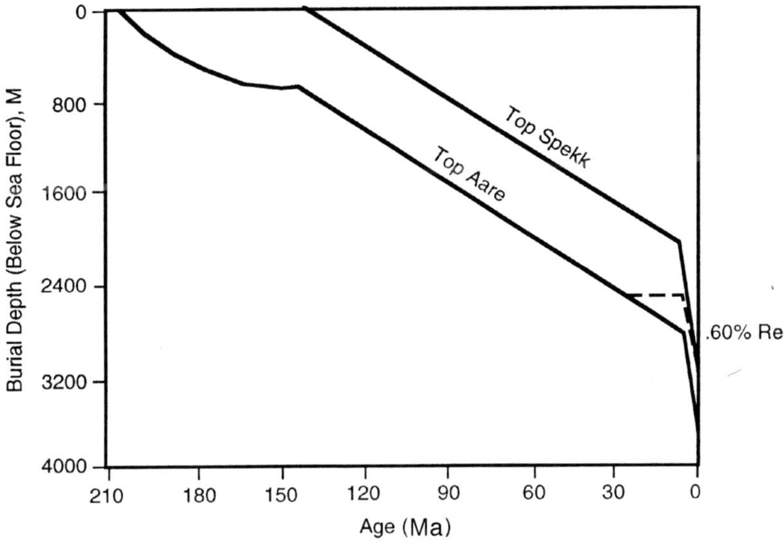

Fig. 4. Burial and maturation histories of Spekk and Aare Formations in well 6407/1-2 (Tyrihans South).

thermal conductivity and interval velocity derived from the data of Goss *et al.* (1975). The relationship is justified by the theoretical functional relationship between thermal conductivity (k), sonic velocity (v) and molecular spacing (δ) based on the liquid phase. Mathematically, the relationship is:

$$k = \frac{(2Rv)}{\delta^2} \qquad (1)$$

where R is the universal gas constant; and $\delta = (m/p)^{1/2}$, where m is molecular weight and π is density (see Brigman 1949; Isaacs 1981). This classical work by Brigman (1949) implies that there is a similar mechanism for the transference of heat energy and of compressive wave through a liquid. Note that according to eqn (1), the thermal conductivity of a solid is greater than that of the corresponding liquid (see Isaac 1981).

In our study we used stacking velocities (interval velocities) wherever sonic velocities were not available. In the study area, the interval velocities based on stacking and sonic log differ by less than 5 % for depth less than 2 km and less than 25 % for deeper section (>2 km). The calculated conductivities were combined within a stratigraphic unit to estimate a weighted average conductivity (thickness-weighted harmonic mean) for a given stratigraphic unit. In the post-drilling study, we used well logs to calculate interval velocity from sonic velocity, and to calculate interval geothermal gradient. The gradient was derived from well-log bottom-hole temperature corrected using the AAPG method for correcting temperature to equilibrium conditions (Kehle 1973). Drill-stem test temperatures were also used in the temperature gradient calculation wherever the data were available.

Geochemical analysis

Well cuttings and cores were collected during the second phase of modelling in the mid 1980s to be described later from five wells in the surrounding area for vitrinite reflectance analysis. The samples were also measured for quantitative fluorescence to determine equivalent Ro values (Thompson-Rizer & Woods 1987). These maturity data were then combined with Ro data measured in the late 1980s to the early 1990s by six other laboratories for samples from nine wells in the study area. We used about four core samples from the Spekk Formation and six samples of Aare coal and sediments from the Heidrun Field for Rock-Eval analysis and for kinetic analysis by our in-house time slice pyrolysis gas chromatography method. These kinetics related data were used for an independent validation of the maturity data (vitrinite reflectance and oil/liquid window). The kinetic analysis is out of the scope of this paper and will be treated in another paper in the future.

Theoretical basis for maturation modelling

Since the introduction of the TTI method by Waples (1980), who derived the time–temperature index from the original empirical scheme of Lopatin (1971), several methods based on the same principle have appeared in the literature (Royden *et al.* 1980; Welte & Yükler 1981; Middleton 1982). This empirical approach is based on the old idea held by physical chemists (see for example, Pauling 1945) that reaction rate doubles with every increase of 10 °C in temperature. The application of this rule of thumb to organic maturation has been questioned by several geochemists (Tissot *et al.* 1987; McKenzie & Quigley 1988; Wood 1988; Larter 1989; Burnham & Sweeney 1989; Sweeney and Burnham, 1990). Recently, Waples *et al.* (1991) also documented the inadequacies of the TTI method and concluded that the TTI method should be replaced with a kinetic-based method.

In 1982 we developed a computerized maturation model based on the Arrhenius equation. This program is called REMOD; its application in various settings was discussed by Leadholm et al. (1985), Dembicki & Pirkle (1985), McLimans (1987) and Daines *et al.* (1990, 1991).

Our kinetic-based model has the following general form:

$$\ln Re = a + b \bullet \ln \int_0^t (\mathrm{EXP}\ (-E/(RT(t)))dt \qquad (2)$$

$$T(t) = To + q \bullet t \qquad (3)$$

$$q = dT/dt \qquad (4)$$

where Re is the estimated (theoretical) Ro value; a and b are constants, in which b is related to the Arrhenius constant); E is activation energy in cal mol^{-1}; R is the universal gas constant (1.987 cal Kelvin^{-1} mol^{-1}); $T(t)$ is the subsurface temperature in Kelvin at time t (in Ma); To is the initial temperature (in Kelvin K); and q is heating rate in K Ma^{-1}. Note from equation (3) that we approximate palaeotemperature by a linear equation instead of quadratic form as proposed by Lerche *et al.* (1984). We prefer the linear approximation because so much uncertainty is associated with palaeotemperature, in particular in a case involving a frontier area.

The time–temperature integral $\int(\mathrm{EXP}(-E/(RT(t)))dt)$ in equation (2) is similar to that published by Wood (1988). However, there are several differences in our approach compared with that of Wood (1988). The primary difference is that we used an average value of activation energy derived from about 80 vitrinite reflectance values instead of using an assumed activation energy based on laboratory experiments, as reported by Wood (1988). These Ro values are based on samples from 13 wells, worldwide, with known burial histories and present subsurface temperatures. The well data (formation age and surface and subsurface temperatures) along with measured Ro values were input to a computer

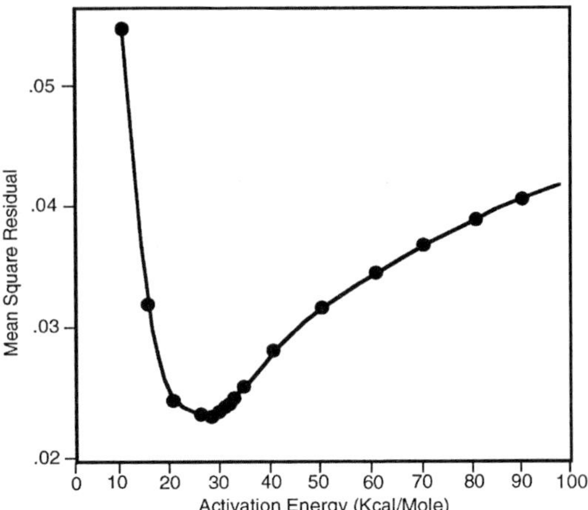

Fig. 5. Numerical optimization of pseudo-activation energy required for the maturation model used in REMOD.

program containing eqs (2) and (3) to search for an optimum value of activation energy by a least squares method that minimized the errors between calculated and observed Ro values. This one dimensional search, shown in Fig. 5, results in an optimum value of activation energy (E) of 30 kcal mol^{-1}. This derived activation energy value is lower than those based on artificial laboratory maturation methods (for example, a value of 52 kcal mol^{-1} was used by Wood (1988)). For this reason, our calculated value of activation energy can be regarded as pseudo-activation energy as suggested by Waples (1985) because it results from the mathematical averaging of several values of activation energy values as discussed by Jüntgen & Klein (1975). We believe our pseudo-activation energy derived from the field data is mathematically and chemically reasonable at this stage of our understanding about the formation of vitrinite. This is because the process involves complex physico-chemical processes that are difficult to represent as multiple parallel reactions proposed by Burnham & Sweeney (1989) (see also Ungerer (1993) and Waples (1994)) The other major difference is: we used two regression equations in the form of equation (2) instead of the one equation used by Wood (1988). The numerical values for the constants in the equations are presented in Table 1. These two equations are used to represent sequential reactions to accommodate different maturation rates during diagenetic (Equation A in Fig. 6) and catagenetic (Equation B in Fig. 6) stages, and to compensate for a difference in Arrhenius factors in the two stages. Thus, our approach is different from

Table 1. *Numerical constants used in the kinetic modelling of vitrinite maturation*

Maturation Stage*	E (cal mol^{-1})	a	b
Diagenetic (Eqn A)	30000	2.198	0.072
Catagenetic (Eqn B)	30000	4.667	0.137

* use eqn 2 when time–temperature integral (in logarithm) is –37.985.

those of Wood (1988), Tissot & Espitaliè (1975), Espitaliè *et al.* (1991) and Burnham & Sweeney (1989) who treat the formation of vitrinite in multiple parallel reactions.

Computed Ro values based on our model were compared with measured valued from several wells in nine basins worldwide. The results showed that the computed values are within ±11 % of the measured values and that the model can predict the top of the oil window to within 120 m of the expected depth. The latter is consistent with the result of Javie (1991) who showed that the analytical error of ±2 kcal mol^{-1} commonly associated with kerogen kinetics leads to ±120 meters error in the estimate for the top of the oil window.

Modelling of thermal and maturation histories on the Halten Terrace

Tissot & Welte (1987) discussed the applications of geochemistry and related basin modelling techniques in oil exploration, and categorized the exploration programme into three stages based on the availability of well data. Similarly, the programme of exploration can be classified according to the applicability of basin modelling (see Ho *et al.* 1992).

The first phase can be called the hypothesis-testing (or concept building) stage. At this stage an explorer has to establish a conceptual model about the geologic and geochemical make-up of a basin based on the testing of plausible models using mainly seismic data, with very limited well control. The second phase of exploration can be identified as the model-calibration stage, because there are sufficient well data at this level of exploration to fine-tune the models established in the first stage.

The third phase of exploration programme can be recognized as the 'problem-solving' phase. This stage of basin modelling in exploration can be used as a trouble-shooting tool to solve some geological, geochemical and fluid flow problems in a well-explored, mature basin using well-established models. In other words, it can be used in a manner analogous to the reservoir simulation models used by engineers. However, since uncertainties in parameters (in particular those related to oil

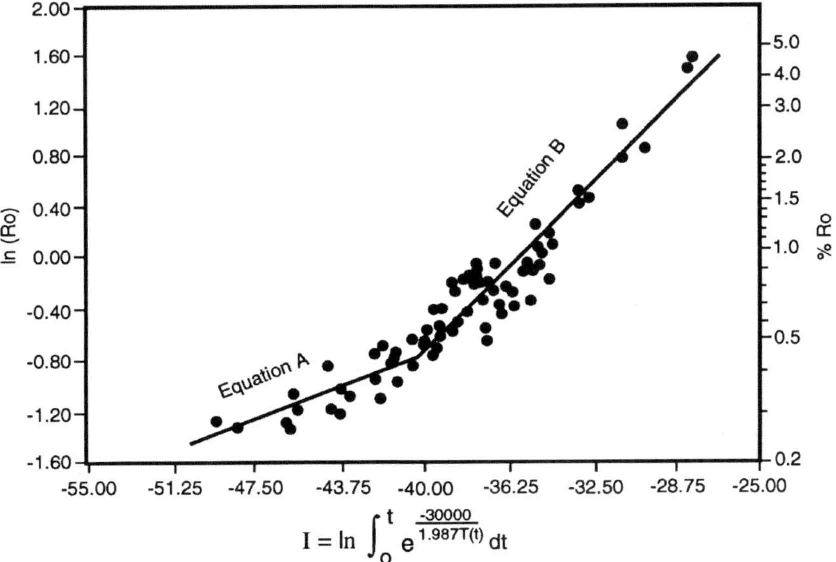

Fig. 6. Ro value as a function of time–temperature integral (I).

migration) still exist even in a mature basin, its use as a problem-solving tool is limited to semi-quantitative or qualitative analysis at present. In spite of this limitation, Ungerer *et al.* (1990) presented a good example of the third stage of basin modelling. Ungerer *et al.* (1990) applied two-dimensional basin modelling techniques to better understand hydrocarbon leakage in the Smørbukk Field in the Halten Terrace. Their modelling results showed that the development of fracture due to overpressuring in the northern structure in this field was responsible for the leakage; hence 6506/12-4 was a dry hole.

In this paper we will give a case history of our applications of maturation modelling in the full cycle of Halten Terrace exploration (frontier, developing and mature stages) that started in the early 1980s. Even with this full cycle, we recognize that our model is not complete, and will continue to undergo transformations as new data and techniques become available.

First stage: hypothesis-testing or concept-building (The 5th Licensing Round)

The maturity of potential upper Jurassic source rocks and the location of the oil window were the major problems encountered in the early 80s by explorers working in the Halten Terrace. This was particularly important because the Midgard gas discovery indicated immature upper Jurassic rocks (Hollander 1982; Ronnevik *et al.* 1983; Campbell & Ormassen 1987). Therefore, maturation modelling became a critical part of the exploration programme, and was a major objective of our basin evaluation during the 5th Round of License Application in 1982.

Prior to the 5th Round, only the Midgard Field had been drilled. The data from this area were, as is common in the industry, kept tight by the operator. Consequently, we had to rely heavily on seismic data during this concept-building stage of basin modelling. To carry out this work we used the seismic–geochemical techniques outlined in Ho & Sahai (1982) and Leadholm *et al.* (1985) to derive geothermal gradients for three prospective structures coded as Prospects A, B and C. These publications show that thermal conductivity can be estimated from interval velocity. Table 2 gives the results of this calculation for Prospect C. These data are significant, because Prospect C was the only structure associated with a mature oil kitchen; the other prospects available for lease at that time were found to be less mature or immature.

As shown in Table 2 an assumed heat flow value of 1.5 HFU was used in the calculation of average geothermal gradients. This heat flow value was taken from what was thought to be the value in an analogous basin, the Central North Sea. The Central North Sea was used for this purpose because the heat flow profile in the Central North Sea is available in the literature (Evans & Coleman 1974; Evans 1977; Toth *et al.* 1981). This approach based on tectonic analogy, is further supported by the similarity of a heat flow value of about 1.1 HFU on the east side of the Central Graben, near the coast, to a heat flow value of 1.14 HFU observed in an analogous area on the east side of Trondelag Platform (Swanberg *et al.* 1974). Additionally, the estimated value is close to the average global heat flow of oceanic crust (Turcotte & Schubert 1982) which, in our opinion, was the best estimate, at that stage of our understanding, for the thermal regime in the Halten Terrace.

Table 2. *Geothermal gradients estimated from the velocity-based thermal conductivity for Prospect C in the 5th Round*

Thermal properties	Location	
	On-structure	Off-structure
Heat flow (HFU)	1.5	1.5
Thermal conductivity ($mcal\ cm^{-2}\ s^{-1}\ °c^{-1}$)	3.97	4.34
Geothermal gradient ($°C\ km^{-1}$)	37.78	34.56

The estimated thermal conductivity data shown in Table 2 are the average conductivity calculated from nine time intervals at Prospect C. They were based on the stacking velocities at nearby shotpoints. Note in the table that the calculated conductivity for the kitchen is larger than that for the structure due to a greater compaction and pressure in the kitchen (off-structure) than *in* the structure. As a result, the geothermal gradient is greater for the structure than for the kitchen.

The geothermal gradients based on several thermal conductivities and velocity-derived depths to the tops of Palaeocene and upper and middle Jurassic were input to the computer program to delineate the possible liquid window in the prospective area. The results of this simplified modelling are shown in Fig. 7. The modelling was done for both the tops of upper and middle Jurassic because they were known to be the source rock and reservoir, respectively, in the North Sea. At Prospect C the liquid window was predicted to be between 2200 and 4120 m below the sea floor. The source rock, off-structure, was predicted to begin oil generation at about 40 Ma (indicated by the calculated Ro value of 0.6 %) and was predicted to be presently at the light oil or condensate gener-

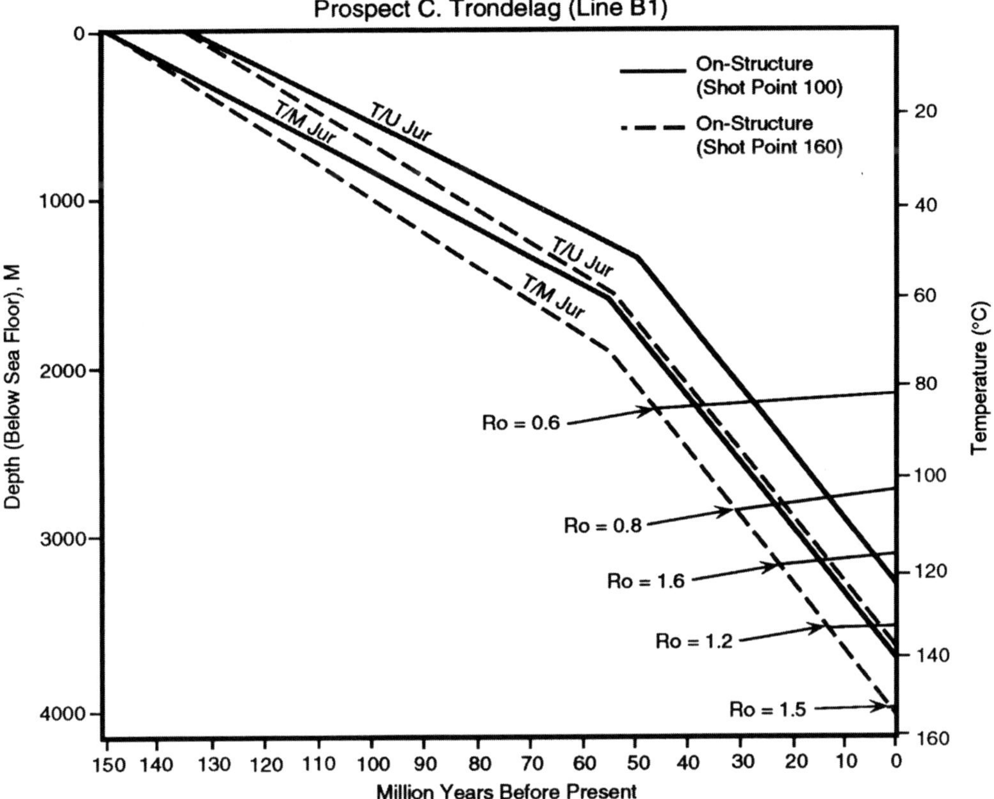

Fig. 7. Burial and maturation histories based on seismic data available in the 5th Round for upper and middle Jurassic horizons in Prospect C.

ating stage as indicated by the calculated Ro value of about 1.3 % for the source rock at this location. Similarly, the source rock at the structure was predicted to be in the oil window (Fig. 7).

Drilling of Prospect C in 1983 resulted in the first oil discovered north of 62nd parallel in the North Sea. This is the present southern branch of Tyrihans Field. The initial drilling report of a condensate discovery (oil gravity of 45 °API) in this structure proved our maturation modelling prediction of a mature oil-prone source rock in this area. It is clear from these results, that it is possible to use only minimum data (four seismic horizons and stacking velocities), to produce valid and reasonably accurate models of maturation.

Second stage: model-calibration (the 8th Licensing Round)

Further validation of our original concept building was carried out during the 8th Licensing Round in 1983 (see Leadholm *et al.* 1985). In this second stage of model calibration multiple gradients derived from velocity-based thermal conductivities were used for stratigraphic units instead of the single average gradient employed in the previous work. Available well data were used to calibrate the maturation model used by our computer program REMOD. Figure 4 represents the modelling results for a well drilled near the prospect C of the 5th Round. Note in Fig. 4 that the well data available at this stage provided us with the data necessary to improve our burial and maturation models. The burial history was significantly altered for the Tertiary, with rapid subsidence occurring in the last five million years, rather than throughout the Tertiary as seen in Fig. 7.

With more data available, the thermal model was also modified to more accurately reflect the thermal difference between sedimentary layers. For example, present-day thermal conductivities of 2.56, 3.81, 4.06 and 5.8 mcal cm^{-1} s^{-1} $°c^{-1}$ were derived for the Neogene, Palaeocene, upper

Jurassic and lower Jurassic sections, respectively, for the well with the modelling results shown in Fig. 4. Similarly, various heat flows that increase westwards were calculated from well data for various areas (see Leadholm *et al.* 1985). The heat flow was calculated by the thermal resistance method similar to that of Chapman *et al.* (1984).

After calibrating the maturation model with well data so that Re (estimated reflectance) matched measured Ro, the measured and calculated values were mapped for the top of Spekk Formation (see also Leadholm *et al.* 1985). A total of 85 measured Ro values from 6 wells and 77 calculated Re values from 79 seismic shot points was used in the mapping (Fig. 8). The mapping shows that the maturity increases basinward towards the west side of the area due to the increase in burial depth and the high heat flow associated with Mesozoic–Tertiary volcanic activity in this area (Bukovics *et al.* 1983; Boen *et al.* 1984). Note also that there is a wide zone of mature source rock (defined by the 0.6 and 1.2 % Re contours) near the centre of the mapped area. This fairway marks the location of the possible oil kitchen on the Halten Terrace.

On the basis of this map, Block 6507/7, containing the structure marked as Prospect B, was selected by Conoco out of nine blocks available for lease during the 8th Round in 1983. The block was selected because mapping suggested that the potential reservoir in Prospect B was fed by a large drainage area to the southwest, therefore a possible trapping of a large volume of oil generated in the kitchen was envizaged. The drilling results in 1985 validated the idea when a large oil column of about 300 m was found in the Heidrun structure (Koenig 1986). It should be noted that the oil-mature zone defined in 1983 also encompasses the oil fields known at that time or discovered later such as Tyrihans and Smorbukk Fields. Furthermore, the kitchen in this zone has been thought to be the source of oil in the Draugen Field discovered later in 1984 (Ellenor & Mozetie 1986).

To obtain a cross-sectional view of the generation window, present depths for the oil and gas

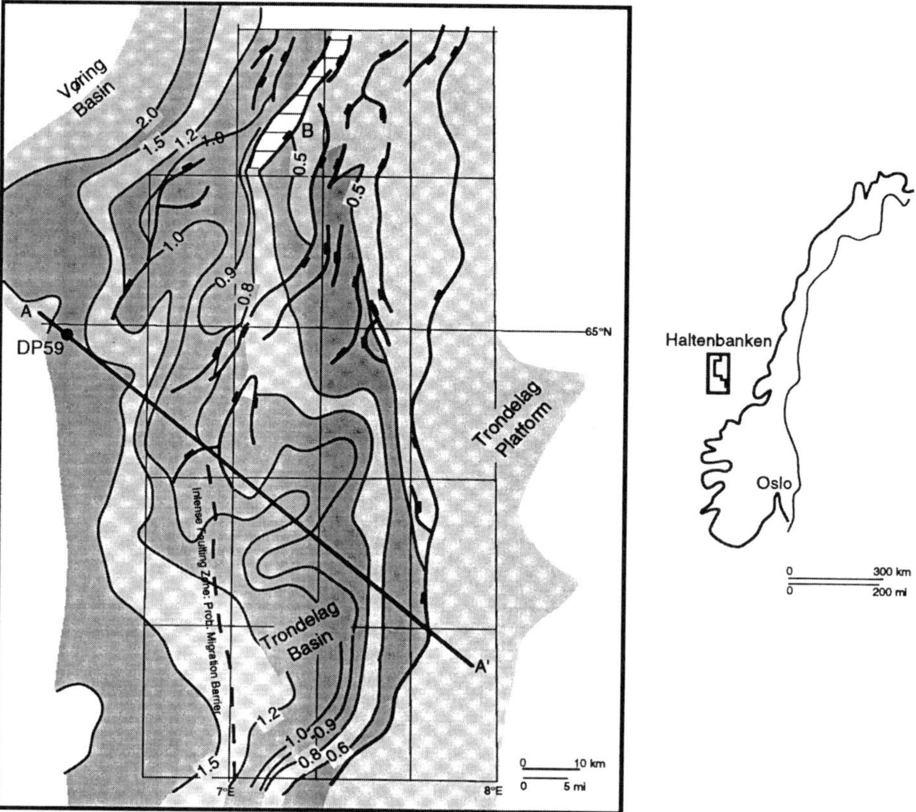

Fig. 8. Present-day maturation map made for the 8th Round for the top of the upper Jurassic (Spekk Formation). Modified from Leadholm *et al.* (1985).

windows, calculated by REMOD for the selected shotpoints, were converted to two-way times and posted on a seismic section, as shown in Fig. 9. Figure 9 shows that the depth to the top of the oil window is deeper in the northwest part of the section than in the southeast part, and that the window gradually increases in width from the northwest to the southeast side of the section. This is obviously due to a higher heat flow in the west than in the east side of the basin as discussed earlier (Leadholm *et al.* 1985).

Third stage: problem solving (The 13th Licensing Round)

Over sixty wells have now been drilled on the Halten Terrace, making this a prime location for the third stage of basin modelling — problem solving. Our increased knowledge has resulted in more sophisticated basin models of temperature and burial history, such as that shown in Fig. 10. The heat flow history in Fig. 10 is based on the stretching theory of McKenzie (1978). We used a stretching factor (beta) of 1.3 to model heat flow during the rifting event in late Jurassic/Early Cretaceous time (see Jensen & Dorè 1993). Note that the maximum heat flow used is 1.23 HFU (see also figs 1 and 8(A) of Jensen & Dorè (1993)).

We are now in a position to use these upgraded models to determine where we might next expect to find either oil or gas. We can also use the available data and models to test or compare the accuracy or performance of various modelling algorithms, or to test the sensitivity of results to various parameters. These subjects will be covered in the next sections.

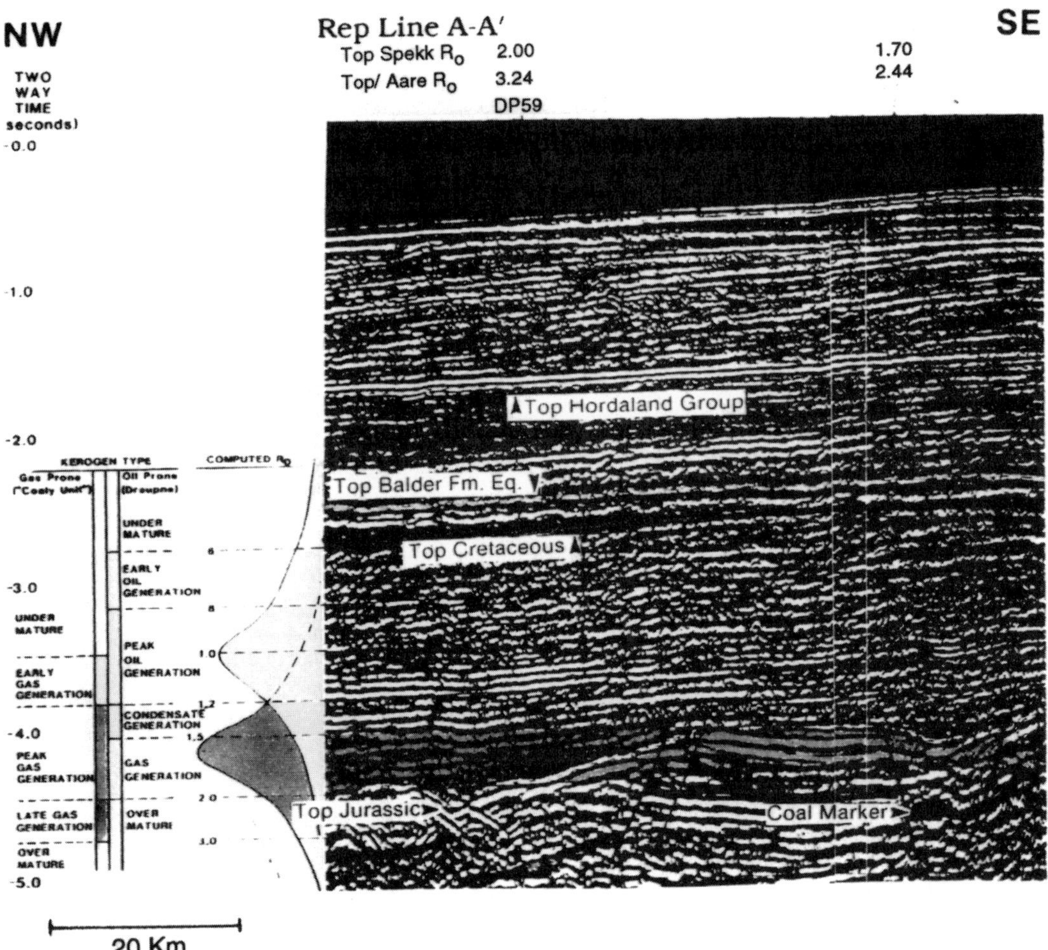

Fig. 9. Seismic section showing present-day hydrocarbon generation window in the Halten Terrace area. This section is located on the northwest end of the seismic line A–A' in Fig. 8.

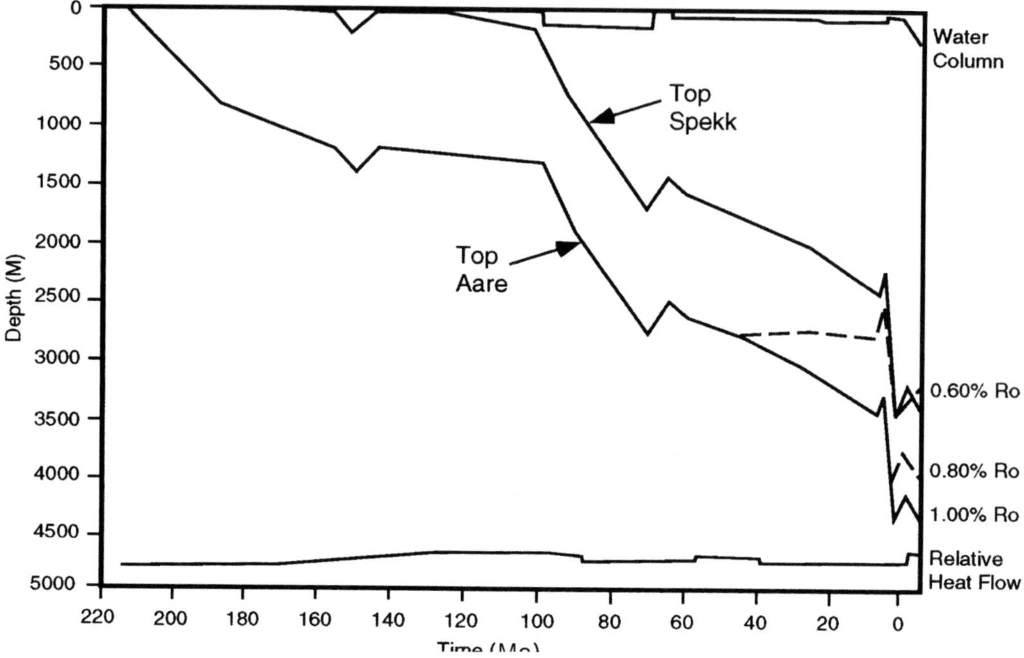

Fig. 10. Detailed burial, heat flow and maturation histories of the Jurassic sediments on the Halten Terrace.

To elucidate the significance of the comparative studies, measured data and modelled data will be compared on the basis of within-well, among-wells (i.e. in map view) and among various basin modelling algorithms.

Comparison of pre-drilling models and predicted data with the post-drilling results

Measured and computed thermal conductivities based on our model: intra-well comparison

Measured values of thermal conductivity for sandstone and shale core samples of sandstones and shales from five wells in the thermal conductivity database of the Rogaland Research Institute of Stavanger, Norway were compared with the velocity-derived thermal conductivities for the same wells and horizons (Fig. 11). There is a good agreement between the computed and observed values. It should be noted that the Heidrun Area sample is the only sample not from the same location, but from a point 5 km away. It should be further noted that the measured thermal conductivity is based on a small water-saturated core of about 1″ (2.54 cm) by 1.5″ (3.81 cm). The differences in measured and computed values for the five wells range from 0.1 to 32.4 % with the overall difference of about 16 %. This difference, which is close to the 15 % analytical error, is geologically insignificant because the available data from well samples indicate that thermal conductivity can vary by as much as 28 % for the same formation in a well in this area. The maximum difference of 32.4 % (see Heidrun, upper Jurassic in Fig. 11) is found in upper Jurassic organic-rich shale. This difference could be due to a pressure–temperature effect that is commonly neglected in laboratory measurements. It could also be caused by a scaling problem, because the measured value represents ‘single grain’ organic-rich shale as compared to bulk conductivity calculated for the whole upper Jurassic section in this location. The higher modelled value could represent inhomogeneous organic (kerogen) distribution in the upper Jurassic section of this location. This inhomogeneity in mineral grains probably explains a relatively large discrepancy in the Midgard data shown in Fig. 11. According to the spider diagram shown in Fig. 1, this maximum difference (+32.4 % deviation) causes an under-estimation of Ro value by 22 % that, as will be discussed later, is geologically insignificant in most cases.

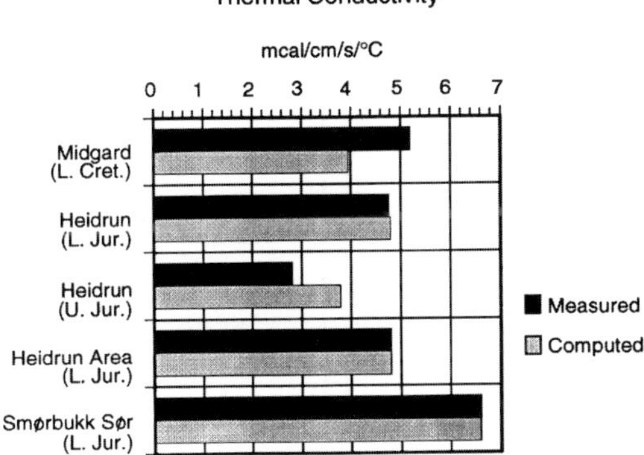

Fig. 11. Comparison of computed and measured thermal conductivities.

Measured computed thermal conductivities based on other models: intra-well comparison

Thermal conductivities calculated by other basin modelling programmes (Platte River Associates' BasinMod and the Yükler programme) were also compared with our modelling results. With the exception of the Midgard sample, all the samples discussed previously (see Fig. 11) were used in the comparison. The results are shown in Fig. 12.

The figure shows that, in general, all the modelled values agree with the measured values, with the exceptions of the values in the Smørbukk Sør location. In this location, the measured value is almost identical to the value calculated by the velocity-based method. However, the values calculated by other models (in particular that based on the Yükler method (Yükler & Welte 1980) for the Smørbukk Sør case are significantly lower than the measured value (see Fig. 12). This range of difference is geochemically very significant as suggested on the spider diagram in Fig. 1, This range of error (–30 % to –70 %) could equate to an increase in calculated Ro (Re) of 40 % to more than 100 %. Ehrenberg *et al.* (1992) and Jensen & Dorè (1993) showed that the Smørbukk Sør Field is located in a transition zone where pore pressure changes from normal to overpressure (see Fig. 13). Therefore, the under-estimated thermal conductivity values based on the conventional modelling techniques mentioned are attributable to over-estimated fluid pressure; hence to over-estimated porosity. These inaccuracies of the conventional modelling techniques (in particular 1-D technique) pertaining to modelling compaction-related porosity and pore pressure in an overpressured area have recently been discussed widely in the literature (Jensen & Dorè 1993; Waples & Kamata 1993; Hermanrud 1994; Wangen & Trondsen 1994).

Measured and computed Ro values based on our model: intra-well comparison

Measured values of vitrinite reflectance of 13 samples from the Spekk and Aare Formation of 13 wells were compared with the computed Ro values for the same depths in the wells. The results are presented in Table 3. The table indicates that the differences (in absolute deviation) range from about 3 % to 16 % with an average of about 8 % that is equivalent to a spread of 320 m. These differences are considered to be geologically

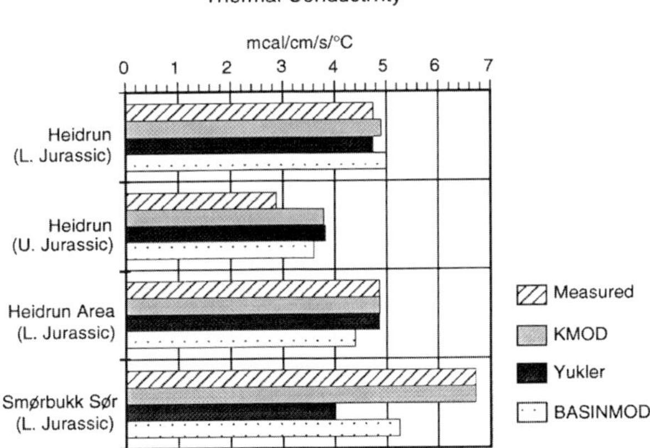

Fig. 12. Comparison of computed thermal conductivities based on various methods with measured values.

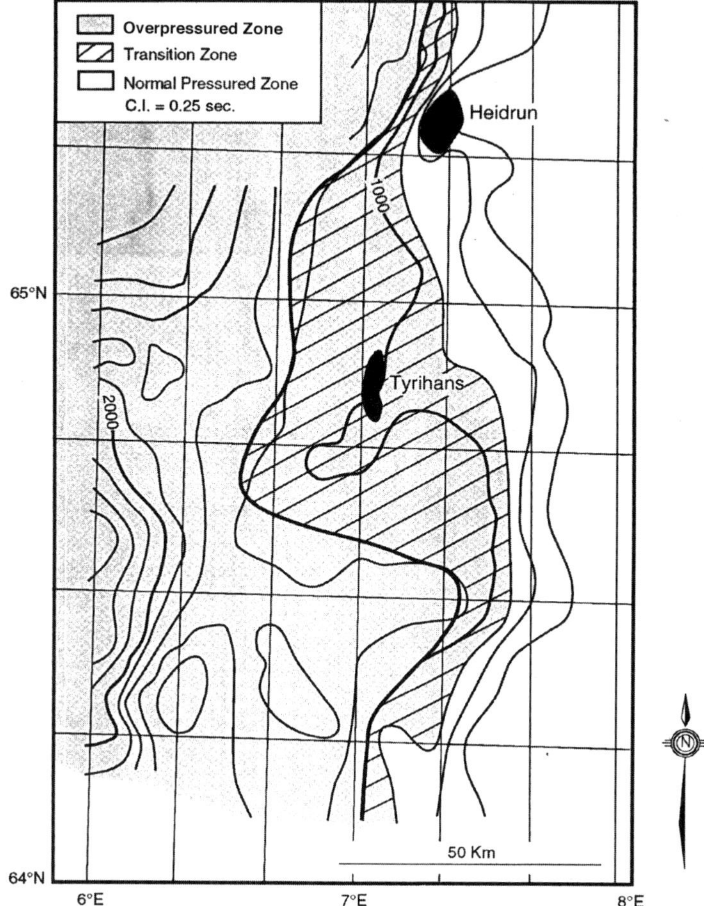

Fig. 13. Jurassic pressure regimes of the Halten Terrace.

insignificant for two reasons:

(i) According to Bostic & Foster (1974) the maximum differences in Ro values can be as high as 20 % for shale and 30 % for sandstone.
(ii) On the basis of Ro trend with depth, Vik & Hermanrud (1991) have recently reported that Ro spreads of 200 to 400 m are normally observed in the Northern North sea.

The conclusion based on a 'point-to-point' comparison (Table 3) is substantiated by the overall trend of increasing Ro value with present depth shown in Fig. 14. This theoretical trend, based on our modelling at location D59 (Figs 8 and 9), fits the trend based on measured data well. Note also that both the REMOD and the LLNL models fit the measured data well (Fig. 14). The measured data are based on about 80 measurements performed by six different laboratories for core and cuttings samples from 11 wells. Considering remarkable inter-laboratory discrepancies in measured Ro values reported by Dembicki (1984), and possible difference in the types of vitrinite measured (Buiskool-Toxopeus 1983; Durand *et al.* 1986), the scatter in the values is remarkably small. The scatter is also partly attributable to variation in geothermal history as implied by present geothermal gradients in the wells studied (see also Jensen & Dorè (1993)). In these wells the present-day gradients range from 33.94 to 38.85 °C km^{-1} (or 1.84 to 2.13 °F 100 ft). The gradients were calculated from well-log and DST (drill-stem test) temperatures. This range of gradients was also predicted previously in the 8th Round by Leadholm *et al.* (1985) using the seismic geochemical technique described earlier.

It should be noted that all the modelled Ro values were also validated by the kinetic modelling

Table 3. *Comparison of pre-drilling modelled Ro values with post-drilling measured Ro values*

Well name	Modelled %Ro	Measured %Ro	Deviation %
	0.60	0.55	+9.1
6407/1-2	0.81	0.77	+5.1
	1.13	1.08	+5.0
6407/1-3	0.80	0.76	+5.3
	1.02	1.05	–2.9
6407/2-1	0.55	0.48	+14.6
	0.62	0.65	–4.6
6407/4-1	0.90	0.86	+4.7
6506/12-1	0.93	0.83	+12.1
	1.27	1.13	+15.5
6506/12-3	0.85	0.82	+3.6
	1.14	1.10	+3.0
6507/7-1	0.97	0.83	+3.6
Average absolute deviation (%) 7.9			

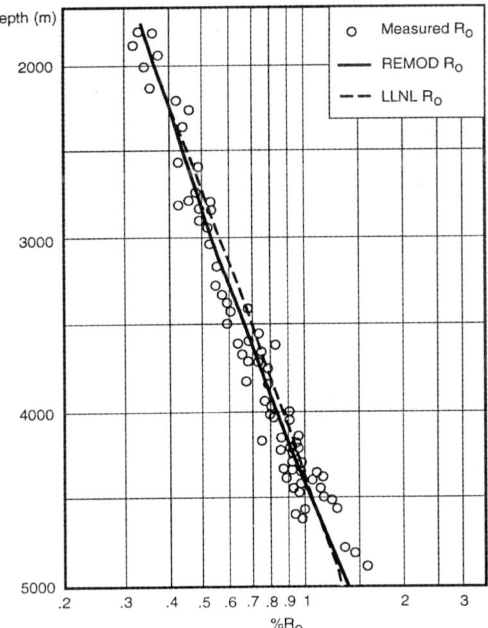

Fig. 14. Depth versus Ro plot based on measured and computed values (derived by REMOD and LLNL methods) for the Halten Terrace.

discussed earlier by checking modelled Ro values with transformation ratios in the manner similar to that reported by Forbes *et al.* (1991) for the Halten Terrace. Detailed accounts are out of the scope of this paper and will be presented elsewhere.

Computed and measured Ro values: inter-well comparison in map view

Our study for the 8th round involved the mapping of calculated Ro values. Now that we have a sufficient number of measured Ro values to compare the early models with the actual data in map view.

Figure 15 shows the oil generative kitchen defined by the 0.6 and 1.2 % Ro dashed contours for the velocity-based thermal conductivity and modelled Re values for present day; the solid contours represent the measured Ro values. As shown in the figure, the maps are remarkably similar, deviating only in the deeper portion of the basin in the west. This deviation is due in part to different method of contouring (i.e. interpolation and extrapolation), and, in part, to a better structural map. This slight deviation may be a result of an insufficient number of data points for measured Ros in the deeper part of the basin (see Fig. 15). In the northwest corner of the study area where both measured and computed 1.2 % Ro contours cross, a high heat flow is suggested by the study of Leadholm *et al.* (1985), while insufficient measured data lends these contours to interpretation of other trends (see Fig. 15).

The modelled Ro map constructed for the third modelling stage (see Fig. 15) was based on the data points throughout the basin. As a result, lower thermal conductivities that should be expected were predicted for areas that became overpressured in the last five million years. These predicted values resulted in higher maturities in the deep areas. Recent work by Jensen & Dorè (1993) indicated that these areas have higher geothermal gradients than the shallower areas. Consequently, As shown in Fig. 15, the shapes of the oil kitchen fairways based on measured and modelled Ros (see also Whitley, 1992) are similar to the overpressure distribution reported by Jensen & Dorè (1993) (compare also Figs 13 and 15). This similarity implies that the velocity-based modelled Ro values can be used, along with measured values to compensate for the uncertainty associated with the paucity of well data, to reduce exploration risk.

A comparison of the oil kitchen maps with the distribution of present oil fields shows the relationships between the oil kitchen fairway and oil/gas fields in this area. As shown in Fig. 16, the oil/gas fields are distributed within or near the oil kitchen fairway. With our increased geologic and geophysical database and the calibrated thermal, maturation and hydrocarbon generation models, we can now address migration questions with more

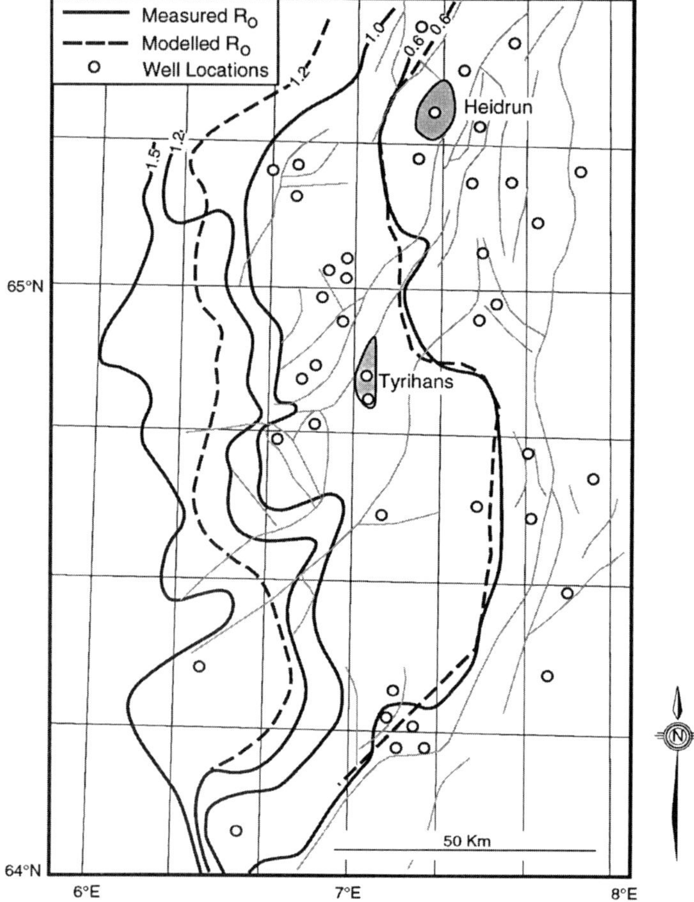

Fig. 15. Maturity map (based on measured and modelled Ro values) for the top of the Upper Jurassic (Spekk Formation) on the Halten Terrace.

confidence than was possible during the earlier stages of exploration.

Discussion

Applicability of maturation modelling in frontier exploration using the 5th Round as an example

In the foregoing discussion we focused our attention mostly on the 8th and 13th Rounds results because we have more field and modelled data for comparison than we did for the 5th Round. From the standpoint of basin modelling and exploration, the 5th Round predictions are interesting and important because the modelling was done solely on the basis of geophysical data. Therefore, it is important to test the reliability of maturation modelling based only on seismic data. The 5th Round predicted value of 2200 m (measured from sea floor), indicated by 0.6 % computed Ro in Fig. 7 for the onset of oil generation, is about 600 m (or 21 %) shallower than the measured data (see Fig. 14). This discrepancy is due partly to an error of up to 25 % in the interval velocity used in our calculation for deeper sections discussed earlier. The difference could also be attributed to the relatively high heat flow value used in the 5th Round — 1.5 HFU as compared to the present value of 1.28 HFU. The spider diagram (Fig. 1) indicates that a 15 % decrease in heat flow value causes a decrease in modelled Ro value of about 17 %, and is acceptable for the level of modelling carried out in the 5th Round. The difference in the

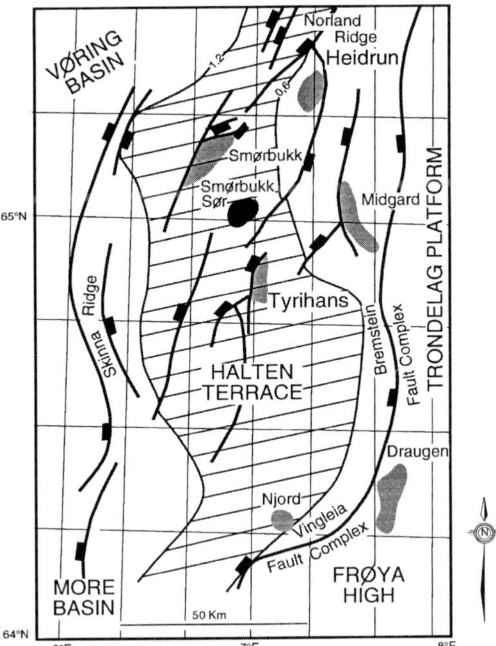

Fig. 16. Relationships between oil kitchen fairway (hatched area defined by 0.6 and 1.2 % Ro contours) and oil fields on the Halten Terrace.

depth to the top of the oil window sounds more alarming considering the reliability of ±400 m error using the kinetic-based method reported recently by Nielsen & Dahl (1991) and an Ro spread of 200 to 400 m at a given depth shown recently by Vik & Hermanrud (1994).

From the standpoint of frontier exploration the critical issue is often the presence or absence of a mature source rock or oil window. Therefore, the exact level of maturity and location of the oil window are relatively unimportant at the concept-building stage. For instance, at the time of our evaluation in 1981 and 1982 the presence of a mature source rock on the Halten Terrace was the key factor in the selection of prospects (see Hollander 1982; Ronnevik *et al.* 1983); in fact, mature source rocks in the deeper western part of the Halten Terrace were speculated (Hollander 1981, 1982, 1984). Therefore, this difference becomes a minor issue because the prognosis based on the 'crude' maturation model available at that time did show (Fig. 7) an oil-mature source rock down-dip from the Midgard Field — the only location with known maturity at that time. The source rocks at Midgard were later shown to be relatively immature (Elvsborg *et al.* 1985).

Reliability and advantages of using velocity-based thermal conductivity

Thermal conductivity of sediments has been known to show a strong correlation with seismic (sonic) velocity (Goss *et al.* 1975; Evans 1977; Houbolt & Wells 1980). Recently, this idea has been expanded by the work of Vacquier *et al.* (1988) and Brigaud *et al.* (1990) who use several wireline log parameters to correlate them with thermal conductivity. All of these well-log based empirical techniques give reasonable conductivity values because they are explicable on the basis of the theoretical mechanism of eqn (1).

According to the recent publication of Brigaud *et al.*(1990) the well-log based technique has an analytical error of about ±20 %. Vacquier *et al.* (1988) indicated an error of ±13 %, which is close to the value of ±10 % reported by Goss *et al.* (1975). Our technique, as mentioned earlier, is based on the correlation of thermal conductivity measured in the laboratory with sonic velocities or interval velocities from well logs reported by Goss *et al.* (1975). These variations in the precision are related to lithologic type, porosity, water content, temperature, etc. According to the study of thermal conductivity carried out by Chapman *et al.* (1984) for three different rock formations in the Uinta Basin (Utah, USA), typical conductivity across a formation is about 30 %. This variation within a formation is very close to the maximum difference between the measured and calculated values shown for the Halten Terrace samples presented in Fig. 11.

Ho *et al.* (1991) compared measured and velocity-based modelled thermal conductivities for 10 samples from the Halten Terrace, and found that about 90 % of the samples studied have a difference of less than 30 %; and 70 % of the samples have a difference of less than about 16 %. The results are very encouraging for using the interval velocity as a parameter for predicting thermal conductivity in an unknown area. Of course the differences mentioned are only based on limited samples. Therefore, we think that more work should be carried out to increase the conductivity database, thus enabling conductivity mapping for better thermal models. More measured conductivity data correlated with well logs, which can be very site-dependent, will therefore help to establish a predictive model for thermal conductivity on the Halten Terrace.

It is well known that the controlling factors of thermal conductivity — lithology, porosity, water content, temperature and compaction — are all correlatable to seismic velocity, as suggested by the Eqn (1). This is why velocity is a good predictor of thermal conductivity for problems in exploration and basin modelling as demonstrated by the results presented in Table 3.

There are three other reasons or advantages for using the seismic geochemical modelling technique in exploration and basin modelling:

(1) It is a good technique for the interpolation of thermal conductivities between wells. Traditionally, the conductivity used in basin modelling has been based on compaction-related modelled porosity, or on wireline logs (such as proposed by Vacquier et al (1988) and Brigaud *et al.* (1990)). These techniques have a disadvantage of being unable to cope with changes in lithofacies. This weakness is not found in the velocity-based technique used in this work because the lateral change in lithofacies or even geopressure between wells is also reflected in seismic velocity at least in a bulk sense. This seismic reflection of lateral change in lithology in Offshore Norway, is also illustrated in a paper by Gonzalez & Tufekcic (1991). In the Halten Terrace the application of the seismic geochemical technique is especially important as the major control of geothermal gradient in this area is due to differences in lithofacies (Vik & Hermanrud 1991).

(2) The velocity-based thermal conductivity can be a complementary parameter for calibrating or checking thermal conductivity based on calculated porosity commonly done by basin modelling computer packages. This is especially important in basin modelling because Waples & Kamata (1991) reported a considerable inaccuracy in calculated porosity based on the simple compaction models found in most computer packages on the market and in the literature. The uncertainty in porosity is not the only problem inherent in this conventional method; there are also problems in duplicating lithologic composition, water flow (see Wangen & Trondsen (1994)), water saturation (see Jensen & Dorè (1993) and Zwach *et al.* (1994)) and compaction within a well using existing basin modelling techniques. These rock properties are also reflected in the seismic velocity; therefore, the velocity-based conductivity represents subsurface conditions relevant to thermal conductivity better than those based on calculated porosity.

(3) The seismic geochemical modelling technique described here can be a link between geochemistry and seismic techniques to enhance the value of organic geochemistry. This is because the technique enables geochemists to translate geochemical concepts into the seismic language used and understood by most explorers. Traditionally, geochemistry has been used as post-drilling technique for trouble shooting or prospect evaluation. Therefore, it is not viewed as being as valuable as the seismic method because the latter has an advantage of being used without well data, thus, allowing explorers to select a prospect ahead of their competitors — a vital exploration strategy in a major oil company.

Geochemistry also has a lesser role in the traditional exploration decision making process because once oil is found, subsequent evaluation based on well data (including geochemical data) becomes a minor task relative to pre- drilling work. Shell Oil Company reports that geochemistry helped to improve the forecasting efficiency by about 35 % over traditional methods of finding structures by the seismic method (Murris 1984; Sluijk & Parker 1986). This reduction of exploration risk due to geochemistry has also being expressed recently by Amoco (Leonard 1989). Hence, inclusion of the seismic geochemical modelling technique in front-line operations for exploration will, without question, further enhance the value of geochemistry by enabling explorers to apply geochemistry in frontier areas where the chance of making a major oil discovery is better than in developed areas as is demonstrated in this work.

Limitation of the seismic method for calculating thermal conductivity

The seismic geochemical techniques described have some limitations. Some of these drawbacks are related to the quality of seismic data. They are more applicable to marine environments land because onshore environments are more prone to static noise problems than offshore environments. The other problem relates to the difference in velocity profiles from different operators. One way to overcome these problems is to use velocities from sonic logs whenever it is possible to do so.

The other limitation of this velocity-based technique is that it is applicable mainly to the modelling of the present-day oil window as it calculates only present-day thermal conductivities. In this connection, this shortcoming is not a problem in the Halten Terrace because, as discussed earlier (see Figs 4 and 10), the rapid subsidence, that is responsible for the major hydrocarbon generation in this area, has continued to the present time for the past 5 Ma. In other words, this technique is most applicable to an area with a late major thermal or subsidence event in its geological history. How to carry the velocity-based thermal conductivities back through time is a research problem to be dealt with in the future.

Conclusion

We have demonstrated in this paper the successful application of seismic geochemical modelling in early stages of oil exploration. This velocity-based calculation of thermal conductivity from seismic stacking velocities, enabled us to predict the presence of a mature source rock on the Halten Terrace. This prediction was later substantiated by drilling and subsequent maturation and thermal conductivity measurements; thus confirming the reliability of our methodology.

Our study shows that seismic geochemical modelling is a useful tool for frontier exploration and is also a potential tool to complement the conventional method of calculating thermal conductivities for basin modelling in both frontier and developed areas. This method provides a means to integrate geochemical and geophysical data used in basin modelling; therefore, it also enhances the exploration value of organic geochemistry and basin modelling as a pre-drilling tool for prospect evaluation and selection rather than just as a post-drilling or *postmortem* tool.

We wish to express our gratitude to Conoco Inc. for permission to publish this paper. We are grateful to Conoco Norway Inc. for providing samples and geochemical and geological data. We also appreciate the access to the thermal conductivity database of Rogaland Research Institute, Stavanger, Norway. The manuscript has been much improved by the helpful review of Mr D. Burggraf of Conoco.

References

BOEN, F., EGGEN, S. & VOLLSET, J. 1984. 17 structures and basins of the margin for 62 to 69 N and their development. *In*: SPENCER, A. M. *et al.* (eds) *Petroleum Geology of the North European Margin.* North European Margin Symposium Proceedings, Graham and Trotman, London, 253–270.

BOSTICK, N. H. & FOSTER, J. N. 1974. Comparison of vitrinite reflectance in coal seams and in kerogen and sandstones, shales, and limestone in the same part of a sedimentary section. *In*: ALPERN, B. (ed.) *Petrographie de la Matiere Organique de Sediments, Relations avec la Paleotemperature et le Potential Petrolier*. C.N.R.S., Paris, 13–25.

BRIGAUD, F., CHAPMAN, D. S. & LE DOUARAN, S. 1990. Estimating thermal conductivity in sedimentary basin using lithological data and geophysical well logs. *American Association of Petroleum Geologists Bulletin*, **74**, 1459–1467.

BRIGMAN, P. W. 1949. *The Physics of High Pressure.* G. Bell & Sons, London.

BUISKOOL-TOXOPEUS, J. M. A. 1983. Selection criteria for the use of vitrinite reflectance as a maturity tool. *In*: BROOK, J. (ed.) *Petroleum Geochemistry and Exploration of Europe*. Blackwell, Oxford, 295–307.

BUKOVICS, C. & ZIEGLER, P. A. 1985. Tectonic development of the Mid-Norway continental margin. *Marine and Petroleum Geology*, **2**, 2–22.

——, CARTIER, E. G., SHAW, N. D. & ZIEGLER, P. A. 1983. Structure and development of the mid-Norway Continental margin. *In*: SPENCER, A. M. *et al.* (eds) *Petroleum Geology of the North European Margin*. North European Margin Symposium Proceedings, Graham & Trotman, London, 407–424.

BURNHAM, A. K. & SWEENEY, J. J. 1989. A chemical kinetic model of vitrinite maturation and reflectance. *Geochimica et Cosmochimica Acta*, **53**, 2649–2657.

CAMPBELL, C. J. & ORMAASEN, E. 1987. The discovery of oil and gas in Norway: an historical synopsis. *In*: SPENCER, A. M. *et al.* (eds) *Geology of the Norwegian Oil and Gas Fields*. Graham & Trotman, London, 1–37.

CAO, S. & LERCHE, I. 1990. Application of sensitivity analysis. *Journal of Petroleum Science and Engineering,* **4,** 83–104.

CHAPMAN, D. S., KEHO, T., BAUER, M. & PICARD, M. D. 1984. Heat flow in the Uinta Basin determined from bottom hole temperature (BHT) data. *Geophysics*, **49**, 453–466.

DAINES S. R., NOPPER, R. W. & HO, T. T. Y. 1990. Consideration of the parameters for the calculation of paleogeothermal gradients (Abstract). *American Association of Petroleum Geologists Bulletin*, **74**, 637.

——, —— & —— 1991. Resolving geothermal gradients: past and present (Abstract). *NPF International Conference on Basin Modelling; Advances and Applications*, 49.

DEMBICKI, H. Jr. 1984. An inter-laboratory comparison of source rock data. *Geochimica et Cosmochimica Acta*, **48**, 2641–2649.

—— & PIRKLE, F. L. 1985. Regional source rock mapping using a source potential index. *American Association of Petroleum Geologists Bulletin*, **73**, 1455–1471.

DORÈ, A. G. 1991. The structural foundation and evolution of Mesozoic seaway between Europe and the Arctic. *Palaeontology, Palaeoclimatology, Palaeoecology,* **87**, 441–491.

DURAND, B., ALPERN, B., PITTION, J. L. & PRADIER, B. 1986. Reflectance of vitrinite as a control of thermal history of sediments: *In*: BURRUS, J. (ed.) *Thermal Modelling in Sedimentary Basin*. Editions Technips, Paris, 441–474.

EHRENBERG, S. N., GJERSTAND, H. M. & HADLER-JACOBSEN, F. 1992. Smørbukk Field: A gas condensate fault trap in the Haltenbanken Province, Offshore Mid-Norway. *In* HALIBOUTY, M. T. (ed.) *Giant Oil and Gas Fields of the Decade 1978–1988.* American Association of Petroleum Geologists Memoir **54**, 323–348.

EKERN, O. F. 1987. Midgard. *In*: SPENCER, A. M. *et al.* (eds) *Geology of the Norwegian oil and Gas Fields*. Graham & Trotman, 403–410.

ELLENOR, D. W. & MOZETIE, A. 1986. The Draugen oil discovery. *In*: SPENCER, A. M. *et al.* (eds) *Habitat of Hydrocarbons on the Norwegian Continental Shelf.* Norwegian Petroleum Society, Graham & Trotman, London, 313–316.

ELVSBORG, A., HAGEVANG, T. & THORNDSEN, T. 1985. Origin of the gas-condensate of the Midgard Field at Haltenbanken. *In*: THOMAS, B. M. *et al.* (eds) *Petroleum Geochemistry in Exploration of the Norwegian Shelf.* Norwegian Petroleum Society, Graham & Trotman, London, 213–219.

ESPITALIÈ, J., MARQUIS, J. F., DROVET, S. & LAFARGUE, E. 1991. Critical study of modelling parameters (Abstract). *International Conference on Basin Modelling. Advances and Applications*. Norwegian Petroleum Society, 17.

EVANS, T. R. 1977. Thermal properties of North Sea rocks. *The Log Analyst*, March–April, 3–12.

—— & COLEMAN, N. C. 1974. North Sea geothermal gradients, *Nature*, **247**, 28–30.

FORBES, P. L., UNGERER, P. M., KUHFUSS, A. B., RIIS, F. & EGGEN, S. 1991. Compositional modelling of petroleum generation and expulsion: trial application to a local mass balance in the Smørbukk Sør Field, Haltenbanken Area, Norway. *American Association of Petroleum Geologists Bulletin*, **75**, 873–893.

GONZALEZ, A. & TUFEKCIC, D. 1991. Interval-velocity inversion 2-D (Abstract). *NPF International Conference on Petroleum Exploration and Exploitation in Norway*, 53. Norwegian Petroleum Society.

GOSS, R., COMBS, J. & TIMUR, A. 1975. Prediction of thermal conductivity in rocks from other physical properties and from standard geophysical well logs. *SPWLA 16th Annual Logging Symposium,* 1–21.

HERMANRUD, C. 1994. Uncertainties in basin modelling — magnitudes and implications (Abstract). *The Geological Society Petroleum Group Conference on Basin Modelling,* November 1 & 2, 1994.

——, EGGEN, S., JACOBSEN, T., CARLSON, E. M. & PALLESEN, S. 1990. Accuracy of modelling hydrocarbon generation and migration: The Egersund Basin Oil Field, Norway. *Organic Geochemistry*, **16**, 389–399.

HEUM, O. R., DALLAND, A. & MEISINGSET, K. K. 1986. Habitat of hydrocarbons at Haltenbanken (PVT modelling as predictive tool in hydrocarbon exploration. *In*: SPENCER, A. M. *et al.* (eds) *Habitat of Hydrocarbons on the Norwegian Continental Shelf.* Norwegian Petroleum Society, Graham & Trotman, London, 259–274.

HO, T. T. Y. & SAHAI, S. K. 1982. Estimation of organic maturation level from seismic and heat flow data (Abstract). *American Association of Petroleum Geologists Bulletin*, **66**, 581–582.

——, —— & THOMPSON-RIZER, C. L. 1991. Seismic geochemical modelling of organic maturity in Haltenbanken area, Offshore Norway: A comparative study of pre- and post-drilling results. (Abstract) *NPF International Conference on Petroleum Exploration and Exploitation in Norway*, 44A-B. Norwegian Petroleum Society.

HOLLANDER, N. B. 1982. Evaluation of the hydrocarbon potential Offshore mid-Norway. *Oil and Gas Journal*, **80**, 168–172.

—— 1984. Geohistory and hydrocarbon evaluation of the Haltenbanken area. *In*: SPENCER, A. M. *et al.* (eds) *Petroleum Geology of the northern Europe Margin.* Graham & Trotman, London, 383–388.

HOUBOLT, J. J. H. C. & WELLS, P. R. A. 1980. Estimation of heat flow in oil wells based on a relation between heat conductivity and sound velocity. *Geologie en Mijnbouw*, **59**, 215–224.

ISAAC, N. S. 1981. *Liquid Phase High Pressure Chemistry.* John Wiley & Sons, New York.

JAVIE, D. M. 1991. Factors affecting Rock-Eval derived kinetic parameters. *Chemical Geology,* **93**, 79–99.

JENSEN, R. P. & DORÈ, A. G. 1993. A recent Norwegian heating event — fact or fantasy?. *In*: DORÈ, A. G. *et al.* (eds) *Basin Modelling: Advances and Applications*, Elsevier, Amsterdam, 85–106.

JÜNTGEN, H. & KLEIN, J. 1975. Formation of natural gas from coaley sediments. *Erdol und Khole-Erdgas-Petrochemie*, **28**, 65–73 (in German).

KEHLE, R. O. 1973. *Geothermal survey of North America 1972 Annual Progress Report*. Project supported by the American Association of Petroleum Geologists.

KOENIG, R. H. 1986. Oil Discovery in 6507/7; an initial look at the Heidrun Field. *In*: SPENCEr, A. M. *et al.* (eds) *Habitat of Hydrocarbons on the Norwegian Continental Shelf.* Norwegian Petroleum Society, Graham & Trotman, London, 307–311.

LARSEN, R. M. & HEUM, D. R. 1988. Haltenbanken hydrocarbon province (Offshore Mid-Norway) (Abstract), *American Association of Petroleum Geologist Bulletin*, **72**, 210.

LARTER, S. 1989. Chemical modelling of vitrinite reflectance evolution. *Geologische Rundschau*, **78**, 349–359.

LEADHOLM, R. H., HO, T. T. Y. & SAHAI, S. K. 1985. Heat flow, geothermal gradients and maturation modelling on the Norwegian Continental Shelf using computer methods. *In*: THOMAS, B. M. *et al.* (eds) *Petroleum Geochemistry in Exploration of the Norwegian Shelf.* Norwegian Petroleum Society, Graham & Trotman, London, 131–144.

LEONARD, R. 1989. *Generation, Migration, & Entrapment of Hydrocarbons on the Southern Norwegian Shelf.* Videotape, International Human Resources Development Company, Boston.

LERCHE, I., YARZAB, R. F. & KENDALL, G. G. ST G. 1984. Determination of paleoheat flux from vitrinite reflectance data. *American Association of Petroleum Geologists Bulletin*, **68**, 1704–1717.

LOPATIN, N. V. 1971. Temperature and geologic time as factors in coalification. *Izvesyiya Akademii Nauk USSR, Seriya Geologicheskaya*, **3**, 95–106 (in Russian).

MCKENZIE, D. 1978. Some remarks on the development of sedimentary basins. *Earth and Planetary Science Letters,* **40**, 319–333.

—— & QUIGLEY, T. M. 1988. Principles of geochemical prospect appraisal. *American Association of*

Petroleum Geologists Bulletin, **72**, 399–415.

McLimans, R. K. 1987. The application of fluid inclusion to migration of oil and diagenesis in petroleum reservoirs. *Applied Geochemistry*, **2**, 585–603.

Middleton, M. F. 1982. Tectonic history from vitrinite reflectance. *Geophysical Journal of the Royal Astronomical Society*, **68**, 121–132.

Murris, R. J. 1984 Introduction. *In*: Demaison, G. & Murris, R. J. (eds) *Petroleum Geochemistry and Basin Evaluation*. American Association of Petroleum Geologists, Memoir, **35**, x–xii.

Nielsen, S. B. & Dahl, B. 1991. Confidence limits on kinetic models of primary cracking and implication for the hydrocarbon generation. *Marine and Petroleum Geology*, **8**, 483–492.

Pauling, L. 1945. *General Chemistry, 2nd edition.* W. H. Freeman, San Francisco, 332–343.

Ronnevik, H., Eggen, S. & Vollset, J. 1983. Exploration of the Norwegian Shelf. *In*: Brooks, J. (ed.) *Petroleum Geochemistry and Exploration of Europe*. Geological Society, London, Special Publications, **12**, 71–93.

Royden, L., Sclater, J. G. & Von Herzen, R. P. 1980. Continental margin subsidence and heat flow: important parameters in formation of petroleum hydrocarbons. *American Association of Petroleum Geologists Bulletin*, **64**, 137–187.

Sluijk, D. & Parker, J. R. 1986. Comparison of pre-drilling predictions with post-drilling outcomes, using Shell's prospect appraisal system. *In*: Rice, D. D. (ed.) American Association of Petroleum Geologists Studies in Geology, **21**, 55–58.

Swanberg, C. A., Chessman, M. D., Simmons, G., Smithson, S. B., Groline, G. & Heier, K. H. 1974. Heat flow generation studies in Norway. *Tectonophysics*, **23**, 31–48.

Sweeney, J. J. & Burnham, A. K. 1990. Evaluation of a simple model of vitrinite reflectance based on chemical kinetics. *American Association of Petroleum Geologists Bulletin*, **74**, 1559–1570.

Thompson-Rizer, C. L. & Woods, R. A. 1987. Microspectrofluorescence measurements of coal and petroleum source rocks. *International Journal of Coal Geology*, **7**, 85–104.

Tissot, B. P. & Espitalie, J. 1975. L'évolution thermique de la matiere organique des sediments: Applications d'une simulation mathematique. *Revue de l'Institut Francais du Petrole*, **30**, 743–777.

—— & Welte, D. H. 1987. The role of geochemistry in exploration risk evaluation and decision making. *Proceedings of the 12th World Petroleum Congress*, John Wiley, New York, **2**, 99–112.

——, Pelet, R. & Ungerer, P. M. 1987. Thermal history of sedimentary basins, maturation indices, and kinetics of oil and gas generation: *American Association of Petroleum Geologists Bulletin,* **71**, 1445–1466.

Toth, D. J., Lerche, I., Petroy, D. E., Meyer, R. J. & Kendall, G. G. 1981. Vitrinite reflectance and the derivation of heat flow changes with time. *In*: Bjoroy, M. (ed.) *Advances in Organic Geochemistry*. John Wiley, New York, 295–307.

Turcotte, D. L. & Schubert, G. 1982. *Geodynamics*. John Wiley and Sons, New York.

Ungerer, Ph, 1993. Modelling of petroleum generation and migration. *In*: Bordenave, M. L. (ed.) *Applied Petroleum Geochemistry*, Editions Technip, Paris, 395–442.

Ungerer, P., Burrus, J., Doligez, B., Chenet, P. Y. & Bessis, F. 1990. Basin evaluation by integrated two-dimensional modelling of heat transfer, fluid flow, hydrocarbon generation and migration. *American Association of Petroleum Geologists Bulletin*, **74**, 309–335.

Vacquier, V., Mathieu, Y., Legendre, E. & Blondin, E. 1988. Experiments on estimating thermal conductivities of sedimentary rocks from oil well logging. *American Association of Petroleum Geologists Bulletin*, **72**, 758–764.

Vielvoye, R. 1984. Haltenbanken heat up. *Oil and Gas Journal*, **82**, 51.

Vik, E. & Hermanrud, C. 1991. Regional heat flow modelling in the Haltenbanken area: Mid-Norwegian Continental Shelf (Abstract). *NPF International Conference on Basin Modelling; Advances and Applications*, 13.

Wangen, M. & Trondsen, T. 1994. A comparison between 1D, 2D and 3D basin simulations (Abstract). *The Geological Society Petroleum Group Conference on Basin Modelling,* November 1 & 2, 1994.

Waples, D. W. 1980. Time and temperature in petroleum formation, application of Lopatin method to petroleum exploration. *American Association of Petroleum Geologists Bulletin,* **64**, 916–926.

—— 1985. *Geochemistry in Petroleum Exploration*. International Human Resources Development Co., Boston, Chapter 9.

—— 1994. Maturity modelling: Thermal indicators, hydrocarbon generation, and oil cracking. *In*: Magoon, L. B. & Dow, W. G. (eds) *The Petroleum System — from Source to Trap,* American Association of Petroleum Geologists, Memoir **60**, 285–306.

—— & Kamata, H. 1993. Modelling porosity reduction as a series of chemical and physical processes and applications: *In*: Dore, A. G. *et al.* (eds) *Basin Modelling: Advances and Applications*, Elsevier, Amsterdam, 303–320.

——, Suzu, M. & Kamata, H. 1991. The art of maturity modelling, Part 2, sensitivity analysis and Alternative models and sensitivity analysis. *American Association of Petroleum Geologists Bulletin,* **76**, 47–66.

Welte, D. H. & Yükler, M. Y. 1981. Petroleum origin and accumulation in basin evolution — a quantitative model. *American Association of Petroleum Geologists Bulletin*, **65**, 1387–1396.

Whitley, P. K. 1992. The geology of Heidrun: A giant oil and gas field on the Mid-Norwegian Shelf: *In*: Halbuty, M. T. (ed.) *Giant Oil and Gas Fields of the Decade 1978–1988*. American Association of Petroleum Geologists Memoir **54**, 383–406.

Wood, D. A. 1988. Relationships between thermal maturity indices calculated using Arrhenius equation and Lopatin method. Implications for petroleum exploration. *American Association of Petroleum Geologists Bulletin*, **72**, 115–134.

YÜKLER, M. A. & WELTE, D. H. 1980. A three-dimensional deterministic dynamic model to determine geologic history and hydrocarbon generation, migration and accumulation. *In*: *Fossil Fuels*. Editions Technip, Paris, 267–285.

ZWACH, C., POELCHAU, H. S., HANTSCHEL, T. & WELTE, D. H. 1994. Simulation with contrasting pore fluids: Can we afford to neglect hydrocarbon saturation in basin modelling? (Abstract). *The Geological Society Petroleum Group Conference on Basin Modelling*, November 1 & 2, 1994.

Aspects of applied basin modelling: sensitivity analysis and scientific risk

R. O. THOMSEN

Saga Petroleum as, Kjørboveien 16, Sandvika, Norway

Present address: Department of Geography & Geophysics, Texas A & M University, College Station, Texas 77843-5309, USA

Abstract: Basin modelling is a powerful quantitative tool and in order to be able to use modelling for the correct purpose and to expand the variety of problems where basin modelling with advantage can be applied, it is necessary to view basin modelling in the context of the general exploration process. Results from basin modelling are often directly used for ranking and risking of exploration targets. Due to the integrated nature of basin modelling it is applied more or less routinely throughout the exploration process for quantitative assessments. Awareness of some fundamental limitations of the models used is therefore vital in order to assess the overall uncertainty associated with the results. It will be shown how, through sensitivity analysis and risking of the modelling results, uncertainty ranges and confidence levels can be attached to the results. Examples of how sensitivity analysis helps in the process of understanding the system behaviour, reveals the critical factors, and provides the resolution limits on specified parameters will be given. Finally, a probabilistic procedure for assessing the uncertainty ranges and the level of confidence associated with modelling results will be given. A case history shows how the procedure can be used for providing risked entry parameters for the general risk assessment.

Basin modelling is used routinely as a quantitative tool in exploration for the purpose of predicting ahead of drilling the volume and type of hydrocarbon trapped in a prospect. In the context of the general exploration process basin modelling is applied as a quantitative tool in basin analysis (Fig. 1). Basin analysis again provides the necessary geological and quantitative input for reserve estimates at play level or at prospect level. Therefore the results from basin modelling can be made to play an important role in the general procedures for risking plays and/or prospects.

Since the first basin models appeared in the late 1970s/early 1980s (Yükler *et al.* 1978; Welte & Yükler 1981; Durand *et al.* 1984; Lerche & Glezen 1984; Bethke 1985) both the models and the way models are used have undergone significant development. Models on the one hand include more processes and have improved the calibration to observed parameters. On the other hand the users have matured and are now becoming aware of the importance of knowing the limitations of a given model. The field of basin modelling itself has developed through three major stages. During the first stage the major effort was put into the description of processes and actually building models that included both dynamical evolution of the sediments and some sort of maturity modelling — the models were forward deterministic models. During the second stage of development calibration was in focus and sophisticated inverse schemes were developed first of all for reconstructing the thermal history from thermal indicators (Lerche *et al.* 1984; Lerche 1988) and later for calibrating the rock parameters determining compaction and pressure build-up (Lerche 1991). Development of inverse models made it possible to investigate two important aspects of basin modelling. First of all the sensitivity of model results to the choice of input parameters and constants could be analysed not only as a response function from changes in the parameters but as a misfit to or departure from observed control information such as Ro, formation pressure, porosity, formation thickness, etc. Second, inverse methods provide a means of directly determining the limits of resolution of any calibration parameter or model constant in a set of control parameters, either as a single parameter or in combination with any number of other parameters. With the development and use of inverse methods in basin modelling came an increasing understanding of the inability of complex basin models to provide a unique answer and basin modelling slowly entered its third stage of development. This stage takes on the uncertainty on basin modelling results, developing the use of

THOMSEN, R. O. 1998. Aspects of applied basin modelling: sensitivity analysis and scientific risk. *In*: DÜPPENBECKER, S. J. & ILIFFE, J. E. (eds) *Basin Modelling: Practice and Progress*. Geological Society, London, Special Publications, **141**, 209–221.

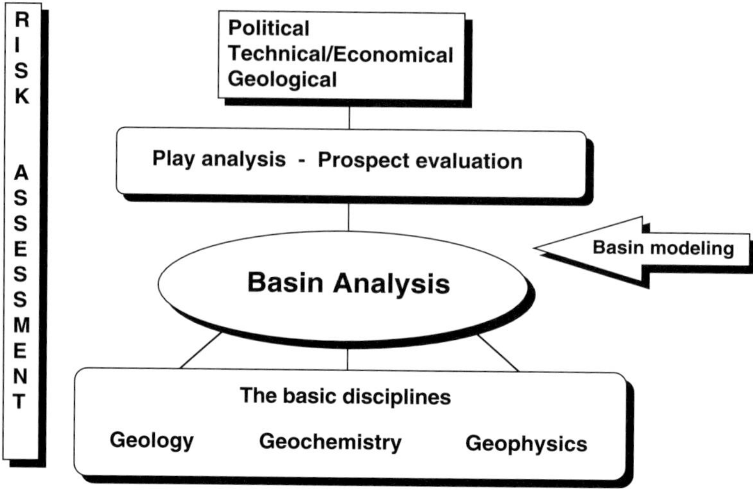

Fig. 1. Basin modelling is a quantitative tool in basin analysis and provides part of the entry parameters for play analysis and prospect evaluation.

probabilistic methods in the evaluation of basin modelling results with the purpose of evaluating the scientific risk associated with results from such models.

This paper deals with the aspects of resolution limits on parameters, sensitivity analysis and assessment of the scientific risk. The theoretical aspects will be covered in some detail where needed, otherwise references will be given to texts with thorough treatment of the theoretical background. Examples are used to illustrate specific points of practical use of the methods described.

Determination of parameters

Ahead of running a simulation using any basin model, a large number of constants and other parameters, describing features of the basin evolution together with the behaviour of the rock–fluid system over the time frame of the basin formation, have to be determined. In the process of calibrating basin models these parameters are varied and eventually fixed in such a way that some acceptable agreement with observed parameters such as porosity, permeability, Ro, formation temperature, formation pressure, etc. is obtained.

There are numerous ways of determining parameters to be used in basin modelling studies. One can simply use the model default values and argue that these are adequate, one can rely on experience and use so called qualified guess, one can use the optimization procedures available in some basin model codes, or one can determine the parameters from observed related entities by applying inverse methods. A few general requirements, however, are that the determination of parameters is done in a consistent way and that the results are reproducible and independent of the user. Furthermore, in order to apply sensitivity analysis and probabilistic evaluation procedures, it is necessary to determine the acceptable dynamic range of variation for each of the parameters. Inverse methods will most likely fulfil all these general requirements. Different inverse methods are available and have been described in detail elsewhere (Glasko 1984; Lerche 1988, 1991; Press *et al.* 1986). In this paper two different inverse schemes are used for the purpose of determining modelling parameters; a linear search scheme valid for one or two parameters and a non-linear tomographic scheme valid for multiple parameters. The linear scheme has the advantage of mapping the misfit of the modelled to the observed quantity over the entire search range. The disadvantage of linear search is that the search is slow especially for two parameters. The advantage of the tomography scheme is that several parameters can be determined simultaneously and the method is fast. The disadvantage being that the misfit function is not mapped in detail and only the path from the initial guess to the minimum is mapped.

In principle there are three classes of parameters to be determined in the process of calibrating basin models. The first class, referred to as dynamical parameters, are parameters associated with the geohistory such as time-rock parameters which describe the behaviour of the physical rock entities with time and burial (compaction, erosion, fluid-flow, porosity, permeability, and pressure etc.). The dynamical parameters impact on the general geometrical reconstruction of the basin. The second class, referred to as thermal parameters, are parameters associated with the thermal history and describe the time–temperature behaviour imposed on the rock system (thermal conductivity and palaeoheat flow etc.). The thermal parameters impact on the reconstruction of the thermal history of the rocks. The third class of parameters is related to the petroleum system and the reconstruction of the hydrocarbon generation history (hydrocarbon kinetics, primary migration, secondary migration, source rock parameters etc.). In the following two sections examples of determination of dynamical and thermal parameters will be given. The third class of parameters will not be treated in detail but rather will be incorporated in the section dealing with risking of model output. For the simulations the 1-D Dynamical and Thermal Indicator Tomography (DTIT) code developed at the University of South Carolina is used.

Dynamical parameters

Sediments undergo major reduction in thickness with burial due to the load of overlying sediments. The reduction in thickness is governed by the ability of the pore water to flow out of the sediment. Hence some calibration of the constants in the equations relating porosity to pressure and permeability to porosity is often necessary in order to obtain an adequate match of observed to modelled thickness, pressure, permeability and porosity. In the following example shale compaction is modelled by:

$$e = e_* \left(\frac{p_f}{p_{f*}}\right)^{-\frac{1}{A}} \quad (1)$$

where e is the void ratio, e_* is the depositional void ratio, P_f is the frame pressure or the part of the pressure supported by the rock frame work, P_{f*} is a scaling frame pressure at deposition, and A is a shale-type specific constant determining the overall level of compaction to a given load. Void ratio, e, is related to porosity, ϕ, by:

$$\phi = \frac{e}{(1+e)} \quad . \quad (2)$$

For sandstone, the compaction is modelled by:

$$e = e_* - C(p_f - p_{f*}) \quad (3)$$

where C is a rock specific constant governing the overall compaction behaviour of the sandstone.

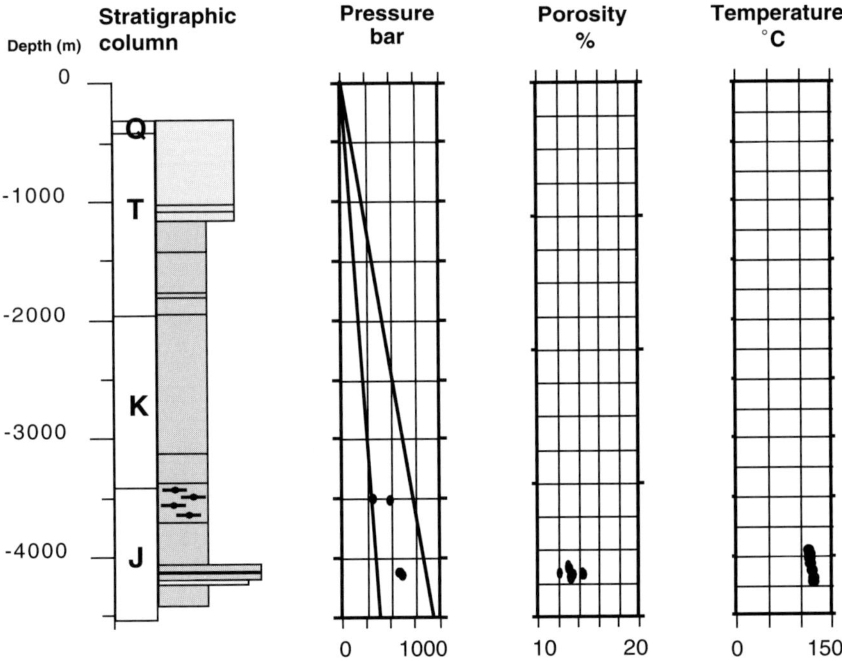

Fig. 2. A typical set of control parameters may consist of formation thickness, fluid pressure, porosity, and formation temperature. Control parameters help determine constants in the equations used in modelling.

The porosity–permeability relation is modelled using the following equation:

$$k = k_* \left(\frac{e}{e_*}\right)^B \quad (4)$$

where k is the permeability, k_* the permeability at the depositional void ratio e_*, and B a rock specific constant determining the change of permeability with loss of porosity.

In the process of calibration the key question is how well constants and other describing parameters can be determined from observations. In the following stratigraphy, measured pressure, and measured sandstone porosity at reservoir level (Fig. 2) is used to illustrate this point. The set of observations (control parameters) available for determining the appropriate rock parameters are formation thickness, total depth, fluid pressure, and sandstone porosity measured in a cored section. Determination of the rock constants is then done by a series of tomographic and linear searches. Multiple parameter tomography is used first to narrow the search range for the parameters then single parameter linear search is performed to evaluate the resolution limits of each of the parameters. As an example of parameter determination and evaluation of resolution limits the sandstone parameters C (C has the dimension of pressure^{-1}) and B (dimensionless) and the shale parameters e_* (dimensionless) and B (dimensionless) is used (Fig. 3). The mis-fit function is the dimensionless mean squared residual (MSR) of the modelled to the measured quantity. By inspection it is seen that the sandstone parameters are found to be resolved only to a level of less than a threshold value. The parameter B which is governing the loss of permeability with compaction must in this case be less than roughly 3.2 limiting how tight the sand can be to around 3 mD at 12 % porosity. The best value of B is seen to be 2.7 giving around 10 mD at 12 % porosity. However, any value of B less than the above mentioned 3.2 gives a reasonable combined fit to all of the control observations. The parameter C which determines the compactional behaviour of the sand as a function of the frame pressure is similarly determined to some threshold

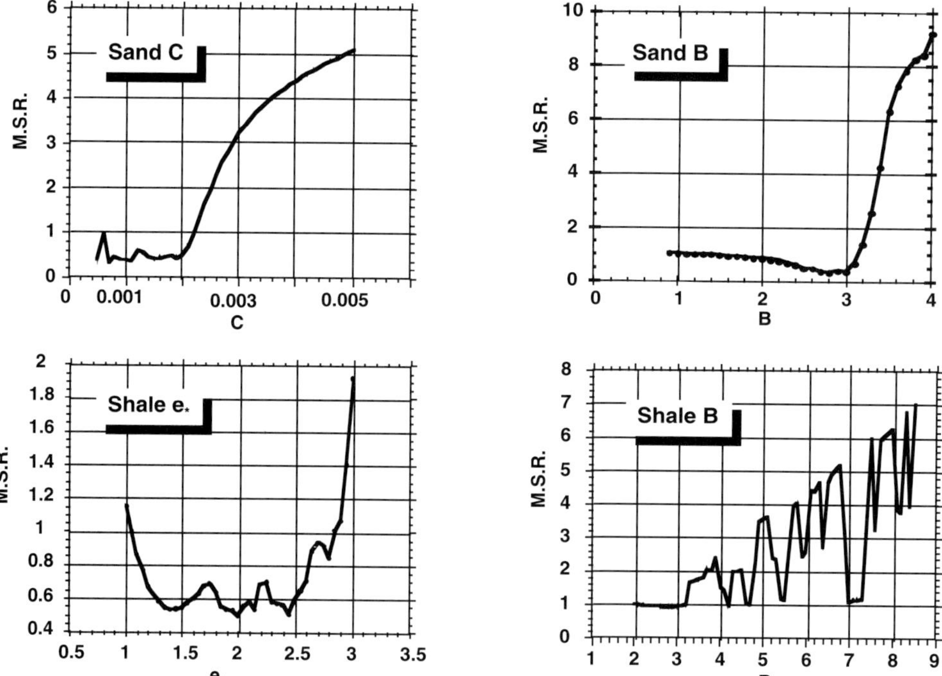

Fig. 3. Parameters are not always well determined. The sand parameters are determined to be less than a threshold value. Once the parameters are set to values below the threshold the system seems insensitive to sand C and B. The shale depositional void ratio, e_*, is determined within a range between 1.2 and 2.6 as indicated by the broad valley in the mis-fit function. The B parameter is determined to less than a threshold value. There are, however, values above the threshold that give an acceptable fit to the observations. This is due to an interplay of low permeability resulting in high overpressure and fracturing of the rock with bleed off of pressure.

value, in this case less than roughly 0.0023 indicating that the sandstone is highly under-compacted.

Thermal parameters

Thermal history can be determined in a similar way (Lerche 1988). Observed vitrinite reflectance from the well together with measured present day formation temperatures will for this purpose be the control data. An appropriate model for heat flow variation with time is chosen; in this case, a rift basin, an exponential decay with time of a heat flow anomaly from the time of rifting to the present day. The relationship is expressed by:

$$Q_t = Q_0 \exp(\beta t) \qquad (5)$$

where $Q(t)$ is the heat flow at the time, t, in million years. Q_0 is the heat flow at the present day and β is a constant determining the rate of decay of the heat flow anomaly. The present day formation temperature helps determine the present day heat flow and the observed vitrinite reflectance profile is used to determine β in eqn (5). In this case there is significant scatter in the measured Ro leading to some difficulty resolving the palaeoheat flow (Fig. 4). The linear search reveals a minimum in the misfit function at $\beta = 0.001$ with the shape of the misfit function around the minimum revealing a distinct lack of resolution of β, which in fact can lie anywhere between –0.0065 and +0.0045.

The previous paragraphs show how it is possible from observations to determine a best value of any parameter bracketed by a minimum and a maximum value. A common procedure is then to evaluate how this uncertainty is reflected in results, by some linear sensitivity plot. As an example, the sensitivity of source rock maturation to uncertainty in determining the maximum heat flow anomaly is shown (Fig. 5). These kind of plots are valuable for illustrating the possible range of results that will satisfy a given set of control observations and for evaluating critical parameters. The plot does not, however, address the question of the likelihood of the answer lying above or below a given threshold value. As will be shown in the next section there are some quite simple methods available for evaluating, in a probabilistic manner, the uncertainties inherited from the limits of resolution of parameters in control data.

Probabilistic evaluation

Knowledge of the resolution limits and sensitivity does not automatically provide a quantitative

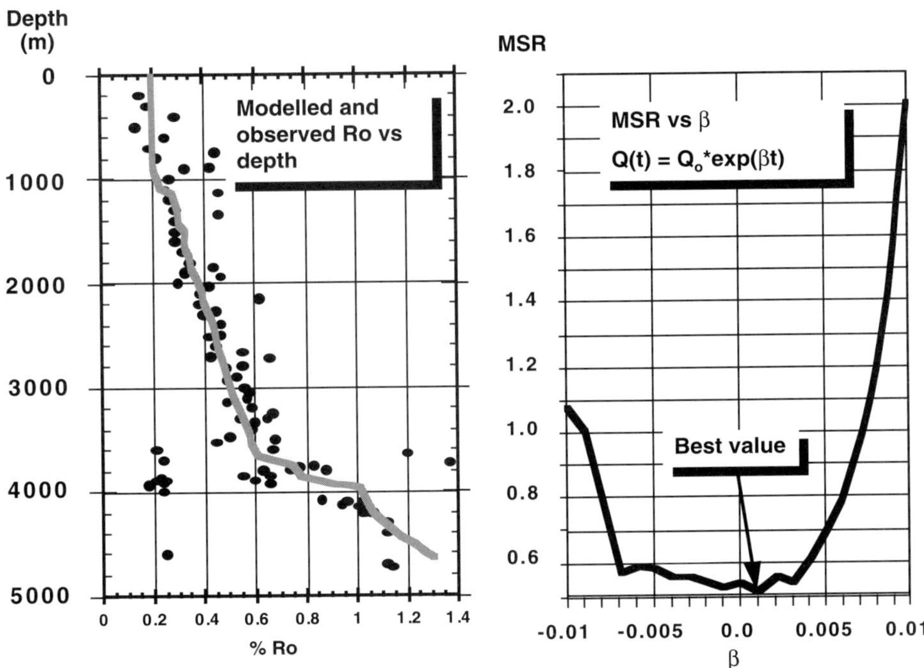

Fig. 4. When the control data are highly scattered the resolution of information in the data becomes poor, indicated by the broad valley in the mis-fit function.

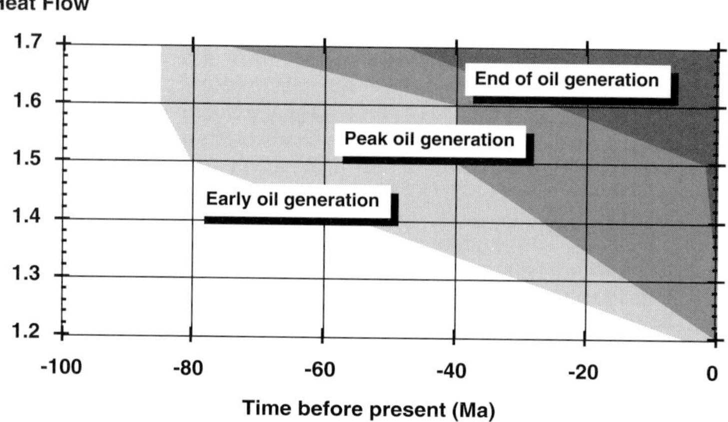

Fig. 5. The consequences of poor resolution of heat flow history can be dramatic as illustrated here where the timing of oil generation is plotted against the maximum palaeoheat flow before decay of the anomaly. Palaeoheat flow is here a critical parameter and without better constraint peak oil generation is predicted to have taken place between 70 Ma and 2 Ma.

measure of confidence that should be attached to model results. In order to evaluate the likely validity of model results, a measure is required of the level of confidence that can be ascribed to any output from such computer models. Quantitative basin analysis deals primarily with measurements of physical properties, interpreted qualities, and equations simulating geological processes of interest. As illustrated in the previous section the major limitation is the accuracy by which any of these quantities can possibly be obtained. As a consequence we have to accept ranges of uncertainty on all parameters entering the equations connecting measurements with the physical quantities being modelled.

One well known procedure for assessing the probability of a particular result being correct, based on ranges of uncertainty of input parameters, is to define ranges of possible values for each parameter, together with a frequency distribution of the occurrence of each value within the range, then perform a series of random Monte Carlo computer runs. While the Monte Carlo method is extremely effective, the computer time needed for Monte Carlo simulation, when dealing with complex systems such as basin models becomes a major problem.

Fortunately, there are other methods available for assessing uncertainty and risk which are less intensive of computer time. The method to be used here is an analytical probabilistic approximation method that has been proposed for use in exploration economics (Lerche 1992). Apart from being useful in exploration economics the method provides us with a generic tool for uncertainty evaluation (Thomsen 1994). For a general introduction to the method the reader is referred to Lerche (1992*a*,*b*,*c*) and the following is merely a short summary of the general points relevant to the arguments used for the application presented in this paper.

The method is based on three intrinsic assumptions:

(i) cumulative probabilities are adequate for the kind of problems we are concerned with;
(ii) a triangular distribution adequately describes the individual frequency distributions;
(iii) the empirical rules for sums and products of distributions (see later) are valid.

The first assumption implies that we are concerned with the probability of a quantity being above or below a threshold ($P(x > y)$ or $P(x < y)$). With $p(x)$ being the differential probability of obtaining x in the range x to x+dx we get:

$$P(x > y) = \int_y^\infty p(x)\mathrm{d}x \tag{6}$$

$$P(x < y) = \int_0^y p(x)\mathrm{d}x \equiv \int_0^\infty p(x)\mathrm{d}x - \int_y^\infty p(x)\mathrm{d}x \equiv 1 - P(x > \mathrm{y}). \tag{7}$$

Later, from the third assumption, we shall see that the two distributions for the differential probabilities necessary to be concerned with are the normal distribution and the log-normal distribution. There are a few statistical parameters needed in order to be able to apply the cumulative probability method: the mean value of x, the mode value and the variance. The mean value or the first moment of x is given by:

$$E^1(x) = x_{\frac{1}{2}} \exp\left(\frac{\mu^2}{2}\right) \quad ; \quad (8)$$

the mode value by:

$$x_m = x_{\frac{1}{2}} \exp(-\mu^2) \quad ; \quad (9)$$

and the variance by:

$$\sigma^2 = E_1(x)^2[\exp(\mu^2)-1] \quad ; \quad (10)$$

where $x_{\frac{1}{2}}$ is the median value (Fig. 6) and μ a scale parameter (see eqn 16).

Due to the fact that most of the parameters being determined in basin modelling are determined only by a likely minimum value, a likely most probable value, and a likely maximum value, the second assumption is necessary and the argument is: because of the inability to exactly determine the distribution of the parameters an approximation by a triangular distribution adequately describes the true distribution of the individual parameters being evaluated (Fig. 7). Using Simpson's triangular rule (Lerche 1992) approximations of the relevant values can be obtained:

$$E_1(x) \cong \frac{1}{3}(x_{\min} + x_{\mathrm{mp}} + x_{\max}] \quad ; \quad (11)$$

$$\sigma^2 = \frac{E_1(x)^2}{2} - \frac{[x_{\min}x_{\max} + x_{\mathrm{mp}}(x_{\min} + x_{\max})]}{6} \; ; \quad (12)$$

and the second moment is given by:

$$E_2(x) = E_1(x)^2 + \sigma^2 \quad . \quad (13)$$

Estimates of μ, $x_{\frac{1}{2}}$, and x_m are then given by:

$$\mu = \left[\ln\left(1 + \frac{\sigma^2}{E_1(x)^2}\right)\right]^{\frac{1}{2}}, \quad (14)$$

$$x_{\frac{1}{2}} = E_1(x)\exp\left(\frac{-\mu^2}{2}\right) \equiv \frac{E_1(x)}{\left[1 + \frac{\sigma^2}{E_1(x)^2}\right]^{\frac{1}{2}}}, \quad (15)$$

and

$$x_m = E_1(x)\left[1 + \frac{\sigma^2}{E_1(x)^2}\right]^{-\frac{3}{2}}, \quad (16)$$

Thus, the cumulative probability, P, takes on the values:

$$P = 0.16 \text{ on } E_1(x)\exp\left(\frac{-3\mu^2}{2}\right) \quad , \quad (17)$$

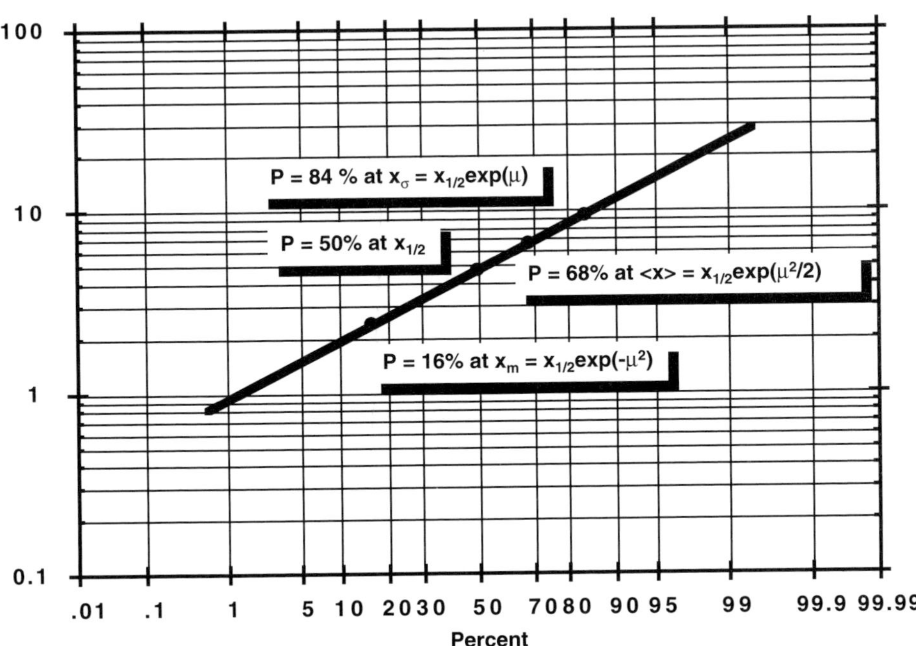

Fig. 6. A log probability plot with important parameters indicated.

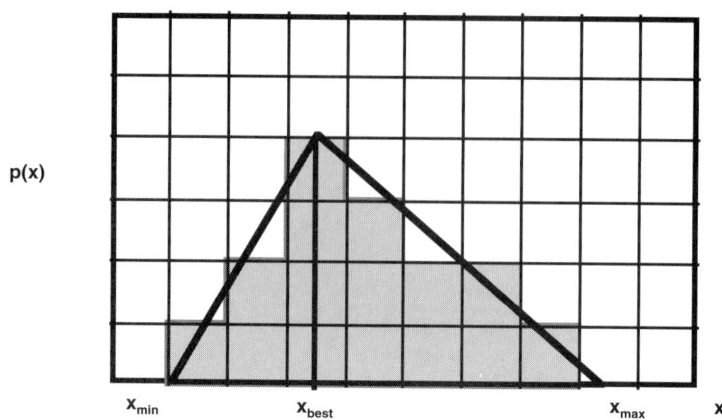

Fig. 7. An illustration of how the determined minimum, most likely (or best), and maximum values are used to describe the probability distribution with Simpson's triangular rule (modified from Lerche 1992).

$$P = 0.5 \text{ on } E_1(x)\exp\left(\frac{-\mu^2}{2}\right) \quad , \tag{18}$$

$$P = 0.68 \text{ on } E_1(x) \quad , \tag{19}$$

$$P = 0.84 \text{ on } E_1(x)\exp\left(\mu - \frac{\mu^2}{2}\right) \quad . \tag{20}$$

The third assumption is that the following two empirical rules are valid: First it appears that N independent random variables from any frequency

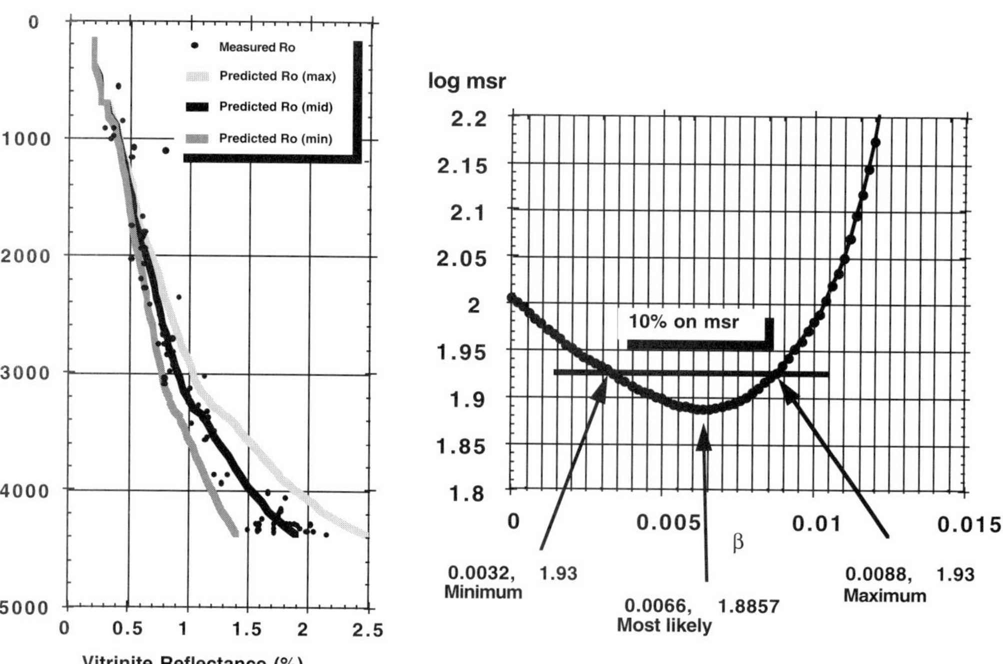

Fig. 8. Determination of heat flow β for a set of vitrinite reflectance data. Accepting a 10 % error in the fit to the data it is seen that the minimum and maximum values bracket the distribution of Ro.

distribution add to give a sum:

$$S_N(\equiv x_1 \pm x_2 \pm x_3 \pm \ldots \pm x_N) \quad (21)$$

which is approximately normally distributed as N becomes large (this is usually obtained for N > 3 or 4), with the mean value:

$$E_1(S_N) \cong \sum_{i=1}^{N} E_1(x_i) \quad (22)$$

and the variance:

$$\sigma(S_N)^2 = \sum_{i=1}^{N} \sigma(x_i)^2 \; . \quad (23)$$

Second, it appears that N independent random variables from any frequency distribution multiply to give a product:

$$P_N(\cong X_1^{a1} X_2^{a2} X_3^{a3} \cdots X_N^{aN}) \quad (24)$$

which is approximately log-normally distributed as N becomes large (this is usually obtained for $N > 3$ or 4), with the mean value:

$$E_1(P_N) \cong \prod_{i=1}^{N} E_1(X_i^{ai}) \quad (25)$$

and the scale parameter, μ, given by:

$$\mu^2(P_N) \cong \sum_{i=1}^{N} \mu^2(X_i) \quad , \quad (26)$$

arguing that the validity of the first assumption is straight forward. For instance once the pore volume available for hydrocarbons in a prospect (HCPV) has been estimated, it becomes interesting to evaluate the probability that at least this volume of hydrocarbons has migrated to the prospect and has been trapped there which is exactly what a cumulative probability distribution can be used for. For the validity of the second assumption the following argument can be used. The shape of the mis-fit function within the acceptable range for each of the parameters being evaluated given by the probable minimum value, the probable most likely value, and the probable maximum value gives an indication of the overall shape of the frequency distribution of the parameter. A triangular distribution can now be fitted to the shape of the mis-fit function to obtain an approximate distribution (compare for instance Figs 7 and 8). For the third assumption the empirical rules can be verified by simply multiplying and adding distributions of truly independent parameters. The validity of the assumption then becomes a question of whether or not the individual parameters being evaluated are truly independent. Arguing that these three assumptions are valid within the limitations mentioned it is possible to use the above formulae to assess the probabilistic uncertainty or scientific risk associated with modelling output as we shall see in the following section.

Risking model output

In this section it will be demonstrated how it is possible, by use of inverse methods for assessing the thermal history from a distribution of vitrinite reflectance values with depth, to obtain a measure of the sensitivity of the results from modelling to the uncertainty in the palaeoheat flow determination. At the same time the determination of the resolution limits of the palaeoheat flow from the distribution of vitrinite reflectance values by linear search is used for assessing the likely most probable, the likely minimum, and the likely maximum heat flow history.

A general model of the geological evolution suggested some constraints on the palaeoheat flow model implying a maximum heat flow anomaly around 130 Ma. Inversion of vitrinite reflectance with the purpose of revealing a general trend for the heat flow history then indicated a decay of the heat flow anomaly from its maximum value at 130 Ma to the present day value at around 60 Ma. Using this to constrain the heat flow history a new linear search was performed in order to obtain a heat flow history most consistent with both the observed vitrinite reflectance and the geological model. In order to comply with the requirements from the geological model, the equation for the heat flow evolution with time was modified from the simple form in eqn (5) to a more complex form allowing both a focus of the heat flow anomaly and a faster decay of the anomaly:

$$Q(t) = Q_0 \exp\left(\beta t + A_1 \sin\left(\frac{\pi t}{t_{\max}}\right) + A_2 \sin\left(\frac{2\pi t}{t_{\max}}\right)\right) \quad (27)$$

where t_{max} is the oldest stratigraphic age (in this case 160 Ma), the present day heat flow, Qo, is fixed at 1.3 HFU ($\approx$55 mW m^{-2}), A_1 at 0.25, A_2 at –0.30, and A_3 at 0.12 for the base case. The linear search for the β parameter in eqn (27) now allowed for an evaluation of the resolution limits of β in the vitrinite reflectance and the determination of the most likely, the minimum, and the maximum value for β (Fig. 8). The resulting heat flow histories show fair agreement for the Tertiary heat flow but an increasing discrepancy towards the peak of the heat flow anomaly indicating an increasing lack of resolution of the heat flow with time before present (Fig. 9).

Having determined the dynamic range of variation of heat flow history most consistent with the observations and the geological indications, the results can be used for evaluating the maturity level

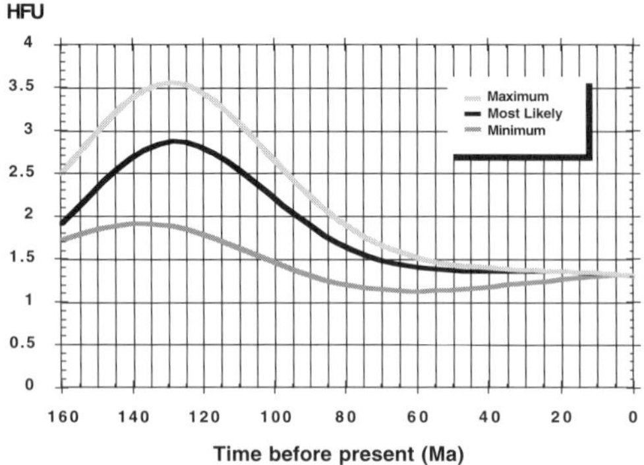

Fig. 9. The resulting heat flow histories acceptable within the resolution limits of the heat flow β in the vitrinite reflectance.

of the basin by extrapolation of the results to pseudo-wells in basin position. With the three heat flow histories the present day depth to an 'oil window' defined to lie between vitrinite reflectance of 0.6 % to 1.3 % can be evaluated using the probabilistic approach. First P(16), P(68), and P(84) are calculated and plotted in a log-normal probability diagram then a straight line is fitted through the three points for extrapolation beyond the P(16) and P(84) (Fig. 10). The diagram represents the probability that the 'oil window' (Ro between 0.6 % or 1.3 %) is located at a particular depth or shallower. For instance, at the depth of 3100 m below the sea floor there is only a 5 % chance that the vitrinite reflectance has reached 0.6 % whereas at 3340 m below the sea floor there is a 95 % chance that a vitrinite reflectance of 0.6 % has been reached. The depth below the sea floor to an iso Ro of 0.6% can then be expressed as 3240^{+100}_{-140} m with a 10 % chance of being wrong. Similarly the depth below the sea floor to an iso Ro of 1.3 % is predicted to be 4380^{+170}_{-180} m with a 10 % chance of being wrong (Fig. 10).

The three heat flow histories will naturally result in different hydrocarbon generation histories for a given source rock. The uncertainty in assessed hydrocarbon generation has a direct impact on the volumetric calculations for undrilled prospects. Using the same procedure as above and allowing for the uncertainty in predicted hydrocarbon generation potential, uncertainty in source rock total organic carbon (TOC), and uncertainty in actual source rock thickness P(16), P(68), and P(84) for the total volume of oil generated can then be evaluated (Fig. 11; Thomsen 1994). By inspection it is seen that the slope of the straight line fit is

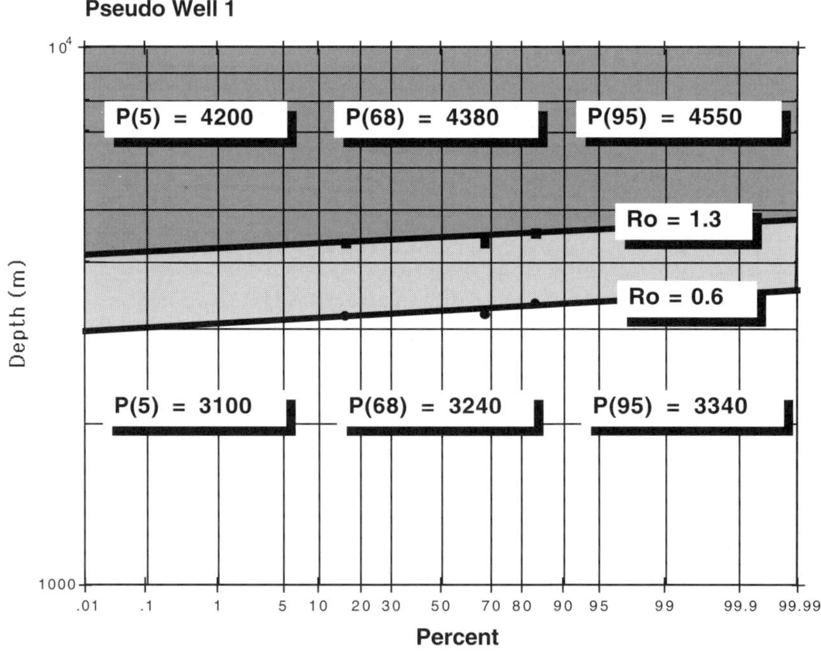

Fig. 10. With the uncertainty in palaeoheat flow, depths to an Ro of 0.6 % and 1.3 % can be predicted for pseudo-wells and risked using the cumulative probability method. Note that depth is below sea floor.

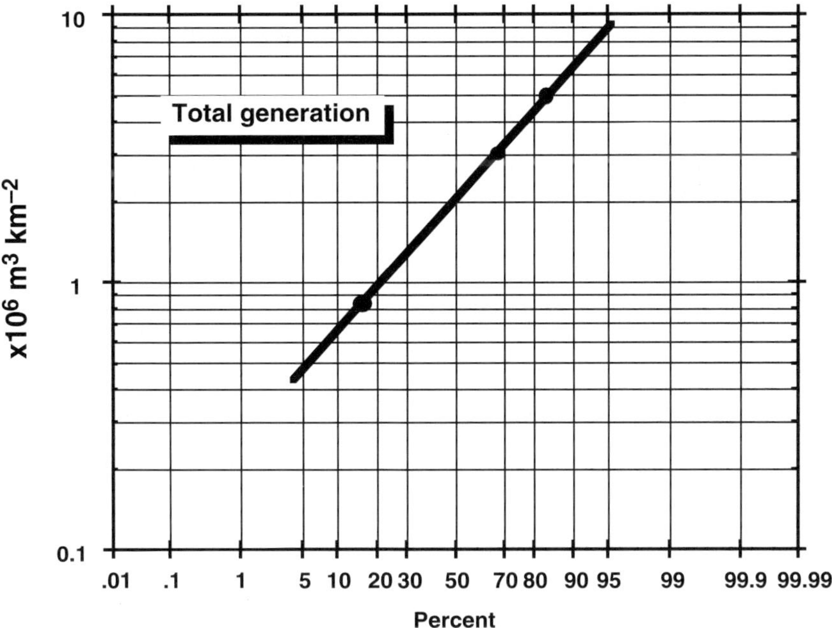

Fig. 11. Taking into consideration all the uncertainty from heat flow history and parameters describing the source rock the total generation can be predicted and the results risked.

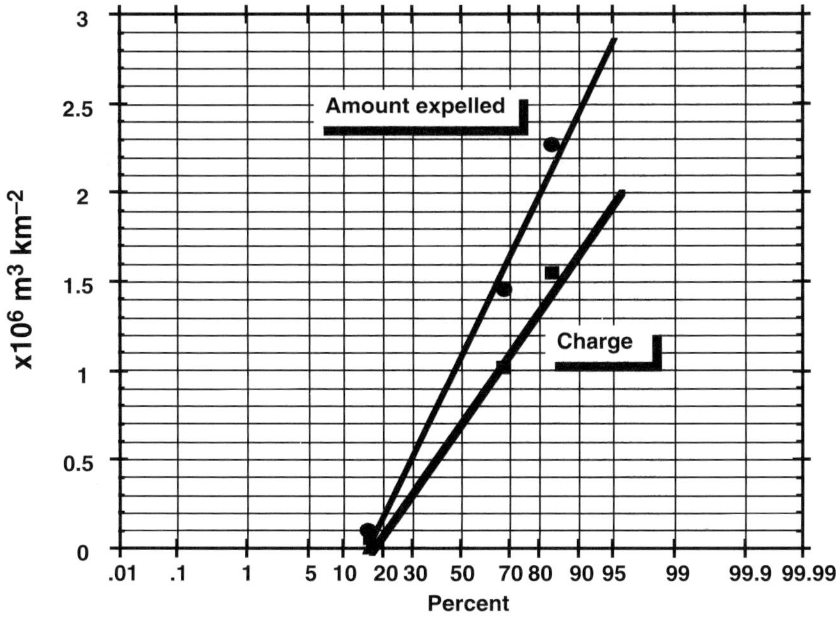

Fig. 12. The risked amounts of expelled hydrocarbons and hydrocarbons migrated to the trap. Note that there is roughly 15 % chance that no hydrocarbons have been expelled and a roughly 17 % chance that none of the migrated hydrocarbons ever made it to the trap.

quite high indicating high uncertainty on the estimated volume. In fact, within the P(5) to P(95) range the total volume generated varies by an order of magnitude from 0.46×10^6 m^3 km^{-2} at P (5) to 8.77×10^6 m^3 km^{-2} at P(95). The amount of expelled and migrated hydrocarbons can be evaluated and risked in the same way taking into account the uncertainty in estimated total generation and uncertainty in predicted source rock porosity and hydrocarbon saturation (Fig. 12). It is easily seen that there is roughly a 15 % chance that no hydrocarbons have been expelled. The mean value, P(68), suggests a most likely expulsion normalized with respect to drainage area of $1.64^{+1.26}_{-1.64} \times 10^6$ m^3 km^{-2} with a 10 % chance of being wrong. The probable charge is evaluated considering residual saturation in the carrier bed and loss through the seal and through fractures. Again, by calculating P(16), P(68), and P(84) and fitting the best straight line through the points it is seen that there is roughly a 17 % chance that none of the expelled hydrocarbons ever made it to the trap. The most likely charge, P(68), normalized with respect to drainage area thus yields a volume of $1.10^{+0.88}_{-1.10} \times 10^6$ m^3 km^{-2}.

The previous paragraphs illustrate the point that if only the 'best' values of all the parameters are used (a typical mean case) the prediction would be a hydrocarbon bearing structure. Furthermore, the risking of the charge will not be directly linked to quantifiable observations but rather reflect some level of qualified guesswork. Following the outlined procedure presented here a direct measure of the risk of a dry structure due to lack of charge is obtained and indeed any number picked off the diagrams has the risk associated with it.

Discussion and conclusion

Predictions of occurrences of hydrocarbons are obviously associated with major uncertainties. Sometimes the uncertainties are overwhelming and one can be tempted to ignore them. The general idea is to try to constrain the results from modelling by calibrating the model both with respect to observations — that is control parameters — and with respect to an underlying geologic model. Calibration using control parameters can be done as a pure mathematical excercise whereas calibration to a geologic model is not a purely mathematical procedure. One of the major uncertainties in evaluation of hydrocarbon generation is the palaeo-heat flow. In the example used in the section 'Risking model output' the heat flow history was constrained both by vitrinite measurements and a geologic model indicating a time for a maximum heat pulse and a general form of heat flux behaviour. In this particular case the individual parameters in the heat flow equation (eqn 27) cannot be allowed to vary independently because of the constrain imposed by the geologic model. Because of the non-linear dependency of the parameters in eqn (27) due to the geologic model, the probabilistic method is not used on the parameters but rather on the resulting heat flow history. The minimum, best, and maximum heat flow histories are determined by the distribution of the vitrinite reflectance measurements and the actual values of the resulting heat flow histories at any instant in time are used for the probabilistic evaluation thus avoiding violation of the intrinsic assumption of independent parameters. If any of the parameters being evaluated show a dependency — whether the relationship is linear or non-linear — the probabilistic method should not be used on the parameters themselves but rather on the results.

Hopefully the examples presented here illustrate the importance that instead of ignoring uncertainties, one should use the uncertainty in sensitivity analysis and reveal critical parameters. Finding the limits of resolution in a set of control observations, of parameters that connect the measured quantities to the processes being active during the formation and evolution of sedimentary basins (including generation, expulsion, migration, and accumulation of hydrocarbons) can be extremely helpful in determining the most critical sources of uncertainty. Acceptable parameter ranges, constrained by observations and geological models can be used to determine the overall uncertainty quantitatively and can eventually be used for assessing the scientific risk.

Evaluating resolution limits and performing sensitivity analysis can help assess model limitations and whether the limitation is in the available data or in the model itself. The procedure presented here illustrates the stepwise path from sensitivity analysis, parameter determination, and probabilistic evaluation to obtaining a risked model output. The probabilistic method provides a way of obtaining a measure of the level of confidence that can be assigned to any result from basin modelling. Basin modelling can thereby provide direct — and reproducible — entry parameters for the overall risking of prospects and plays.

The author wish to thank Saga Petroleum a.s. for allowing the publication of this work. The thoughts and ideas presented here are those of the author and should not be viewed as the general philosophy of Saga Petroleum a.s.. A special acknowledgment goes to I. Lerche for his fruitful discussions concerning the probabilistic method presented here. I also wish to thank the referees for their helpful and constructive comments.

References

BETHKE, C. M. 1985. A numerical model of compaction-driven groundwater flow and heat transfer and its application to paleohydrology of intercratonic sedimentary basins. *Journal of Geophysical Research*, **90**, 6817–6828.

DURAND, B., UNGERER, P., CHIARELLI, A. & OUDIN, J. L. 1984. Modèlisation de la migration de l'huile. Application à deux examples de bassins sèdimentaires. Proceedings of the 11th World Petroleum Congress, London, 1983.

GLASKO, V. B., 1984. *Inverse problems of mathematical physics*. American Institute of Physics, New York.

LERCHE, I. & GLEZEN, W. H. 1984. *Deposition, compaction and fluid migration: Time-dependent models in one and two dimensions*. Gulf Research and Development Co., Pittsburgh, Pa.

——, YARZAP, R. F. & KENDALL, C. G. ST. C. 1984. Determination of paleoheat flux from vitrinite reflectance data. *American Association of Petroleum Geologists Bulletin,* **17**(11), 1704–1717.

—— 1988. Inversion of multiple thermal indicators: Quantitative methods of determining paleoheat flux and geological parameters. I. Theoretical development for paleoheat flux. *Mathematical Geology*, **20**, 1–36.

—— 1991. Inversion of dynamical indicators in quantitative basin analysis models: Theoretical Considerations. *Mathematical Geology*, **23**, 817–832.

—— 1992. *Oil Exploration: Basin Analysis and Economics.* Academic Press Inc., San Diego.

PRESS, W. H., FLANNERY, B. P., TEUKOLSKY, S. A. & VETTERLING, W. T. 1986. *Numerical Recipes. The art of Scientific Computing.* Cambridge University Press, Cambridge.

THOMSEN, R. O. 1994. Dynamical models in geology: Sensitivity analysis and scientific risk. *Energy Exploration and Exploitation*, **11**, 329–356.

WELTE, D. H. & YÜKLER, A. 1981. Petroleum origin and accumulation in basin evolution — a quantitative model. *American Association of Petroleum Geologists Bulletin,* **65**, 1387–1396.

YÜKLER, A., CORNFORD, C. & WELTE, D. H. 1978. One-dimensional model to simulate geologic, hydrodynamic and thermodynamic development of a sedimentary basin. *Geologishe Rundschau*, **67**, 960–979.

A novel approach for constraining heat flow histories in sedimentary basins

K. GALLAGHER[1] & D. W. MORROW[2]

[1]*T. H. Huxley School of Environment, Earth Sciences and Engineering, Imperial College of Science, Technology and Medicine, Prince Consort Road, London, SW7 2BP, UK*

[2]*Geological Survey of Canada, Institute of Sedimentary and Petroleum Geology, 3303 -33rd Street N.W., Calgary, Alberta, T2L 2A7, Canada*

Abstract: We examine the application of a novel inversion procedure to determine a heat flow history from down-hole thermal indicator data from 2 wells in northern Canada. The approach is based around Occam's razor in that we attempt to constrain the simplest heat flow history consistent with the observed data. Simple models are defined in terms of the variation about the present day heat flow and the inverse problem is to minimise a combined function of the data misfit, weighted by the individual error on each observation, and the model complexity.

The results obtained for the 2 wells, Pan Am Beaver River YT G-01 and Imperial Island River No. 1 in the Liard Basin, agree with heat flow histories determined from forward modelling. However, the inversion approach discussed provides useful estimates of model resolution and sensitivity. The effect of variations in the data errors and the nature of the forward model are also considered. As we expect intuitively, the quality of the solution depends directly on the quality of the input data and forward model.

Thermal history modelling is one of the major components of modern integrated approaches to sedimentary basin analysis. This aspect is particularly important for assessing maturation levels and timing of hydrocarbon generation — probably the key aims in exploration-related basin modelling. In detail, the thermal evolution of a sedimentary basin will be determined by a variety of physical factors and processes. Generally, however, the most significant factor that must be considered is the heat flow from the deep Earth up into the basin sediments. Heat flow is estimated from the product of an observed temperature gradient and some measure of the integrated thermal conductivity over the depth that the temperature gradient has been measured. The advantages of considering thermal evolution in terms of heat flow rather than temperature gradients are well known. For example, temperature gradients will vary with depth because of variability in thermal conductivities due to lithological differences and compaction. Moreover, heat flow has a direct interpretation in terms of the physical processes involved in basin formation and is one of the first-order predictions of many basin formation models.

Conventionally, most applications of thermal history modelling use a forward approach, where the temperature history of a given sedimentary layer is determined by combining the burial history with an *a priori* specified heat flow history and adopting an appropriate form of the energy conservation equation. The validity of a given thermal model is assessed by comparing various types of observed thermal indicator data (e.g. vitrinite, fission track, organic geochemistry) with model predictions. If a model is deemed unsuitable (i.e. does not fit the data), then the heat flow history is adjusted manually. This process is repeated until satisfactory agreement between the observed data and model predictions is achieved.

Unfortunately, this ad-hoc approach has several major drawbacks. Firstly, it is usually time consuming. Secondly, and more importantly, it does not explicitly allow the observed data to determine the form of the heat flow history. Thirdly, such approaches generally do not explicitly incorporate the errors inherent in the data. Finally, it is not possible to quantify the resolution of the model readily. Therefore, we cannot address how much useful information on the thermal history is contained in the observed data, let alone how this maps into the preferred model.

In this paper, we consider a methodology recently proposed by Gallagher & Sambridge (1992) specifically to address some of these

GALLAGHER, K. & MORROW, D. W. 1998. A novel approach for constraining heat flow histories in sedimentary basins. *In*: DÜPPENBECKER, S. J. & ILIFFE, J. E. (eds) *Basin Modelling: Practice and Progress*. Geological Society, London, Special Publications, **141**, 223–239.

problems. The approach is based around a robust inversion method which has been applied to other geophysical modelling problems, such as electromagnetic sounding (Constable *et al.* 1987) and 3-D seismic velocity structure (Sambridge 1990). Here, we wish to concentrate on the philosophy and application of the approach to thermal history modelling, rather than the mathematical details. The latter are summarized in the original paper and references therein. In essence, the aim of the inversion is to find the simplest model (in our case we are interested in the heat flow history) which predicts the observed data adequately (i.e. to within the known uncertainties). The motivation for seeking the simplest heat flow history is that variation, or structure, in the simple model should also occur in the true heat flow history. Moreover, if we fit the observed data to within their known, and often variable, errors, a more complicated heat flow model is not required nor indeed justified by fitting the data. Thus, we want to fit our observed data well but, with the approach described here, we can avoid over-interpreting data, heat flow and the thermal history. The approach has been referred to as Occam's Inversion (Constable *et al.* 1987), after William of Occam's razor — 'it is vain to do with more what can be done with less'.

Inversion methodology

Generalities

The general approach used in inversion problems requires an appropriate forward model (in our case a form of the energy conservation equation to calculate temperature), some parametrization of the unknown model (here, the heat flow history) and a measure of how well the observed data (e.g. vitrinite reflectance, fission track data) are matched by the values predicted for a given heat flow model.

We adopt a relatively simple forward model to predict temperature as a function of heat flow into the base of the sedimentary basin. More involved thermal models can be incorporated into the inversion scheme described in this paper, but we consider the simple model presented here is adequate for most situations. Thus, we neglect transient, advective and internal heat source contributions. The first two terms can generally be neglected as a consequence of the short thermal time constant associated with a sedimentary column a few kilometres thick — we are only considering the value of heat flow at the base of the sedimentary basin, not on the lithospheric scale. Internal heat sources are generally of second-order importance relative to the deep heat flow. With these assumptions, the heat flow is constant through the sedimentary column at any given time, and the energy balance is expressed as

$$\frac{\mathrm{d}}{\mathrm{d}z}\left[k_z \frac{\mathrm{d}T}{\mathrm{d}z}\right] = 0 \tag{1}$$

where T is temperature, z is depth and k_z is the thermal conductivity, the subscript indicating this property is a function of depth. The solution to this equation is given as

$$T_t(z) = T_t(0) + Q_t \int_0^z \frac{1}{k_z}\,\mathrm{d}z \tag{2}$$

where Q is the heat flow into the base of the sedimentary column, $T(0)$ is the surface temperature and the subscript t refers to time.

The heat flow history, or heat flow as a function of time, is the desired solution to our inverse problem, so we need to specify, or parametrize, the form of this function in terms of the model parameters we wish to determine. As discussed by Gallagher & Sambridge (1992), we want to use a parametrization that imposes the least constraint on the heat flow, allowing it to be as flexible as required to fit the data. Given these requirements, a suitable form is then one where the heat flow history is defined in terms of a constant value over a small time interval, i.e.

$$Q_t = Q_i \quad t_i > t \geq t_{i-1},\ i = 1,N \tag{3}$$

with the present day (t_0) heat flow given as Q_1. More complex functions could be defined. For example, we could use linear or quadratic variation over each small time interval, with the additional condition that these local functions are continuous across successive time intervals. However, the simple discrete value function readily allows an interpretation of the role of each model parameter and is also of a form amenable to impose the simplicity constraints we discuss later. The last general requirement is a measure of how well we fit the observed data with our predicted values. This is usually referred to as the misfit or objective function, and in any inversion scheme the aim is to find the model that yields the minimum value of the misfit. A standard misfit function is the weighted least squares, or L_2 norm, measure. For N observed data (d_i, $i = 1,N$), this is given as

$$\phi = \sum_{i=1}^{N} \frac{\left(d_i^{\text{observed}} - d_i^{\text{predicted}}\right)^2}{\sigma_i^2} \tag{4}$$

where σ_i is the error associated with the i-th datum. It is important to recognize that we do not expect to obtain a misfit of zero because of the errors in the data. However, given certain assumptions, we can define an expected value for the misfit. If we obtain a value of ϕ equal to this expected value, ϕ^*, then we can infer that we have obtained an adequate fit to the observed data. If the data have independent Gaussian errors, then the appropriate value of ϕ^* is N, the number of data. It can be seen from equation (4) that we will obtain a value $\phi \leq \phi^*$ if the difference between each observation, d_i^{observed}, and the predicted value, $d_i^{\text{predicted}}$, is less than its associated error, σ_i, on average.

Defining a simplicity : least rough or smoothest models

Assuming that we know the burial history, the other important constraint we can impose is the present day heat flow. This is generally determined from down-hole temperatures, and some assumed, or inferred, thermal conductivity profile in the sediments. It is straightforward to obtain an internally consistent heat flow estimate, and it is not crucial that this value represent the true present day heat flow very accurately. This is because the observed data will be sensitive to variations about the estimated, rather than the absolute value. We shall return to this aspect later. Thus, we are interested in heat flow variations with time, and we want to keep these variations as simple as possible, while still achieving a satisfactory fit to the data.

Two types of simple model can be defined in terms of the rate of change of heat flow with time. Recalling that we have parametrized the heat flow history in terms of discrete values over small time intervals, the first derivative of the heat flow with respect to time can be written as a first-order finite difference approximation,

$$\left.\frac{\partial Q}{\partial t}\right|_i \approx \left(\frac{Q_i - Q_{i-1}}{\Delta t}\right) \tag{5a}$$

and the second derivative with respect to time can be written as

$$\left.\frac{\partial^2 Q}{\partial t^2}\right|_i \approx \left(\frac{Q_{i+1} - 2Q_i + Q_{i-1}}{\Delta t^2}\right). \tag{5b}$$

The first derivative represents the rate of change of heat flow over time, while the second derivative represents the rate at which the rate of change in heat flow varies. Thus, the sum of the derivatives is a measure of how much variation there is in the heat flow history. For each type of derivative we define this measure as a sum of squares, so all terms are positive,

$$R_1 = \sum_{i=2}^{N} (Q_i - Q_{i-1})^2 \tag{6a}$$

and

$$R_2 = \sum_{i=2}^{N-1} (Q_{i+1} - 2Q_i + Q_{i-1})^2 . \tag{6b}$$

The choice of the symbol R reflects that this function is a measure of model roughness or 'wiggliness' about the present day heat flow. Thus, if we minimize R_1, equivalent to the sum of squares of the first derivatives, we are trying to keep the heat flow constant for the whole thermal history. Similarly, if we minimize R_2, equivalent to the sum of squares of the second derivatives, we try to keep the rate of change (increasing or decreasing) in heat flow constant for the whole thermal history. In both cases, the smaller the sum of the relevant derivatives, the smoother (or less rough) the heat flow history will be. Obviously, a constant heat flow over the total thermal history is as simple as we can get and this situation would lead to zero sum of derivatives for both the first and second derivative cases.

We can now specify our inversion philosophy more precisely, bearing in mind a comment attributed to Albert Einstein, 'Everything should be made as simple as possible, but not simpler'. Thus, while we want to keep the heat flow history as simple as possible (by minimizing the sum of the derivatives), this simple heat flow history needs to fit the data adequately (i.e. to a misfit level of ϕ^*). Thus, we can define a combined objective function, which incorporates the data misfit and the smoothing constraint. This is given as

$$U = (\phi - \phi^*) + \mu R \tag{7}$$

and the aim is to find the heat flow history which minimizes U. The parameter μ is a number which controls the trade-off between fitting the data and keeping the model simple or smooth. A large value of μ means that the smoothing term dominates the overall objective function. As μ decreases, the data fit becomes more important and the model will tend to exhibit more variation or structure in attempting to provide a better fit to the data. Therefore we want to find the largest value of μ associated with a model which predicts the data misfit as ϕ^*. This is equivalent to finding the simplest (i.e. least complex) heat flow history that predicts the data adequately. However, there is no guarantee that we will predict the observed data equivalent to a value equal ϕ^*, but still we are interested in heat flow that corresponds to the minimum value of U.

The significance of the simplest heat flow model is this : any variation in this heat flow history is the minimum required to satisfy the data. Thus, the true heat flow history must have at least the same amount of variation over time. Furthermore, while the true heat flow may be more complicated than our simplest model, the data alone cannot distinguish between the two. Therefore, we require additional information to justify the inclusion of more complexity in the heat flow model. In other words we are addressing how much information on heat flow variations back in time is actually contained in the observed thermal indicator data. This is obviously a very important part of any interpretation procedure.

As discussed by Gallagher & Sambridge (1992), the inverse problem we are interested in is non-linear, although the mathematical formulation we use is based on the solution to a linear problem. This means firstly that we need to adopt an iterative scheme, where we assume an initial heat flow model and successively update it until we reach convergence (i.e. the data misfit is ϕ^* or as close as we can get). Secondly, the nature of the problem is such that we do not expect to always improve the misfit as we allow more complexity by decreasing μ. The non-linearity generally results in the situation where the data misfit can become worse for small values of μ. The general algorithm is to start with a large value of μ, and progressively decrease it until we find a minimum in the overall objective function. This may or may not lead to a data misfit less than or equal to ϕ^*. If not, the best model is used as the starting model for the next iteration and the process repeated. If we find a model which has a data misfit less than ϕ^*, then we assume that we are over-fitting the data. In this case we search for the largest value of μ which yields a data misfit equal to ϕ^*, recalling that a larger μ means a simpler model.

In practice, we can use either first or second derivative smoothing to obtain a solution. The heat flow history departs from the appropriate smoothing constraint only where it is required to fit the data. An advantage of using both smoothing models in 2 separate inversions is that they diverge when the heat flow becomes poorly constrained, allowing us to assess which parts of the heat flow history are well resolved (Gallagher & Sambridge 1992).

The mapping of data information into the heat flow history can be assessed by examining the partial derivatives used in the numerical inversion scheme. The partial derivatives are the rate of change in a predicted thermal indicator with respect to heat flow as a function of time. We can then see how different data values are influenced by the heat flow at different times. Thus, we can write the partial derivative of the i-th predicted datum with respect to the j-th heat flow parameter as

$$\frac{\partial d_i}{\partial Q_j} \approx \frac{\Delta d_i}{\Delta Q_i} \qquad (8)$$

then

$$\Delta Q_j \approx \Delta d_i \left(\frac{\partial d_i}{\partial Q_j} \right)^{-1} . \qquad (9)$$

If we replace Δd_i with $d_i^{\text{observed}} - d_i^{\text{predicted}}$, then we can approximate ΔQ_j using the partial derivatives calculated at the best solution. This is very much an approximation, however, as we implicitly assume a linear relationship between the data misfit and heat flow. Furthermore, changing the heat flow for one observation will generally influence the data fit for other observations. Consequently, it is not possible to use this information to find a better data fitting solution — indeed we hope to have already found this. However, plotting some estimate of ΔQ_j for each observation can be qualitatively useful to illustrate the sensitivity of each datum to each heat flow parameter. In practice, we have found it informative to examine $Ln(|\Delta Q_j|)$, such that small values of this function indicate high sensitivity.

Another approximate integrated measure of model sensitivity over all the data or resolution for each can be obtained from the following expression:

$$Ln\left(\sum_{i=1}^{N} \frac{\Delta d_i}{\sigma_i} \left(\frac{\partial d_i}{\partial Q_j} \right)^{-1} \right) . \qquad (10)$$

Again, the absolute value of this function has no particular useful interpretation. Rather it is the relative magnitudes that indicate where the model parameters influence the overall data-fit most significantly.

Application of methodology to real data

To illustrate the practical application of the methodology to real data, we have selected data from 2 wells from the Liard Basin region in southeastern Yukon Territory and northeastern British Columbia. This basin has been defined on the basis of its anomalously thick upper Palaeozoic sequence and contains Canada's northernmost producing gas fields : the Beaver River, Kotaneelee and Pointed Mountain fields. Morrow *et al.* (1993) undertook a study of organic maturation of these wells incorporating suites of new vitrinite reflectance measurements as constraints for forward thermal history modelling, the aim being to improve understanding of the source and maturation history of the gas fields. The vitrinite reflectance values range up to nearly 5 %, indicating extreme heating relative to the generally accepted oil window values of 0.5–1.2 %. However, results obtained with the inversion scheme presented here depend on the ability of the data to resolve heat flow variations over time. Consequently, the approach is suitable for any range of reflectance, provided the predictive or forward vitrinite reflectance model is applicable over that range.

The 2 wells we have selected, Pan Am Beaver River YT G-01, and Imperial Island River No. 1, were both included in the study of Morrow *et al.* (1993). The former well is in the Liard Basin proper, with a thick Upper Palaeozoic sequence, while the latter lies ≈100 km to the east of the basin, in the Interior Plains, where the Upper Palaeozoic sequence is thin. Figure 1 illustrates the burial history of these 2 wells, demonstrating the difference in Upper Palaeozoic burial. A summary of the basic data used for these calculations is given in the Appendix and Morrow *et al.* (1993) discuss the information used to construct these burial histories in more detail. Although it is important to note that the estimates of eroded section are probably not definitive, we accept these as the optimum burial scenarios to date for the purposes of this paper. The details of the data we use in this paper are summarized in the Appendix.

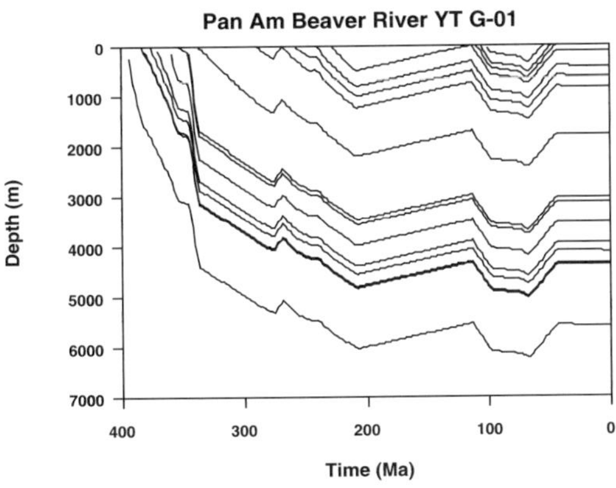

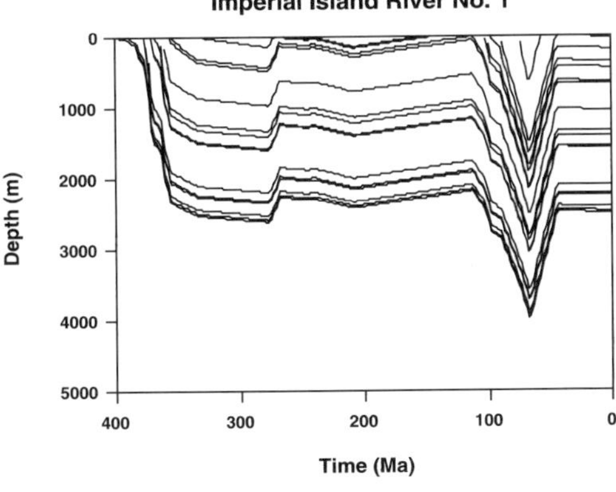

Fig. 1. Burial histories for Pan Am Beaver River YT G-01, and Imperial Island River No. 1, Liard Basin, Western Canada (after Morrow *et al.*, 1993).

Pan Am Beaver River YT G-01

The modelling study of Morrow *et al.* (1993) used the EASY%Ro algorithm, based on the Burnham & Sweeney (1989) kinetic model, to predict vitrinite reflectance having specified the heat flow history *a priori*. Initially, Morrow *et al.* (1993) specified time invariant heat flows of between 40 and 140 mW m^{-2} at intervals of 20 mW m^{-2}. None of the individual simulation results provided a satisfactory fit to the data, although the observations were bounded by the predictions with heat flow between 60 and 120 mW m^{-2}, the deeper vitrinite observations being more consistent with the higher values of heat flow. Thus, Morrow *et al.* (1993) inferred that heat flow was not time invariant, and produced a refined model where heat flow was high during the Palaeozoic (135 mW m^{-2}) and lower in the Mesozoic and Tertiary (80 and 60 mW m^{-2} respectively).

We used the same raw stratigraphic and thermal data adopted by Morrow *et al.* (1993) as input for the inversion scheme, although the details of the forward modelling (e.g. decompaction procedures, calculation of thermal conductivities) differ to some extent. The calculated standard deviation of the individual vitrinite reflectance distributions were adopted as the input errors for each observed value. However, some of the individual means actually reflect only one observation. In these cases, the associated error is assigned as 20 % of the sample value, compared to an average relative error of about 10 % for the more well-defined mean values. The present day heat flow was calculated using the bottom hole temperature (BHT) (227 °C @ 5590 m) and surface temperature of 10 °C. The estimate of ≈77 mW m^{-2} is in agreement with the regional heat flow data reported by Majorowicz *et al.* (1988). The present day heat flow was constrained at this value, and the initial heat flow model was constant over time.

The results of the inversion using both first and second derivative smoothing are given in Fig. 2, where we also show the best models found during successive iterations. For a given iteration, the starting model is taken as the best model found from the previous iteration and the inversion scheme attempts to update this model such that the misfit decreases. The second derivative smoothing solution converged to the expected misfit after 7 iterations, while the solution from the first derivative smoothing effectively stalled within 5 % of the convergence misfit criterion. Such a

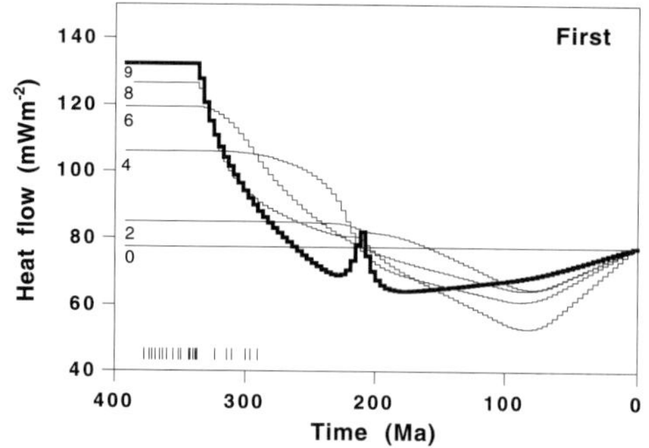

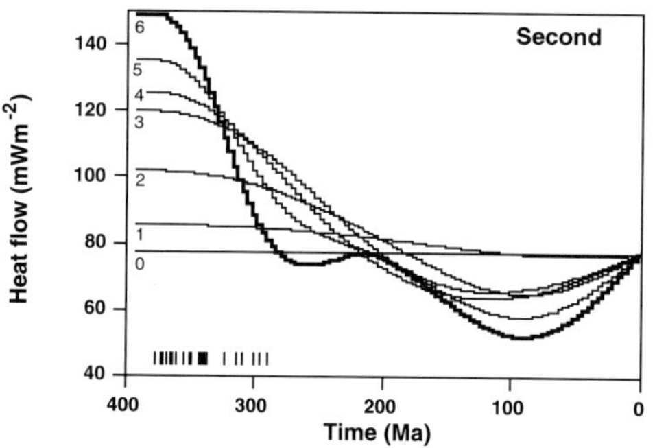

Fig. 2. Heat flow solutions for Pan Am Beaver River YT G-01, using first and second derivative smoothing. The initial heat flow model was constant and set equal to the present day estimate, and successive iterations are labelled accordingly. The final models are shown as heavier lines. The equivalent stratigraphic age of each vitrinite sample is indicated by the short vertical dashes.

difference occurs because of the different paths each approach takes through the model space because of the linearizing approximations involved in the scheme, combined with the influence of the different smoothing constraints.

Both smoothing models give broadly similar best solutions, with a heat flow minimum occurring between 200 and 80 Ma, thereafter increasing to a value higher than the present day value. The timing of this heat flow low is consistent with that inferred by Morrow *et al.* (1993) based on the relatively low vitrinite reflectances (≤0.5 %) preserved in the Early Carboniferous Mattson Formation. Both inversion solutions have a local heat flow maximum around 200 Ma, corresponding to the time of near maximum burial. At this time, the first derivative smoothing solution has a small spike. The extra complexity is introduced in an attempt to reduce the misfit, and is also compensated for by a change in the form of the heat flow minimum. However, the improvement to the misfit between this final solution and the prior iteration is <0.5 %, while the model roughness increases by 100 %. Although an igneous intrusion could be responsible for a short-lived and local heat flow anomaly, there is no other evidence for such an event and the amplitude of this feature is probably too low to be attributed to such a cause. Furthermore, the first derivative solutions prior to the final iteration look very similar to the final second derivative heat flow in terms of the timing of the heat flow minimum, and a small difference in the errors assigned to the data would have resulted in convergence on one of these simpler solutions.

The heat flow for both smoothing solutions is essentially the same between 250 and 350 Ma. After this time the first derivative model heat flow tends to stay at a constant value, while the second derivative model heat flow keeps increasing, although both models fit the data equally well. As shown by Gallagher & Sambridge (1992), this divergence illustrates a lack of resolution of this part of the heat flow history. Thus, the first derivative smoothing will keep the heat flow flat or at a constant value unless the data fit requires additional structure. Similarly, the second derivative smoothing tries to keep the rate of change of heat flow constant. Therefore, if the heat flow needs to increase back in time to fit the data, then the second derivative smoothing will tend to keep the heat flow increasing, while the first derivative smoothing will keep the heat flow constant if no other structure is required. Thus the two solutions are compatible overall, particularly if we consider the lack of convergence in the first derivative model which introduced additional complexity without significantly improving the data fit.

Looking at the fit to the data (Fig. 3), we can see how the predicted values from both of the best solutions differ from the constant heat flow case (iteration 0). While the two solutions predict very similar values, the first derivative model predicts slightly higher values for the deeper samples. A constant heat flow over-predicts relative to the shallower observations and under-predicts the deeper values. Qualitatively, then, we would infer that we need a higher heat flow early on, and a lower heat flow more recently, precisely the form of the inversion solutions.

In Fig. 4 we show the minimum of the data misfit and the relationship between the data misfit and the model roughness for each iteration. During each iteration, the model roughness (R_1 or R_2) correlates inversely with μ such that for decreasing μ, the model roughness increases (or smoothness decreases). It can be seen that successive iterations require progressively more complex models, as measured by increasing R, to reduce the data misfit.

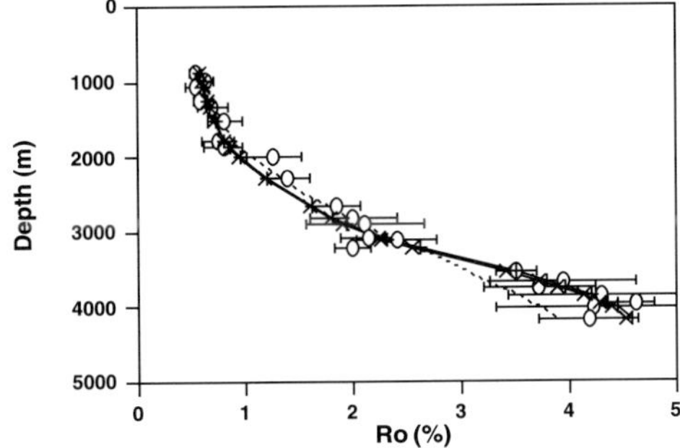

Fig. 3. Predicted and observed vitrinite reflectance for Pan Am Beaver River YT G-01; + — first derivative smoothing; x — second derivative smoothing; o — observed. The predictions from a constant heat flow model (equal to the present day value) are shown as a dashed line.

This figure clearly illustrates the lack of significant improvement over the last few iterations of the first derivative smoothing model, while the model roughness begins to increase rapidly.

The temperature histories for each solution are shown in Fig. 5. Here we can see that the timing and value of maximum temperature is similar for both models, reflecting the general wisdom that vitrinite reflectance is dominated by the maximum temperature experienced by the host sediment. There are two main episodes where maximum temperatures are achieved. For the deeper samples, the time is about 340 Ma, corresponding to the time where the rate of burial drops off (see Fig. 1). For the shallower samples, the maximum temperatures occur just before 200 Ma, reflecting a time of near maximum burial depth. This timing corresponds to the place where the evolving first and second derivative smoothing models diverge rapidly, as if the solutions pivot around this time (Fig. 2). Gallagher & Sambridge (1992) showed how the heat flow solutions for the two type of smoothing tend to agree where the heat flow is well resolved and diverge elsewhere. While this is broadly true for our solutions here, the similarity between the two solutions in terms of the increasing heat flow after 250 Ma is not really an indication that this is well resolved. Rather, this path is the simplest (in

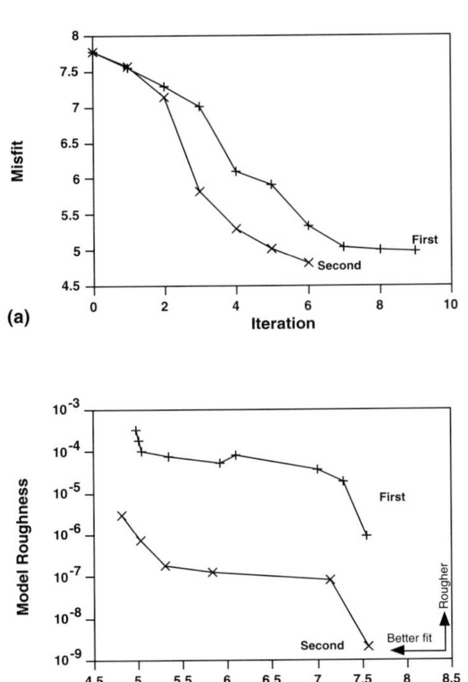

Fig. 4. (a) Misfit as a function of iteration; + — first derivative smoothing; x — second derivative smoothing. (b) Model roughness as a function of data misfit; + — first derivative smoothing; x — second derivative smoothing.

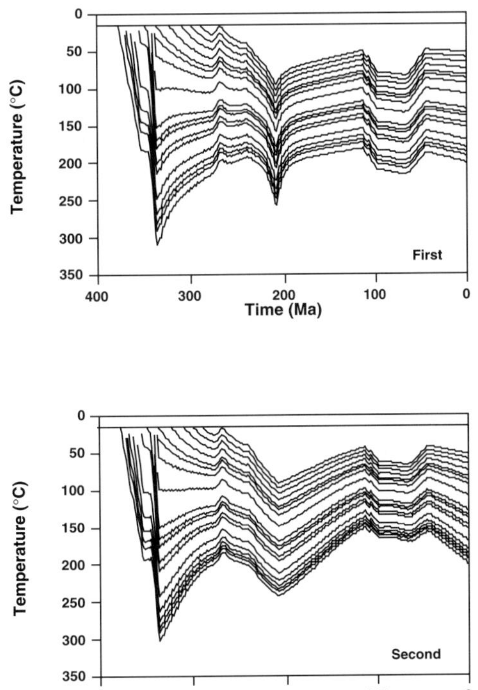

Fig. 5. Temperature histories for the stratigraphic layers containing the vitrinite reflectance observations corresponding to the best heat flow models for Pan Am Beaver River YT G-01.

both senses of our smooth models) to achieve the two timings of maximum temperature for the shallower and deeper samples.

In Fig. 6, we show the evolving vitrinite reflectance, illustrating when individual vitrinite values reach their maximum values. As we expect, this timing reflects the maximum palaeotemperature for a particular sample, and there are two significant times, ≈340 and ≈200 Ma. If we examine the sensitivity plot, $Ln(|\Delta Q_j|)$ as a function of individual data points d_i, for the first derivative smoothing solution, (Fig. 7), we can see how the different data points are influenced by the heat flow at different times. An advantage of this plot over just examining the evolving vitrinite reflectance (Fig. 6) is that the latter will plateau at its maximum temperature while the sensitivity plot, as well as revealing the more well-resolved parts of the heat flow history, can show structure after the time of maximum palaeotemperature. In this particular example, it is clear that the heat flows around 340 and 200 Ma are the dominant influences on the data fit, with a secondary influence around 100 Ma, corresponding to the time of maximum burial. The darker horizontal bands in

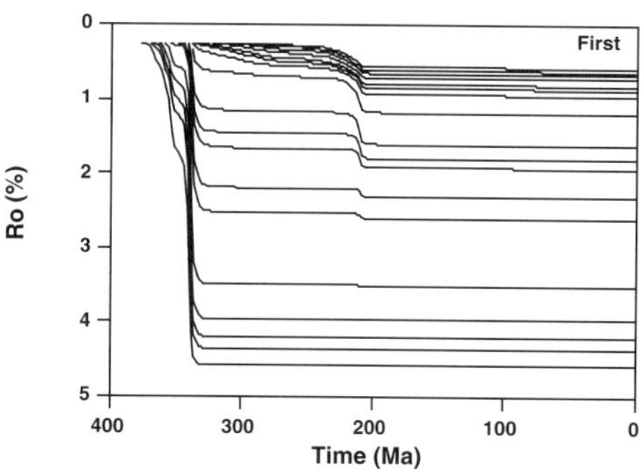

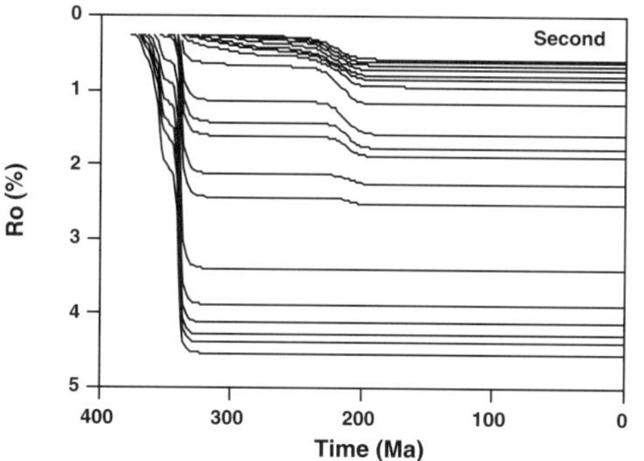

Fig. 6. Evolution of vitrinite reflectance for the best heat flow models for Pan Am Beaver River YT G-01.

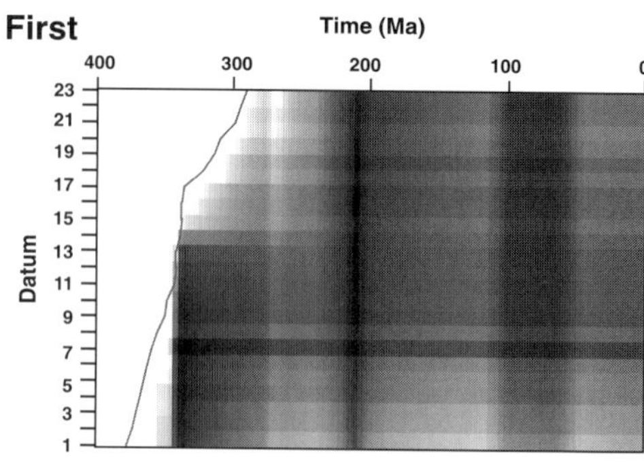

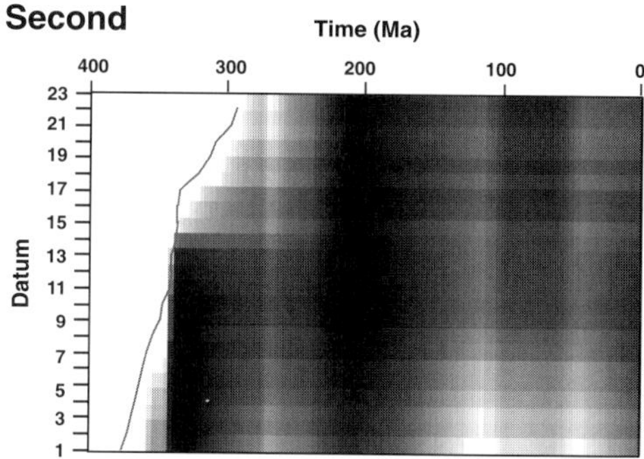

Fig. 7. Model sensitivity or resolution as a function of the individual data points for the best heat flow models from Pan Am Beaver River YT G-01. A darker shading indicates a greater sensitivity of a datum to the heat flow at that time.

this plot correspond to data points that are well predicted by the data (i.e. a small individual misfit as achieved for samples 7 and 9), while the darker vertical bands correspond to times where the heat flow is well resolved. Again, as we would expect intuitively, the heat flow around the times of near-maximum burial is generally the most important, or equivalently, the most well constrained from the data. Thus, the heat flow around 100–80 Ma cannot be much higher, otherwise maximum palaeotemperatures will occur at this time.

The integrated model sensitivity (Fig. 8) shows a similar trend to the previous plot, although lacks the detail of the influence on individual data points. The dominance of the burial history is reflected in the symmetry of this plot and the burial history for the deepest sample. We can easily see that the first derivative model is dominated by the heat flow spike around 200 Ma, while the second derivative model is less extreme. The relative importance of

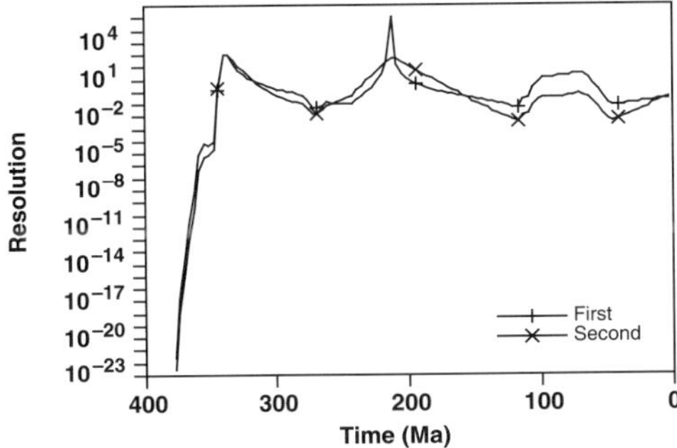

Fig. 8. Integrated model sensitivity or resolution for the best heat flow models from Pan Am Beaver River YT G-01; + — first derivative; x — second derivative. The absolute value of the resolution has no intuitive interpretation, but a more positive value indicates more sensitivity in terms of the data prediction.

the heat flow at 340 and 100–80 Ma is also apparent in these plots.

Thus, the various illustrations of model sensitivity and resolution provide different insights into the nature of the heat flow solutions obtained from the inversion procedure. As we expect, these solutions are strongly influenced by the timing of maximum palaeotemperature but can also reflect potentially well-resolved heat flow lows (particularly if these correspond to periods of maximum burial).

Imperial Island River No. 1

Morrow *et al.* (1993) examined the same range of time invariant heat flows (40–140 mW m^{-2}) to model the vitrinite reflectance from the Imperial Island River No. 1 well, and concluded that the data were adequately explained by a constant heat flow between 80 and 100 mW m^{-2}, provided that the Cretaceous subsidence (and subsequent erosion) was considerably more than the 700 m of section preserved in local synclines nearby. The preferred thickness of this missing section was 2100 m, a value consistent with an independent estimate inferred from apatite fission track data (Arne 1991). As noted by Morrow *et al.* (1993), this late stage burial/erosion event should dominate the maturation history, with the pre-Cretaceous thermal history having a relatively minor influence. We have used the calculated standard deviation of the individual vitrinite reflectance distributions as the input errors, as for the Pan Am Beaver River YT G-01 well. The data quality for the Imperial Island River No. 1 well is variable. In particular, the deepest mean value is based on only one measurement. An additional limitation of this data set is the restricted spread in the stratigraphic age of the vitrinite host sediments. Thus, of the 16 data points, all except one are hosted in sediment between 355 and 377 Ma. We would expect all these samples to have experienced their maximum palaeotemperatures at the same time, so the potential for resolving variations in heat flow over time becomes limited.

The present day heat flow was calculated using the BHT (122 °C @ 2499 m) and a surface temperature of 10 °C, leading to a value of ≈78 mW m^{-2}, again consistent with the regional heat flow (Majorowicz *et al.* 1988).

The results of applying the inversion scheme with the Imperial Island River No. 1 input data are shown in Fig. 9, for both types of smoothing. These results confirm the conclusion of Morrow *et al.* (1993) in that both first and second derivative smoothing produce heat flow models in which maximum palaeotemperatures are dominated by the late Cretaceous burial episode. The best solutions show a heat flow maximum around

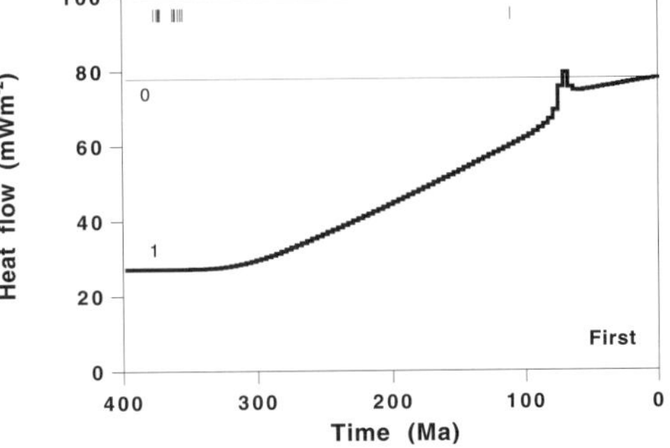

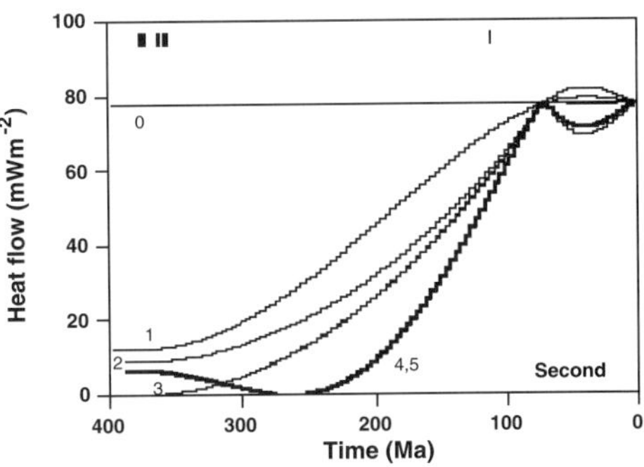

Fig. 9. Heat flow solutions for Imperial Island River No. 1, using first and second derivative smoothing. The initial heat flow model was constant and set equal to the present day estimate, and successive iterations are labelled accordingly. The final models are shown as heavier lines. The equivalent stratigraphic age of each vitrinite sample is indicated by the short vertical dashes.

75 Ma, then a progressive decrease in heat flow back in time. The general inference of Morrow *et al.* (1993) is supported in that the heat flow back to the time of maximum burial is more or less constant. As this episode dominates the data fit, we would expect the heat flow prior to this time to be constant for the first derivative model and vary at a constant rate for the second derivative model. The latter does so until it approaches zero around 270 Ma, then starts to increase. This structure is an artefact of the constraint that we do not permit the heat flow to become negative. The apparently anomalous structure in the first derivative model is attributable to the lack of convergence, combined with the iterative approach to deal with the non-linearity of the problem. However, we do not consider the solution in either case to closely represent the 'true' heat flow, and the unrealistic structure in both solutions clearly points to this.

These problems are further demonstrated by the fact that the constant heat flow predicts the data reasonably well, although there is a tendency to over-predict the observed values (Fig. 10). Thus, we would not expect the inversion heat flow solution to depart too significantly from a constant value. It is also significant that both best solutions did not converge to the desired misfit criterion. The best solution misfits were about 100 % higher than this level, although the misfit of the two best solutions differed by less than 0.5 %, the first derivative solution having the slightly better fit. This lack of convergence in misfit is attributable in part to the 2 deepest data points, which both appear anomalously high. As nearly all the data have the same effective stratigraphic age and the data with reflectance values between 1 and 2 % are separated by only a few hundred metres, it will be difficult to achieve a good fit with any heat flow model.

The other contribution to the lack of convergence comes from points which have very low associated errors. For example, 3 data points have errors less than 0.03, and these account for ≈80 % of the total misfit. Equation (4) shows that smaller associated errors (the σ term) will result in a larger value of misfit for a given difference between the observed and predicted values. Obviously, this is a desirable property as we want the less error-prone data to have more influence. The convergence requirement is to fit the data to within error, at least on average. Therefore, the larger the error, the less well we need to fit the actual value and the converse is true. It is also important to appreciate that we do not expect to fit all the data to within the individual errors. Some of the data may be inconsistent, or outliers. The fitting process attempts a best fit through the data, not each individual observation. Thus, if the error terms on the three data points mentioned above were 0.1 (which is less than the mean error of 0.12), we would have achieved numerical convergence with these heat

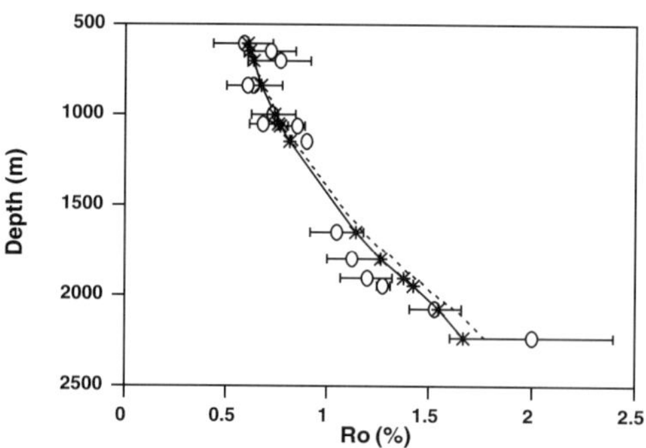

Fig. 10. Predicted and observed vitrinite reflectance for Imperial Island River No. 1; + — first derivative smoothing; x — second derivative smoothing; o — observed. The predictions from a constant heat flow model (equal to the present day value) are shown as a dashed line.

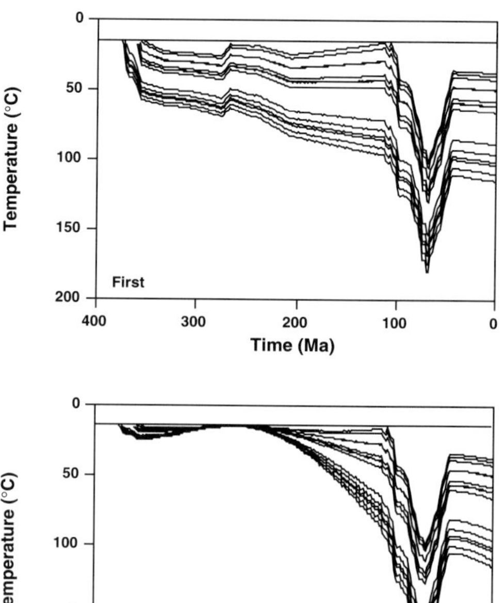

Fig. 11. Temperature histories for the stratigraphic layers containing the vitrinite reflectance observations corresponding to the best heat flow models for Imperial Island River No. 1.

flow solutions. Obviously, when assessing the convergence criterion and indeed the applicability of this inversion methodology at all, it is also very important to understand not only how good the data are, but also how well reliable the error terms are. We return to this aspect later.

The thermal histories associated with the 2 best heat flow solutions are shown in Fig. 11 and the dominance of the Cretaceous burial episode is apparent. The individual data sensitivity plot (not shown) and the integrated model resolution (see Fig. 15 later) also emphasize this feature showing relatively little structure outside this time. In this paper, we are only interested in demonstrating the application of the inversion methodology. In practice, however, these results warrant a reassessment of the burial history for this well. This does not mean that the scenario adopted here is wrong — rather the strong dependence of the thermal history on the amount of burial and subsequent erosion means that we would want to determine these parameters as reliably as possible. It is possible to specify the amount of burial and erosion as a parameter to be found and include this in the inversion scheme. Generally, we expect some trade off between the heat flow and amount of burial and the stability of the inversion will be sensitive to this relationship. A more straightforward option would be to vary the amount of burial and subsequent erosion and look for the value that leads to the simplest heat flow history.

Parameter sensitivity

The discussion of the model results above assumes that the details of the forward model are correct. That is, not only have we assumed that the vitrinite reflectance and burial history models are appropriate, but also that parameters such as the down-hole thermal conductivity structure and the estimate of present day heat flow (which obviously depends on the thermal conductivity) are accurate. Furthermore, we have also briefly discussed the effect of the errors assigned to the observed data in determining the nature of the best fit heat flow solution. A detailed analysis of the model and parameter sensitivities is beyond the scope of this paper, but it is appropriate to examine some additional situations. It was stated earlier that it is not too important that the present day heat flow represent the absolute value very accurately, provided that we have obtained an internally consistent heat flow estimate. For the situation with one BHT measurement at depth z, then the heat flow is determined as

$$Q = K\left(\frac{T(z) - T(0)}{z}\right) \qquad (11a)$$

where K is an integral of the thermal conductivity over depth,

$$K^{-1} = \int_0^z \frac{1}{k(z)}dz \ . \qquad (11b)$$

As the observable data in equation (11) are the down-hole temperature(s), these need to be honoured. This is achieved by keeping the ratio Q/K constant. Thus, when we vary $k(z)$ (and therefore K), then the calculated value of Q will vary sympathetically to predict the present day temperatures adequately.

In practice, we specify the matrix thermal conductivity of particular lithologies or formations and calculate a lithology and porosity dependent thermal conductivity structure. Therefore, we do not expect a 1:1 correspondence between the heat flow history (or indeed the present day heat flow) and our specified thermal conductivity parameters. However, to demonstrate that the solutions do not change wildly, we have determined heat flow solutions for Pan Am Beaver River YT G-01 specifying the matrix thermal conductivity (k_m) for all lithologies as 2 W m^{-1} K^{-1} and also as 4 W m^{-1} K^{-1}, resulting in present day heat flow estimates of 67 and 123 mW m^{-2}, respectively, for a BHT of 227 °C @ 5590 m and surface temperature of 10 °C. The calculated down-hole thermal conductivity and temperature gradient for the original model and the 2 constant values of matrix conductivities, using the same porosity–depth model, are shown in Fig. 12. We have only used the BHT to determine the heat flow, with no intermediate temperature constraints. As can be seen the predicted present day temperatures differ between the 2 constant matrix thermal conductivities and the original model, generally being somewhat higher at greater depth in the latter. This is a reflection of the high thermal conductivity in the deepest stratigraphic unit.

The inversion generated heat flows, using first derivative smoothing, and the predicted vitrinite reflectances are shown in Fig. 13, together with the original best fit solution. Obviously, the absolute values of heat flow differ significantly, but the general form of all 3 solutions is broadly similar, showing a slight decrease in heat flow back to about 200 Ma, then a relatively rapid increase to about 330–340 Ma.

The main difference between the heat flow models is the prediction of the deepest (>3500 m) samples, although these data have relatively large errors and are likely to be a relatively poor fit anyway. The difference between these solutions in terms of the actual thermal histories is greater than the difference between the original first and second derivative smoothing models described earlier, with maximum palaeotemperatures differing by up

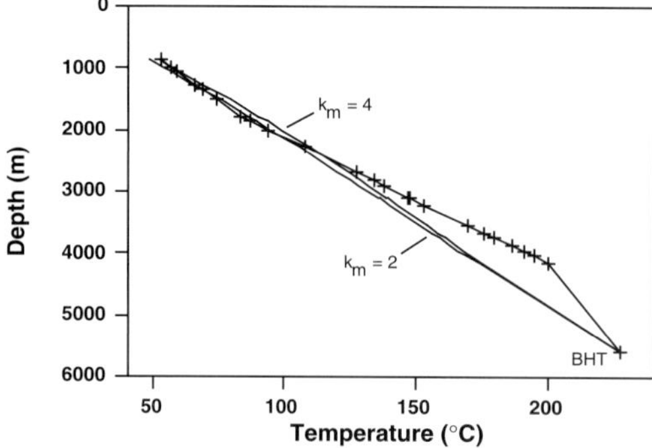

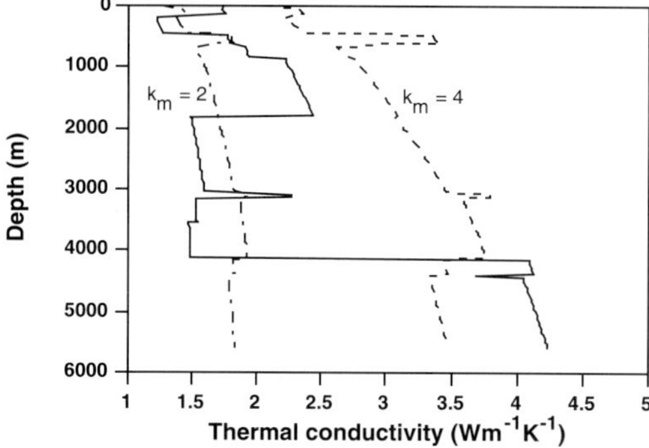

Fig. 12. Down-hole thermal conductivity structure and predicted temperatures (at the depth of each vitrinite datum) for Pan Am Beaver River YT G-01 in the earlier models (solid line) and also using constant matrix thermal conductivities (k_m) of 2 and 4 W m^{-1} K^{-1}.

to 40 °C for the deepest samples (equivalent to about 15 %). However, this discrepancy is similar to the differences in the predicted present day temperatures. The model incorporating a matrix thermal conductivity of 4 W m^{-1} K^{-1} tends to result in the lowest temperatures. This is manifested in the under-prediction of the deeper samples relative to the heat flow models. The general timings of maximum palaeotemperatures are much the same, with the shallower samples reaching peak temperatures around 200 Ma, and the deeper samples peaking around 300–340 Ma. Overall, these results are reassuring in that it is probably not necessary to obtain detailed thermal conductivity down-hole measurements for individual lithologies or formations, provided we are interested in relative heat flow variations.

Much of the difference in these solutions is attributable to the fact that the estimated heat flow is based only on the present day BHT. In reality, we can generally assume thermal conductivities for particular lithologies which would provide a far better approximation to the true value than the constant values we adopted here and the heat flow solutions would be less variable. However, the absolute heat flow values and reliability of the temperature histories will depend on the quantity and quality of the down-hole temperature measurements. Good temperature data is very important, not least because we can use this to assess the suitability of the inferred down-hole thermal conductivities.

As another example of model dependence on input data, we re-ran the inversions for Imperial Island River No. 1, but assigned an error of 0.05 to all the observed vitrinite reflectances. This value is less than half of the mean error originally adopted for this data. Thus each datum is weighted equally and as these errors are relatively small, we are essentially looking to overfit the data, relative to the original solutions. The heat flow histories and predicted vitrinite reflectances are shown in Fig. 14. Both the first and second derivative smoothing

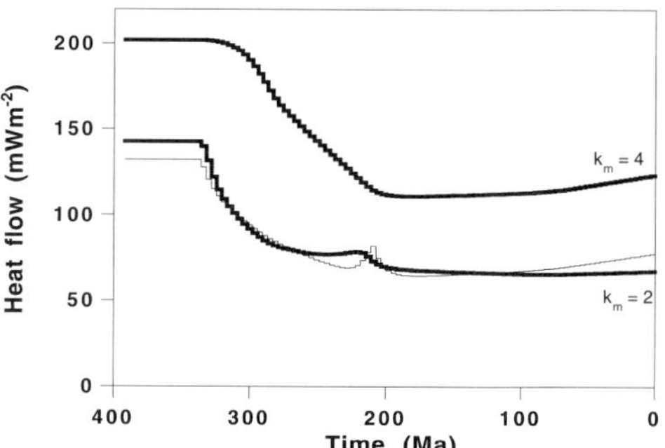

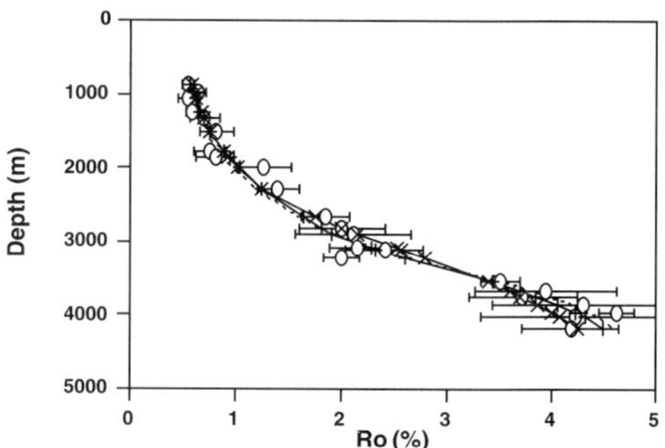

Fig. 13. Heat flow solutions and model predictions for Pan Am Beaver River YT G-01, using first derivative smoothing and constant matrix thermal conductivities of 2 (+) and 4 (x) W m^{-1} K^{-1}. The original first derivative smoothing solution is also shown as a lighter line and the predicted values as a dashed line.

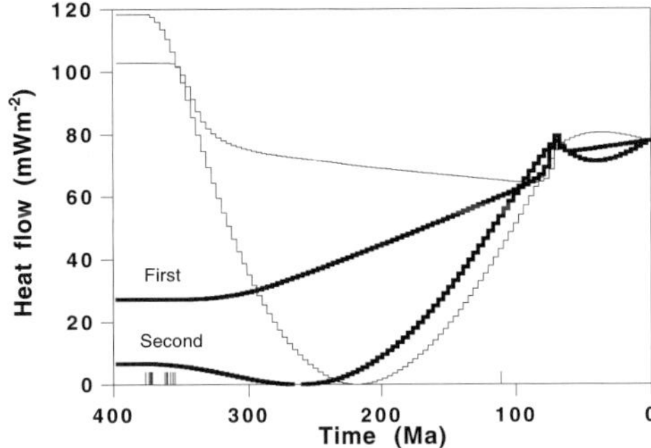

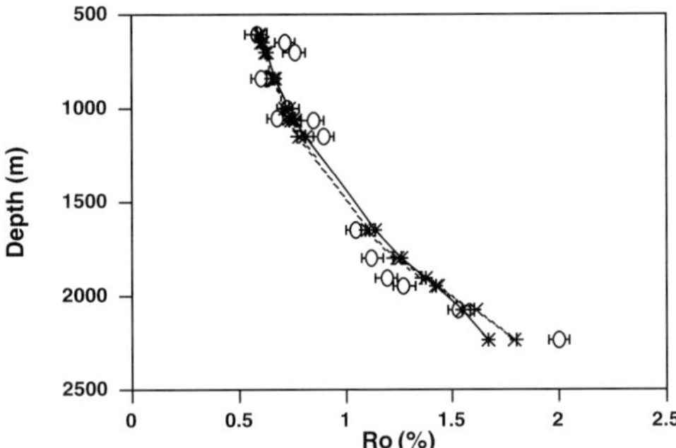

Fig. 14. Heat flow solutions and predicted vitrinite reflectances for Imperial Island River No. 1, using first (+) and second (x) derivative smoothing with the errors on all the data points set to 0.05. The solutions and predictions using the original errors are also shown as lighter and dashed lines, respectively.

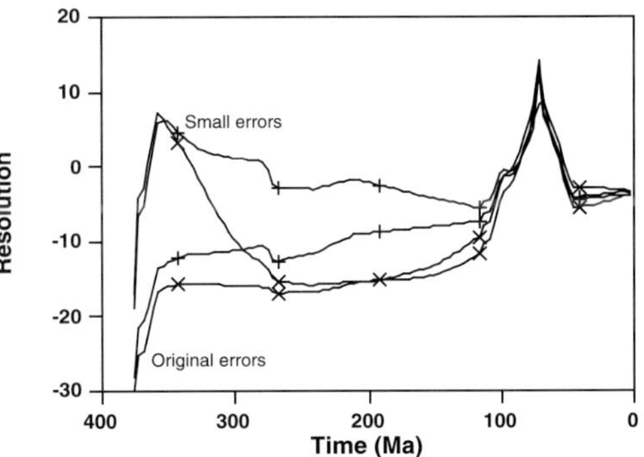

Fig. 15. Integrated model sensitivity or resolution for the best heat flow models shown in Fig. 14; + — first; x — second derivative.

solutions using the equal errors have similar misfits, and lead to a better fit than the original solutions, but the weighted misfit is numerically much the same, and we have not achieved convergence. As we expect, the two solutions obtained with the small errors are also less smooth (by a factor of ≈2, as measured by eqs 6a and 6b) than the original solutions as we introduce more structure to try and improve the fit to the data. The two sets of solutions are similar back to the timing of maximum burial (≈70 Ma) and then tend to diverge, in much the same way as the original first and second derivative solutions do, reflecting a lack of resolution. However, the heat flows obtained with the small errors tend to converge much later and increase up to a peak value around about 350 Ma. Using the criterion that overlap between the first and second derivative solutions indicates that good resolution would lead us to conclude that this part of the thermal history is well resolved, in contrast to our earlier conclusion. However, this reflects the fact that the solution now provides a better fit to the deeper samples, particularly the deepest which had a relatively large error in the initial inversions. Recalling that all except the shallowest sample have stratigraphic ages between 350 and 370 Ma, the heat flow needs to be initially high, and decline rapidly in order to achieve this, without unduly over-predicting the other samples. The evolution of shallowest sample is determined by the heat flow at the time of maximum burial at ≈70 Ma. The integrated heat flow sensitivity plot (Fig. 15) clearly shows the increased contribution of the early heat flow to the overall data fit, particularly for the second derivative smoothing. The individual data resolution plots, not shown here, also show the influence of this earlier heat flow high on the prediction of the deeper observations.

Although we can improve the absolute level of agreement between the observed data and predicted values by reducing the magnitude of the errors, the weighted misfit has not altered significantly and the final solutions contain considerably more structure. As we suggested earlier, this particular data set is problematic in that firstly we do not achieve a satisfactory range in the effective stratigraphic age of the vitrinite and secondly the burial history is such that the Cretaceous burial–erosion episode dominates the solution. While the solutions obtained with the inversion procedure are unrealistic, these are not totally negative results. We consider that it is important to show that there are severe limitations when using this data set and that the palaeoheat flow/thermal history cannot be well resolved. The inversion approach discussed here demonstrates these problems clearly.

Summary

There are two basic philosophies that can be adopted in thermal history modelling — forward and inverse modelling. The two approaches are to some extent complementary in that forward modelling is a convenient method of testing whether data are consistent with a given hypothesis, while the inversion approach allows the thermal indicator data to determine the form of the thermal history. Regardless of whether a preferred hypothesis exists, it is clearly important to understand the information (on thermal histories for example), contained in the observed data. Then, the inversion procedure is the most appropriate approach. However, given that data are error prone, the non-linear inversion problem is non-unique, i.e. more than one solution will exist. In this case, the recommended approach is to try and bound some property of the family of all possible solutions (e.g. Parker 1977). The property of thermal histories we have examined in this paper is the roughness or complexity of the heat flow.

The inversion procedure as applied to vitrinite reflectance provides heat flow solutions that reflect the timing of maximum palaeotemperatures. This is because, although vitrinite reflectance is an integrated measure of the thermal history for an individual data point, the time–temperature integral in the Burnham & Sweeney model is dominated by maximum temperature. Other vitrinite reflectance models have a more significant or different time-dependence (see Gallagher & Sambridge 1992), as can other thermal indicators (e.g. apatite fission track analysis, isomerisation and aromatisation reactions). The inversion scheme we have discussed here is set up to incorporate multiple data types in one pass, although we have not yet obtained sufficient quality data to proceed in this direction. However, individual thermal indicators and their predictive models need to be considered in terms of their ability to resolve temperature histories, rather than just maximum temperatures. In specific situations, some data types may provide more information than others. Of course, an advantage of parametrizing the thermal history in terms of heat flow means that thermal indicator data from different stratigraphic horizons can provide multiple constraints to reduce the potential lack of resolution. The 2 examples considered here illustrate how the quality of data influences the heat flow solution. The Pan Am Beaver River YT G-01 well has reasonable quality data. Consequently, we achieve reasonable inversion results, although the somewhat restricted spread in effective stratigraphic age of the vitrinite data does lead to some lack of resolution. In contrast, the quality of the Imperial Island River data is considerably worse. The effective stratigraphic age of the vitrinite is very restricted, more or less equivalent to 2 data points. Furthermore, the nature of burial history results in maximum temperatures occurring more or less at the same time for all horizons. The quality of the heat flow solution reflects these limitations.

Having solved the problem of generating a heat flow model consistent with the observed data, it is important to examine the model sensitivity and resolution. Certainly a direction for future work is the development of more directly quantitative or probabilistic estimates of model resolution and data sensitivity (Gallagher 1998). Another refinement is to allow the present day heat flow to vary within a specified range, rather than setting it equal to a constant value. For example, the uncertainties in the down-hole temperatures and thermal conductivities can be mapped into upper and lower bounds on the heat flow estimate. These constraints are readily incorporated into the inversion scheme by a transformation of the relevant model parameters (Gallagher 1998) and the procedure is invoked in the same way described in Gallagher & Sambridge (1992).

We would like to thank M. Sambridge and D. Issler for useful discussions on various aspects of the problems addressed here and F. Brigaud and K. Nakayama for reviews of an earlier version of the manuscript.

Appendix

Basic data for thermal history calculations.

Table A1. *Stratigraphic and lithological details of Pan Am Beaver River YT G-01 (after Morrow* et al. *1993)*

Formation name	Depth (m)	Age (Ma)	Lithology fractions 1	2	3	4	5	6	7
Quaternary	15	44	0.5	0.5	0	0	0	0	0
Wapititi	−140(4)	75	1	0	0	0	0	0	0
Kotanalee	−45(4)	88	0.2	0.3	0.5	0	0	0	0
Dunvegan	−45(4)	98	1	0	0	0	0	0	0
Sully	−140(4)	101	0	0	1	0	0	0	0
Sikanni	−25(4)	102	0.5	0.5	0	0	0	0	0
Lepine	−163(4)	106	0.15	0.15	0.7	0	0	0	0
Scatter	−70(4)	109	0.1	0.10	0.8	0	0	0	0
Garbutt	−90(4)	111	0	0	0.5	0	0	0	0.5
Garbutt	137	113	0.5	0.25	0.25	0	0	0	0
Tod Grayling	−503(3)	228	0.1	0.1	0.8	0	0	0	0
Tod Grayling	434	240	0.10	0.1	0.8	0	0	0	0
Fantasq	−11(2)	246	0	0	0.5	0	0	0	0.5
Fantasq	623	258	0	0	0.5	0	0	0	0.5
Kindle	843	268	0.5	0.25	0.25	0	0	0	0
Mattson	−255(1)	289	0.65	0.2	0.15	0	0	0	0
Mattson	1788	337	0.65	0.2	0.15	0	0	0	0
Golata	3042	344	0.1	0.15	0.75	0	0	0	0
Prophet	3131	354	0	0	0.25	0.5	0	0	0.25
Besa above Exshaw	3542	360	0	0.1	0.85	0	0	0	0.05
Besa below Exshaw	4113	376	0	0	0.95	0	0	0	0.05
Nahanni	4360	383	0	0	0	0	0.95	0	0.05
Headless	4390	384	0	0	0.05	0	0.95	0	0
Sub-Headless	5590	396	0.05	0	0	0	0.95	0	0

A negative depth indicates lost section and the integer in brackets refers to the erosion event (see Tables A5 and A6)

Table A2. V*itrinite reflectance values for Pan Am Beaver River YT G-01*

Depth (m)	Ro (%)	∂Ro (%)
871.73	0.55	0.05
981.46	0.65	0.07
1057.7	0.55	0.11
1261.9	0.58	0.06
1341.1	0.71	0.14
1514.9	0.82	0.16
1780.0	0.76	0.15
1865.4	0.80	0.19
1999.5	1.27	0.25
2279.9	1.40	0.20
2671.6	1.85	0.23
2810.3	2.01	0.40
2892.6	2.11	0.55
3087.6	2.15	0.26
3104.1	2.41	0.37
3221.7	2.00	0.17
3544.8	3.51	0.19
3663.7	3.94	0.67
3739.9	3.72	0.52
3861.8	4.30	0.86
3953.3	4.67	0.17
4020.3	4.24	0.92
4164.8	4.18	0.46

Table A3. *Stratigraphic and lithological details of Imperial Island River No. 1 (after Morrow* et al. *1993)*

Formation name	Depth (m)	Age (Ma)	Lithology fractions 1	2	3	4	5	6	7
Quaternary	20	44	0.5	0.5	0	0	0	0	0
Wapititi	−700(4)	75	1	0	0	0	0	0	0
Kotanalee	−700(4)	88	0.2	0.3	0.5	0	0	0	0
Dunvegan	−140(4)	90	1	0	0	0	0	0	0
Dunvegan	183	98	1	0	0	0	0	0	0
Sully	360	101	0	0	1	0	0	0	0
Sikanni	445	102	0.50	0.5	0	0	0	0	0
Buckinghorse	643	113	0.15	0.15	0.7	0	0	0	0
Tod Grayling	−140(3)	240	0.1	0.1	0.8	0	0	0	0
Kindle	−20(3)	258	0.5	0.25	0.25	0	0	0	0
Fantasq	−20(2)	268	0	0	0.5	0	0.5	0	0
Mattson	−140(1)	333	0.9	0	0.1	0	0	0	0
Rundle	−200(1)	349	0	0.1	0.5	0.4	0	0	0
Rundle	−75(3)	354	0	0.1	0.5	0.4	0	0	0
Rundle	678	356	0	0.1	0.5	0.4	0	0	0
Banff/Exshaw	1047	361	0	0.1	0.75	0.15	0	0	0
Kotcho /Tetcho	1346	364	0	0.1	0.5	0.4	0	0	0
Trout/Kakisa	1417	369	0.55	0.05	0.35	0.05	0	0	0
RedKnife	1560	370	0.05	0.05	0.8	0.1	0	0	0
Jean Marie	1576	371	0	0	0	1	0	0	0
Fort Simp /Muskwa	2106	374	0.15	0.1	0.75	0	0	0	0
Slave Point	2224	376	0	0	0	1	0	0	0
Watt/Sulphur	2247	378	0	0	0.1	0.9	0	0	0
Keg River	2412	384	0	0	0.1	0.8	0.1	0	0
Chinchaga	2471	390	0.05	0.05	0.05	0	0.45	0.4	0
Sub-Chinchaga	2499	401	0.7	0	0	0	0.2	0.1	0

A negative depth indicates lost section and the integer in brackets refers to the erosion event (see tables A5 and A6)

Table A4. *Vitrinite reflectance values for Imperial Island River No. 1*

Depth (m)	Ro (%)	∂Ro (%)
607.9	0.58	0.15
650.6	0.72	0.13
704.3	0.76	0.15
837.5	0.64	0.14
841.5	0.61	0.02
998.5	0.73	0.11
1056.4	0.68	0.07
1061.6	0.85	0.04
1154.6	0.90	0.02
1649.1	1.05	0.13
1798.5	1.13	0.13
1902.4	1.19	0.13
1951.2	1.28	0.03
2077.4	1.53	0.13
2229.6	2.00	0.40

Table A5. *Compaction constants, ($\phi(z) = \phi_0 exp[-cz]$), and matrix thermal conductivity*

Lithology	ID	Surface porosity (ϕ_0)	Compaction constant (c; m^{-1}, x 10^{-4})	Matrix thermal conductivity (k_m, W m^{-1} K^{-1})
Sand	1	0.29	3.53	4.2
Silt	2	0.46	3.99	2.7
Shale	3	0.51	5.30	1.5
Limestone	4	0.51	5.18	2.9
Dolomite	5	0.30	2.17	5.0
Anhydrite	6	0.00	1.20	5.5
Chert	7	0.00	1.20	2.9

Geometric mean model was used, with k_{water} = 0.607 W $m^{-1}K^{-1}$.

Table A6. *Timing of erosion events*

Erosion event	ID	Begin (Ma)	End (Ma)
Sub Permian	1	276	268
Sub Triassic	2	245	240
Sub Cretaceous erosion	3	208	113
Tertiary erosion	4	67	44

References

ARNE, D. C. 1991. Regional thermal history of the Pine Point area Northwest Territories, Canada from apatite fission track analysis. *Economic Geology*, **86**, 428–435.

BURNHAM, A. K. & SWEENEY, J. J. 1989. A chemical kinetic model of vitrinite maturation and reflectance. *Geochimica Cosmochimica Acta,* **53**, 2649–2658.

CONSTABLE, S. C., PARKER, R. L. & CONSTABLE, C. G. 1987. Occam's inversion : a practical algorithm for generating smooth models from electromagnetic sounding data, *Geophysics,* **52**, 289–300.

GALLAGHER, K. 1998. Inverse thermal history modelling as a hydrocarbon exploration tool. *Inverse Problems*, in press.

—— & SAMBRIDGE, M. 1992. The resolution of past heat flow in sedimentary basins from non-linear inversion of geochemical data : the smoothest model approach, with synthetic examples. *Geophysical Journal International*, **109**, 78–95.

MAJOROWICZ, J. A., JONES, F. W. & JESSOP, A. M. 1988. Preliminary geothermics of the sedimentary basins in the Yukon and Northwest Territories (60°–70° N) — Estimates from petroleum bottom hole temperatures. *Bulletin of Canadian Petroleum Geology*, **36**, 39–51.

MORROW, D. W., POTTER, J., RICHARDS, B. & GOODARZI, F. 1993. Palaeozoic burial and organic maturation in the Liard Basin Region, northern Canada.*Bulletin of Canadian Petroleum Geology*, **41**, 17–31.

PARKER, R. L. 1977. Understanding Inverse Theory. *Annual Review of Earth and Planetary Science*, **5**, 35–64.

SAMBRIDGE, M. S. 1990. Non-linear arrival time inversion: constraining velocity anomalies by seeking smooth models in 3-D, *Geophysical Journal International*, **102**, 653–677.

Index

Page numbers in *italics* refer to Tables or Figures